FOREST SERVICE MEMORIES

Past Lives and Times in the United States Forest Service

More Than 300 Stories, Letters, Poems, Biographies and Essays From 1891 to the Present

Edited by
Gilbert W. Davies and **Florice M. Frank**

HiStory ink Books
P.O. BOX 52
HAT CREEK, CA 96040

Front cover: Built in 1906, the first constructed ranger station on the Klamath National Forest, California, was located about three and one half miles upriver from the town of Orleans along the Orleans-Happy Camp trail. Left to right: Kate (mule), Ranger Bill Hotelling, Gypsy (horse), Assistant Ranger Albert Wilder – c1911.

Back cover: Bald Mountain Lookout, Klamath National Forest – 1922.

Second Printing: 1998

Library of Congress Catalog Card Number: 97-70410

ISBN 1-887200-03-7

Printed in the United States of America

CONTENTS

The Forties

The Fifties

The Sixties

---

*R-5 FSX Club Newsletter.

**R-6 Thirty Year Club Timberlines.

PREFACE

This is primarily a book of history, though not in the traditional text book format. It is a history of an agency, a people and the land they took care of. It is written by those who lived the experiences and experienced the living of the rich panorama of Forest Service life.

From these stories one may learn how the agency developed, what it covered and where it was going. It is a history book of people who had a burning desire to do their best. From the beginning, the pioneering passion and creative spirit carried on through decades as the agency grew from a few remarkable and tough individuals to today's scientific studies, computerization and social programs.

The diversity of stories show the wide spectrum of activities the Forest Service performs in its stewardship of the land. Bucking a sometimes unfriendly public of individual landowners in the beginning to the present day wants of large businesses and preservationist groups, the Forest Service has often been in the middle of controversy.

This history book does not have a scholarly look. It has no footnotes or bibliography. It does not delve into causes and effects or summaries and conclusions. It does not make a study of various functions and programs, nor does it use charts, tables, appendices or guides. There is no historical revisionism present. The words of the writers are offered intact. Generally, the only changes from the original accounts involve spelling, punctuation and capitalization. Beyond that, some stories were abridged, but not at the expense of changing the basic meaning. And others were slightly rearranged so that the time sequence would be clearer to the reader. Titles were created where necessary.

The stories are often humorous, sometimes tragic or sad, usually dramatic, frequently poignant. A few are suspenseful and a handful are unbelievable. A recurring theme that seem to run throughout is that the contributors found their work a grand experience and they loved what they did.

Stories from dozens of locations throughout the country are included. We limited the number about one subject in a decade and we shied away from autobiographies. We did, however, include a few short biographical sketches of some fascinating people. And we found some poems and essays worth sharing.

We believe the preservation of Forest Service history is vitally important and that much of it is in danger of being lost. Our antidote has been to solicit voluntary remembrances which we offer here from those who lived that history. Let us hope that someday this history will be given the wide attention it so richly deserves. Perhaps no other Federal agency has had as many employees share a combination of lengthy service, pioneering attitude and personal adventures as the United States Forest Service.

Statistically, there are 14 main head chapters, 254 sub-chapters, 30 photographs, 144 contributors and references to more than 90 specific national forests. The photographs are from the U.S. Forest Service.

GWD & FMF

ACKNOWLEDGMENTS

The 144 contributors to this book (many deceased) are the first in line for a huge thanks. Without them, of course, there wouldn't be a book. They are the folks who lived the stories and wrote the records. Contributions came from all over the United States. They were provided by retirees, spouses, current employees and a couple of anonymous sources.

Many of the stories have been reprinted (some abridged and edited) from two Forest Service retiree newsletters: Region 5, with its FSX Club Quarterly Newsletter and Region 6 with its Thirty Year Club Timberlines (Timber Lines). Permission was granted to use these stories. An attempt was made to contact each living author for his or her approval.

Lyle Klubben has been the editor of the R5 FSX Club Newsletter for many years. Before Lyle, the editor was John Hall. The late Everett Jensen was the editor of the "I Remember" stories. The stories have a short, creative and entertaining introduction, ending with the initials E. J. Some of those introductions are included.

The R6 Thirty Year Club Timberlines originated in the last of the 1940s. Starting in 1957, past editors of those stories were Foster Steele, Frank Flack, Carroll E. Brown, Merle S. Lowden, Albert Arnst and Dick Deleissegues. The current editor of the Thirty Year Club Newsletter is Robert C. McQuown.

Contributions from the above two publications are identified by asterisks after the title in the Contents section, as follows:

* – R5 FSX Club Newsletter
** – R6 Thirty Year Club Timberlines

Several stories are taken from other books with permission from their authors. A majority of these books are still available and all are excellent reading. Most of the books were written entirely by the authors listed alphabetically by last name.

Robert W. Cermak: Two stories, "The Matilija Fire of 1932" and "An Old Time Forest Supervisor," are from his book *Fire In the Forest*.

Jay H. Cravens: Three stories are from his book *A Well Worn Path*, published by University Editions, 1994.

Bob Gray: Four stories are from his *Forests, Fires, and Wild Things*.

Les Joslin: "Two Historic Ranger Stations" is from his book *Uncle Sam's Cabins*, published by Wilderness Associates, 1995. "A Campfire In a Catface" is from his book *Toiyabe Patrol*, published by Wilderness Associates, 1993.

Jackie Johnson Maughan: "The White Pickup" is from her book *Go Tell It On the Mountain*, published by Stackpole Books, 1996. "Fire Watch" first appeared in *Women's Sports and Fitness*, August, 1996.

James R. Pratley: Three stories are from his book *Shorty*, about a dog's adventures in the U.S. Forest Service.

Stan Tixier: Two poems are from his book *A Good Lookin' Horse; Cowboy Poetry and Other Verse*.

James D. McNeill's two stories first appeared in the Klamath National Forest newsletter *The Klamath Kurrents*.

All of Don Bauer's stories are from five oral interviews conducted by Elliott L. Graham on the San Bernardino National Forest during 1982 and 1983.

The two stories by Roy H. Blood and Dick Chatham are taken from the old *California Ranger* publication.

Several other people have helped in a major way with the publication of this book. Everett Jensen allowed us to use his complete collection of FSX Newsletters and wrote several letters of encouragement and ideas. Harvey Mack, the founder and guiding light to the Forest Service History Center currently located on the Cleveland National Forest, also supplied copies of the FSX Newsletters and stories, as well as some photographs. Ron Johnson was our contact for the copies of Timberlines. His complete collection was read and he graciously made copies of the hundreds of pages we requested, plus other valuable information and some photographs.

Dan Davies, retired city editor of the *Oregon Statesman* in Salem made suggestions and did much of the proof reading.

However, the main kudos go to those of you who supported this effort by purchasing our other Forest Service books. Without that, there would have been no future publications. Thank you all. We hope this current book is worthy of your support.

"A land without memories is a land without history"

Abram Joseph Ryan
(1834-1886)

DEDICATED TO THOSE EARLY FOREST RANGERS WHOSE INDOMITABLE SPIRIT, PASSIONATE DRIVE AND UNWAVERING CONVICTION IN THEIR MISSION, PIONEERED THE WAY AND LAID THE FOUNDATION FOR THOSE WHO FOLLOWED.

"The Ranger then was a doughty cuss, who chewed up nails and spit out rust"

Tom McArdle

INTRODUCTION

"On March 30, 1891, with little fanfare, the United States took its first step toward a national forest system with the establishment of the Yellowstone Park Forest Reserve. Before then, Government policy was to dispose of public lands. Concerned citizens, many of them scientists and sportsmen, convinced the Congress and President Harrison of the benefits of public ownership of natural resources managed for the public good. This turned out to be what the history books have called the watershed event in American Conservation history, the Creation Act of 1891.

"After that, the wholesale disposition of the public domain stopped, and the complex of national forests, wildlife refuges, BLM lands, and vastly enlarged national parks system was begun. Of equal importance was the acceptance of the idea that the Nation's natural resources should be managed, an idea brought to fruition with the Organic Act of 1897."

The above was written and issued in 1990 by the Chief's Office in Washington, DC, for the forthcoming 100th anniversary of the National Forest System.

The following quotes are the beginning paragraphs from Foster Steele's Forest Service memoirs, written 50 years after his 1905 entrance into the Service.

"America is blessed by a kind Providence which has furnished inspired leadership in times of great crisis. George Washington in the birth of a republic; Abraham Lincoln in the birth of a nation; Gifford Pinchot in the birth of conservation, which saved for America its priceless natural resources.

"History has recorded the exploits of these great leaders but time alone reveals their stature in the light of the results of their accomplishments.

"America is only now beginning to realize the stature of Gifford Pinchot. They all appear greater in retrospect than they did in their present. So too will those who served with Pinchot in the Forest Service come to be recognized for their contribution to a great crusade which saved for America a heritage of inestimable wealth.

"To have had a part in that crusade in the formative era when the preservation of the nation's forests hung in the balance is a reward in itself—perhaps the greatest satisfaction these men and women will ever know. Certain it is that the financial return measured by the value of the present day dollar would be no incentive for dedicating a lifetime of service to such an undertaking."

Grover C. Blake also writes after 50 years from his early start in the Service.

"The ranger force of that early day were, for the most part, raw recruits. They were woodsmen, yes, but foresters, not yet. They were to become foresters through the tough school of experience and study. Their chief assets were health, stamina, and a capacity for learning. The ranger went forth equipped with a badge, a Use Book, map, compass, marking hatchet, and a few other articles and a good supply of blank forms, cloth notices, and posters, also a month's food supply, and pack and saddle horses. There were no telephone lines so the supervisor would likely not hear from the ranger for several weeks. In the meantime he had worked on a large assortment of jobs. His duties were many and his job was rather hard but he was deeply interested and he loved his work."

And finally, Fred Wehmeyer presents an essay of early pioneering.

"Forestry under the supervision of the Department of Agriculture! I have been a part of it from its formation, either by active participation, close association with others, or in retirement. The service has been the victim of excessive progress the same as every other activity in this rapidly developing age. Down through the years a job was obsolete almost before completion. Game trails were replaced with graded pack horse trails, then by motor ways, and later have given way to landing fields. Communication has changed from the heliograph, to telephone, to radio. All phases of the profession have seen equal, startling changes.

"No doubt someone with the ability to paint a graphic picture in writing will some day make a complete and adequate record. He should dwell a bit on the early day personnel, recruited from the western frontiers of that day—real cowpunchers, miners and lumberjacks who rose above mediocrity through loyalty, zeal and honor. As Mr. Churchill would express it, 'Never was so much accomplished by so few and with so little.' There were areas as large as some states, impenetrable except along the meandering blazes of some trapper or prospector. At the turn of the century a friend who was a Ranger of that period said he wrote in for a classification of his duties and after many weeks got the answer from Washington, DC, that, inasmuch as he was signed up as a Ranger, he should get out and range.

"The early days were a historical period in forestry development. Those who took an active part can well remember the early pioneering—the seemingly endless lodgepole thickets; the bottomless swamps of the river and creek bottoms; the miles of debris left through snow slides; the yellow jackets; the devil clubs; the Swede killers; the month-long pack horse trips with the many privations, miseries and frustrations. After being almost sucked dry of blood by clouds of mosquitoes, no-see-ums, gnats, deer flies and horse flies all day and night, grown men have been seen to set down and bawl, then grit their teeth, get up and resume whatever they were doing with added vigor. Members of the service seemed to have a fierce determination of accomplishment, and Uncle Jimmy Wilson's instruction 'to carry on, keeping in mind the greatest good to the greatest number in the long run' took its place in the scheme of things as the Eleventh Commandment.

"Life also had its lighter moments. To those looking for a bit of humor, there was always sufficient time for an occasional chuckle. One of the early Guards woke up and thought he heard some kind of machinery running. He had been clearing trail on the head of Ransay Creek on the Upper Methow. Worn out from a heavy day, he lay there and listened and conjectured on what he heard. Then he got a whiff of an odor that was especially fetid and opened his eyes to see a large puma standing straddle his body with its head a matter of inches from his own. The animal, evidently pleased to have found company, was purring but had evidently failed to use a mouth wash for its breath was even beyond the halitosis stage. Archie, for that was the Guard's name, reached up under his pillow and grabbed his revolver and in his excitement pulled the trigger, blowing most all the feathers out of his pillow. The puma threw large quantities of pine needles, cones and twigs over the bed as it left the area in a series of great leaps.

"Similar stories and yarns were a part of every evening campfire where two or more were gathered. I have often thought if I could relive one day of my life, I would select one when and where the rangers were gathered together to do some construction job. When a forest had to get a job completed in those far off times, the Supervisor called all his men together and, standing shoulder to shoulder with them, would swing an axe, dig holes for telephone poles, or take an active part in whatever labor was needed and also gorge himself on a good grouse mulligan at meal time along with the best or worst of his men."

Perhaps no further comments are necessary to explain the reasons for the production of this book. This history is now a bound record. The old-timers have their magnificent stories in print. Their legacy is secure and preserved. They can smile and rest in peace. They were true "pioneers."

THE BEGINNINGS

IN THE BEGINNING AT THE WASHINGTON OFFICE

Robert L. Campbell

In the fall of 1904, I qualified for government service as a stenographer following a Civil Service examination. About December 1, 1904, I received a telegram offering a position in the Bureau of Forestry, Washington, D. C., at a salary of $720 per annum. I declined it. A second offer, this time at $900, was accepted. Up to that time I had never heard of the Bureau of Forestry. Thus, by chance, my name on the employment register and a vacancy in what was to become the Forest Service came up at the same time. I remained in the Service for an overall period of 42 years. In retrospect, I think it was a fortunate circumstance.

Compared to the city of today, Washington was then an "overgrown village." The New Willard was the leading hotel, with the Raleigh nearby, not to mention the Weiser Oyster House in the rear of the latter where one could stand before a counter on a sawdust floor and enjoy Chesapeake Bay oysters on the half shell at 15¢ the dozen! The old New National was the leading playhouse, where big names in drama, comedy and tragedy filed past in rapid succession. The Post Office Building was the only imposing government structure on Pennsylvania Avenue, and the lower end below the Capitol was rather dilapidated. Near the upper end, not far from the White House, was the "restricted district" through which passed interurban cars bound for Alexandria.

Working hours for government employees were 9:00 a.m. to 4:30 p.m.—very convenient for baseball fans who could catch the 7th Street streetcar and be out at the park in time for the first pitch. The National Cathedral was but a cornerstone sunk in the ground, with a small wooden church nearby. The steamer Jamestown plied the Potomac on summer evenings—tariff for two, including dancing, 50¢. The Atlantic Building stood on F Street, near the corner of 7th Street, and from its rear windows one could look down upon the roof of the old Ford Theatre (then used as a warehouse) and into the alley where Spangler held a horse for John Wilkes Booth on the night President Lincoln was assassinated. Just beyond the elevator, on the ground floor of the Atlantic Building, was one of the old style "bucket shops" with its stock quotation board, now legally extinguished. On the upper floors, the Bureau of Forestry was located, one of the smaller rooms being occupied by "Judge" Harry Towers, who administered the oath, and I became a full-fledged "bureaucrat."

Immediately to the left of the elevator on the sixth floor was a room in which an atmosphere of clerical precision was maintained under the direction of George G. Anderson. It was my first assignment. Here were prepared typographically perfect letters conveying to the recipient, the satisfying information that in accordance with his request, Bulletin No. 15, The Woodlot, or the like, was being forwarded under separate cover. A copy was sent over to the Superintendent of Documents and that was that. Thus was beginning the trickle of public service which would later swell into a mighty flood. One afternoon about closing time, a tall man with piercing eye, flowing mustache and springy step hurried in to cleanse his hands in the basin which stood in a corner. It was my first glimpse of Mr. Gifford Pinchot.

The fiscal department was located in a large room on about the seventh floor and about this time, George E. King was acting as fiscal agent. He was a kindly man, as I recall, and not devoid of humor. One day he called my attention to a voucher which was crossing his desk. It contained the entry "Berth in Pullman - $5.00." Of course everything is higher now!

The top flight at this time, in addition to the Forester and Associate Forester Overton W. Price, was Wm. L. Hall in forest extension; Thos. H. Sherrard in timber management; Dr. Herman von Schrenk in forest products; and F. E. Olmsted in reserve boundaries, with Geo. B. Sudworth as dendrologist. The office force was small and incidentally, the female segment was almost entirely single, the few married ones being widows. Joe Santucci, as chief messenger, hustled papers about the place, while down on the fifth floor outside the office of reserve boundaries, to which I was soon transferred, "Captain" Markwood kept lonely vigil, broken only by a rare excursion to one of the floors above. He was, I believe, a veteran of the then not too recent Civil War, enjoying the GI benefits of his time.

Ranging the Western States was an enthusiastic and dedicated group of young men, recently out of forest school who were busily engaged in prospecting for new forested areas. Their names—E. T. Allen, Coert Dubois, Smith Riley, H. J. Tompkins, R. E. Benedict, Clyde Leavitt, W. H. B. Kent and others are well known. Nothing escaped them. I recall one letter which recited the results of a timber reconnaissance from a Pullman car covering the Bridger Mountains in Montana! Be that as it may, the results of their labors culminated in the office of reserve boundaries where, after type and title maps and proclamation diagrams had been prepared, F. E. Olmsted dictated a letter to "The Honorable, The Secretary of Agriculture, Sir" recommending the creation of another unit. President Theodore Roosevelt did the rest by proclamation, as this was before the Congress had deprived the Executive of such authority. Writing letters of this sort was my chief occupation while on this assignment.

It was in this section of reserve boundaries that a historic letter was first placed on paper which was to keynote the conception of what the relationship between the new regime and the public should be. I refer to the letter dated February 1, 1905, signed by James Wilson, then Secretary of Agriculture, which contained the following:

> "In the administration of the forest reserves it must be clearly borne in mind that all land is to be devoted to the most productive use for the permanent good of the whole people and not for the temporary benefit of

> individuals or companies..... You will see to it that the water, wood, and forage of the reserves are conserved and wisely used..... When conflicting interests must be reconciled, the question will always be decided from the standpoint of the greatest good for the greatest number in the long run."

This letter was dictated to me by F. E. Olmsted and typed for the signature of the Secretary on his official letterhead, after being initialed by the Forester. Whether or not this letter as it finally appeared was the same as that dictated by Mr. Olmsted I cannot say, but I think it safe to conclude that the wording above quoted was his idea of the future responsibility of the Forest Service.

In that first year of 1905, when the changes authorized by the Act transferring the forest reserves from the General Land Office to the Bureau of Forestry became effective, the course of the Forest Service was charted. Standards of conduct and procedures were set up which have persisted unto this day. The "Eager Beavers," who had been so busily engaged in examining lands for the creation of new forests, now became inspectors whose visits to the field resulted in major upheavals. The knife cut deep. Dismissals, suspensions, demotions and enforced resignations came thick and fast. There were one or two cases of criminal prosecution. Action was quick and sure. It was a survival of the fit rather than of the fittest.

There were also revolutionary changes in the office procedure. At the start, administrative matters fell into the lap of the section of reserve boundaries. We moved upstairs. We made a field program. It consisted of a single sheet of paper with the names of the forests and their supervisors, with mail and telegraph addresses of each. Captain J. B. Satterlee, who had been in charge of timber sales, and Geo. Leland, who handled grazing affairs, came over from the General Land Office, but very few other personnel.

It had been Land Office procedure to fold letters for filing in vertical files, with an identification briefed on the back. Also when replying to a letter it was the practice to first summarize at length the contents of the letter to which reply was being made before getting down to the reply itself. One day Harry Hill, who was then secretary to the Forester, said the Chief wanted to circulate two contrasting letters written on the same subject, one in the old Land Office style and the other in the newly appointed Forest Service style. We prepared them with some care, one very verbose upwards of two pages in length and the other short and to the point on a single sheet. Next day Harry returned and, indicating the General Land Office style, remarked: " G. P. says it is not half rotten enough."

Thus the elaborate training programs of today had their lowly beginnings in 1905. I recall one instance where a "trainee" underwent rather a harrowing experience. He was a new arrival in Washington with a good record in the field but unschooled in paper work—he incidentally later rose to a position of eminence and retired at the close of a successful career. However, shortly after his arrival, the task of composing a rather important letter for the signature of the Secretary, after approval by the Chief, was thrust upon him. After long cogitation he prepared the necessary reply and with some misgivings I typed it for him and sent it on its way. It was promptly returned. Across the face was written in the bold stub pen hand used by Mr. Pinchot, the notation "Could not be worse –G. P." I recall a similar comment on another occasion which was perhaps even more cutting: "I have written another reply and I hope a better one –G. P." Nevertheless, despite the severity of such methods, I think all, then in the Service, had much respect for the "first schoolmaster."

Although it was not in operation until after 1905, mention might be made of what became known in the Washington office as the "Bull Pen." In a large front room in the Atlantic Building facing on F Street, six desks were placed. They were occupied by six supervisors, one detailed from each district. In addition I was provided with a desk and there was one for a stenographer (Mr. Bechtold). There were also two other stenographers assigned to the project (Miss Dow and Miss Chappel). The supervisor personnel rotated every 30 days, as new men were brought in from the field and others returned to their stations. Mail on administrative matters as received was turned over to the supervisor from whose district it came. It was his job to prepare a proper answer. It was mine to edit it. The resultant letters were signed by Clyde Leavitt as Acting Chief of Operation (later succeeded by Geo. H. Cecil, as Chief). That is how the "Bull Pen" was originally organized. The atmosphere provided by these men drawn together from all parts of the West, most of whom had never been East, much less visited the Capitol city, was quite refreshing and something to be remembered.

THE REMINISCENCES OF A STUDENT ASSISTANT

A. O. Waha

Shortly after attending the first summer school session in 1901 of the Yale Forest School at Milford, Penn., I was directed to report to Washington.

When after several weeks in Washington I was told that I was to have an assignment in the field, my gratification was quite complete. There were three in our party headed by John Keach, an Amherst man, and Ferris Miller, another student assistant from Ohio. With a map, compass, tally board, chain and calipers we set out for western Maryland for the purpose of securing data on the timber resources of Garrett County. Our monthly wage was reduced to $25, which seemed altogether reasonable since the government reimbursed us for our expenses. While we were working out of Oakland, Maryland and living in a "family style" hotel, everything went well, but after that it was a case of "catch as catch can." We would start out in the morning with no idea of where we would be at night. Many days our search for food and shelter involved a farm to farm canvass. But it was all splendid experience. We stayed with families whose standards of living were most primitive—the simplest of fare along with feather beds. Sleeping in a different bed each night and meeting new people and conditions every day gave us a zest and an experience which I feel sure has always been helpful to me.

The student assistants were a most congenial set, all of whom were very much interested in their work and eager to learn. Weekly meetings after work were held at which various topics were discussed. I recall that the first topic assigned to me was on the characteristics of bark of various trees and their uses. Then the Thursday evening meetings of the so-called "Baked Apple Club" at Gifford Pinchot's house on Rhode Island Avenue were indeed a bright spot in our lives. The best talent available was brought in to address us after which there would be informal discussions, followed by a feed of baked apples, cream and ginger bread. Our most notable speaker was President Theodore Roosevelt who gave us a "bully" talk in March 1903. What a crusader he was. Enthusiasm fairly oozed from him!

After a few months of office work, I was sent to McKeever, New York, for the purpose of making a rather intensive study of hardwoods.

Cleland, from Maine, and I were the only employees assigned to this project. We lived at a lumber camp where all of the lumberjacks were French Canadians and I soon learned how temperamental they can be when drunk. One minute they would be trying to gouge each other's eyes out while the next minute they would be sobbing and kissing one another. One evening at supper for no apparent reason so far as I could see, one of the loggers who had just returned from town and was still very drunk, grabbed hold of the board of the table on which dishes of pea soup had been placed and turned it upside down. Fortunately, the cook was persuaded from committing murder.

My work carried over into January 1904, so I got a taste of real winter in the north woods. At times the temperature reached a low of 40 degrees below zero.

The logs that had been cut, both spruce and hardwood, were hauled out to the river bank for the spring drive. Since my next work on the project was at the sawmill in McKeever and the logs would not be available until after the spring drive, I was ordered to Washington where I remained for a month or so on office work before being again assigned to the deep south. This time it was Alabama, south of Tuscaloosa. At Tuscaloosa I had several hours to wait for a train which would take me to a town called Moundville.

I shall always remember Moundville but for a quite different reason. On reaching there in the late afternoon, I learned that the camp to which I had been assigned was located about 10 miles from town and that Franklin Reed who was in charge of the party would be in town the following morning. So I stayed in Moundville that night at the only hotel and in leaving the following morning I arranged with the owner of the hotel to store my trunk along with a new derby hat which I could not get in my trunk. The first night in camp we had a terrible wind and rain storm, but as we learned the following day, it was nothing in comparison to the tornado that had struck Moundville, killing 40 persons and leveling to the ground all buildings in town with the exception of the church. It was a month or more before I got back to what was left of the town. In the meantime I had learned that my trunk had been found about one quarter mile from the hotel and had been brought back to the church where it had been stored. Everything in the trunk was intact but what a mess my clothes presented, being thoroughly mildewed due to water having come in through a large ragged hole that had been made in the bottom of the trunk. The derby hat, of course, had gone with the wind as had also many other things. Needless to say, I have always been thankful that circumstances were such that I did not have to remain in Moundville two nights. Had I left Washington a day later, the chances are more than good that I would not now be writing these reminiscences.

In returning to Washington upon the completion of our work, I stopped off at Asheville, North Carolina, with my companion Bill Farnum, to visit the Biltmore Forest School. Farnum had been a student there and had told me so much about it that I was greatly interested in seeing the layout. I attended one of Dr. Schenk's lectures and even though I had been told previously that the students could smoke during classes, it did seem strange to actually see it occur. That night one of the famous beer parties was thrown and by midnight there was more than one student resting peacefully under the table. A horseback trip (all of the students kept horses) was made to Vanderbilt's Estate where Dr. Schenk was the resident forester and I was duly impressed with the forestry work involving thinnings and plantations which was pointed out to me.

TWO HISTORIC RANGER STATIONS

Les Joslin

ALTA RANGER STATION

Alta Ranger Station, a small log cabin deep in the Bitterroot National Forest, is generally recognized as the first U.S. Forest Service ranger station even though it predates creation of the Forest Service by almost six years. Built in the spring of 1899 by newly appointed U.S. Department of the Interior forest rangers N.W. "Than" Wilkerson and H.C. "Hank" Tuttle, this historic one-time ranger station gives visitors a good feel for the rigors and hardships faced by the earliest rangers.

Rangers Wilkerson and Tuttle were among the first men in the United States appointed as forest rangers on the forest reserves then under Department of the Interior custodianship. In June 1899, these pioneer rangers were posted along the West Fork of the Bitterroot River to patrol some 300,000 acres of rugged Montana back country in the newly created Bitter Root Forest Reserve. Their main jobs were to prevent timber theft and fight forest fires. They needed shelter and, out of necessity, built the rustic, sod roofed, one room Alta Ranger Station cabin. It took rangers Wilkerson and Tuttle, who borrowed a workhorse to skid logs to the site, fifteen days to build the cabin. As Ranger Wilkerson wrote in his diary:

> *Ranger Hank Tuttle and myself constructed the Alta Cabin on our own initiative.... we went down into our own pockets for every cent. . . . Uncle Sam wasn't putting out a good cent for a pair of hill-billy rangers to live while they took care of his woods.*

And so the two rangers, whose pay was fifty dollars per month, bought the hardware and one glass window for the log cabin with their own funds. But, as Michael Frome put it in *Whose Woods These Are*, "when they erected a flagpole and hoisted a flag aloft, this *was* a ranger station." On July 4, 1899, the two rangers dressed up in their "Sunday best" and had their picture taken with the new cabin and the flag they also had purchased with their own money. Alta's flying of Old Glory that day began the custom of flying the flag at ranger stations.

At the time Alta Ranger Station was built, the only other habitation on the West Fork of the Bitterroot was a lone prospector's cabin. As mining intensified in the area, the community grew. An assay office, post office, store, inn, and saloon were built during the opening years of the twentieth century and, in 1909, a telephone line connected Alta to the outside world. For some reason, the ranger station site was not withdrawn from entry as a homestead or mining claim under the public land and mining laws, and the land was patented as a mining claim in 1904. The cabin was sold to a private owner and moved to Darby that same year. As a result, the Forest Service—which took over the forest reserves in 1905 and renamed them national forests in 1907—withdrew twenty-seven acres adjoining the original site and built another ranger station in 1908. It no longer exists.

Rangers Wilkerson and Tuttle couldn't have known in 1899 that their hastily built 12 foot by 16 foot log cabin would become an historic site—listed on the National Register of Historic Places seventy-six years later as the nation's first ranger station. It's not much to look at, but its significance has long been recognized. Back in 1941, the Hamilton, Montana, Lions Club purchased the site of the original Alta Ranger Station, moved the restored cabin to the site, and deeded both to the Forest Service for preservation and interpretation as an historic landmark. But the cabin again fell into disuse and neglect, and was in poor shape when Peyton Moncure wrote about, "The First Ranger Station" in the October 4, 1959, issue of *The Salt Lake Tribune Home Magazine*:

Only the pack rats scurry among the old newspapers and magazines that litter the floor now, and a few old window panes are broken out, but the day may come when this old landmark will be something of an official monument to pioneer foresters who had such a personal interest in protecting the nation's forests that they were willing to sacrifice much of their own time, comfort, and meager funds.

That day came. Renewed interest in historic building preservation in the 1960s led to a second restoration. Historic Alta Ranger Station, which now receives regular Forest Service and private financial support, remains to tell its tale.

WAPITI RANGER STATION

America's first national forest claims one of America's first ranger stations. Shoshone National Forest, carved out of lands set aside by President Benjamin Harrison as the Yellowstone Timberland Reserve in 1891, to be the first unit of the National Forest System, is the site of Wapiti Ranger Station, the first ranger station built with government funds and still in service at its original location.

Other ranger stations were built before Rangers Harry Thurston and Milton Benedict erected Wapiti in 1903. Alta Ranger Station was built in 1899 on the Bitter Root Forest Reserve, but with the rangers' personal funds. And West Fork Ranger Station, built in 1900 on the San Gabriel Forest Reserve, was the first ranger station in California built with federal funds, but has been moved almost ten miles from its original location and reconstructed at a visitor center.

At any rate, it was October 1903, when Ranger Thurston rendezvoused with Ranger Benedict and his sixteen packhorse loads of supplies at the confluence of the North Fork and Elk Fork of the Shoshone River to construct quarters for W.H. Pierce, Supervisor of the Shoshone Division of the Reserve, who had decided to relocate his headquarters from Cody. "The site," retired Forest Supervisor Thurston wrote in 1941, "Was delightfully located in a broad stretch of mountain valley. Water was near, wood abundant, and fish and game plentiful." The two rangers built a three room dwelling and a separate office cabin of logs previously "cut, peeled, and adzed by Ranger Sherwood." The space between the two buildings was later enclosed. In 1904, the first telephone line built by the government for administration and protection of forest reserves linked Wapiti with Cody, and the stable and corrals were built. Supervisor Pierce and Ranger Thurston lived and worked there until 1907. Distinguished visitors to Wapiti during those first years included Colonel William F. "Buffalo Bill" Cody and Chief Forester Gifford Pinchot. When, in 1907, the divisions of the Reserve were separated and designated "national forests." Shoshone National Forest headquarters moved to Cody and Wapiti Ranger Station became headquarters of its North Fork Ranger District.

After the North Fork and Canyon Creek districts were combined to form the Wapiti Ranger District in 1921, Ranger Clifford Spencer and his wife, Helen, moved to Wapiti Ranger Station from flood ravaged Canyon Creek Ranger Station. Wapiti had been unoccupied for some time when the Spencers moved in during the summer of 1923. After cleaning, repairs including placing a new foundation under the building, and adding a bedroom, the Spencers were at home.

Helen Spencer, "practically an assistant ranger" to her husband, recalled her life as a Forest Service wife at Wapiti Ranger Station in a 1979 interview. In addition to chopping wood, carrying water, providing information to visitors, lining out fire crews, and myriad other tasks, Mrs. Spencer's specialty seemed to be catching and saddling the horses for Ranger Spencer. "If he wanted the horses in a hurry," she explained, "I (would) rattle the oat can. They would come tearing to the house. When he did it, the horses decided he wanted to use them. so they wouldn't come." Ranger Spencer did a lot of his work from horseback, and Mrs.. Spencer accompanied him on many of his pack trips.

> *I was his trail crew. The Forest Service didn't get any money in those days to hire a trail crew, so I went along to help. He always took along a two handled saw so I could get on the other end of it.*

Often, when the Spencers crossed the road between Cody and Yellowstone National Park with a pack string, tourists stopped to take their pictures.

And, as was Forest Service custom, Mrs. Spencer fed all comers—from Washington, D.C., big shots, to a Civilian Conservation Corps crew on a bug job, to Great Depression victims who would "stop in and want a meal." Among the Spencers' better known Wapiti guests were Yellowstone National Park Superintendent Horace W. Albright, who became National Park Service director, and Department of Agriculture entomologist Dr. F.C. Craighead and his twin sons, Frank and John, who became famous grizzly bear researchers. Otherwise, social life was limited to dances at Wapiti Schoolhouse at which ranger station and neighboring ranch couples celebrated birthdays and wedding anniversaries.

Although she summed up her Wapiti Ranger Station years "as an enjoyable time," Mrs. Spencer recalled some tense moments. Once, on a day after Christmas when she was alone at the station, she answered the knock of a drunken man who claimed, "he had just shot a man at the sawmill" and was on the run from the sheriff. "You're in a helluva fix, aren't ya?" he menaced. "I could hurt you if I wanted to . . . but I won't." He borrowed a can opener, ate his canned sardine lunch in the garage, returned the can opener, and helped himself to oats for his horses before he left. "He was so drunk," Mrs. Spencer recalled, "That when Clifford came home he wanted to know what kind of party I'd been having in the garage. It smelled so terribly of moonshine." The sheriff caught up with and arrested the fugitive, who had shot two fingers off his victim's hand during a Christmas Day quarrel, the next day.

The original Wapiti Ranger Station building was improved over the years. When, toward the end of the Spencer's years at Wapiti, the CCC added a long hoped for fireplace and a bathroom, Mrs. Spencer "took a bath about three times a day ... it was such a treat to have a bathtub and play in hot water." After Ranger Spencer was transferred to Colorado in 1937, two bedrooms were added in the early 1940s and a utility room was added in the early 1950s.

Wapiti Ranger Station remained a ranger district headquarters until 1957, when the Wapiti and South Fork districts were combined into a larger Wapiti Ranger District and the office moved to Cody. The historic ranger station continues in use as a guard station and work center from which campgrounds are maintained and at which pack animals—horses and mules—are kept.

FOREST SERVICE MEMOIRS

Foster Steele

At the beginning there were two fields of endeavor. One was the sphere of the professional forester whose job it was to introduce the principles of scientific forestry in all of its ramifications. The other was the domain of the nonprofessional members of the team whose duty it was to gain acceptance of these principles and practices by industry and the general public.

It was in the domain of the nonprofessional group that I landed in July 1905, when Homer E. Fenn, a practical forester and son of the late Major Frank A. Fenn, was sent to St. Anthony, Idaho, to build an organization and put the newly created Targhee Forest Reserve under the administration of the government. In 1907, I assisted Jno. T. Wedemeyer, another nonprofessional but a man with considerable experience in ranging livestock, in organizing and placing the Caribou Forest Reserve under administration. Our headquarters were at Idaho Falls, Idaho. In between times I was assigned to the job of working out boundaries for proposed additions to reserves already established in that area from the official plats of surveys made by the GLO and filed in the General Land Office at Blackfoot, Idaho. These boundaries were accepted and became a part of President Theodore Roosevelt's final proclamation made just before he signed the bill taking that power from him. These experiences taught me the value of the rustic and his place on the team of college bred professional foresters.

The name “Forest Reserves” was soon changed to “National Forests,” but the fact that old-timers, after fifty years, still call them “reserves” is proof of the lasting effect the proclamations of “Teddy” Roosevelt had upon the local forest users.

In the Intermountain Region, from early spring to late fall, the job required many long hours in the saddle over country that knew no trails. I have ridden horseback thousands of miles with my grub, blankets and working tools diamond hitched to a pack horse—or merely with what I could tie on behind my saddle for short trips of only a few days. Fighting fire with what tools we could muster for days on end trying to persuade timber operators to comply with rules laid down by Washington and the more hum drum business of coaxing sheep headers to “bed out,” and the general public to be careful with fire. Laying out timber sales, marking and scaling also occupied much of my time.

I experienced some exciting adventures during that early period of forest administration. Perhaps the most exciting was when a misguided character took a couple of shots at me with a high power rifle. His bullets clipped the heads off of the tall rye grass in which I was standing, not more than a foot in front of me. I have never figured out whether he was trying to scare me or was just a poor marksman. I didn’t stick around long enough to learn. I do know he succeeded in the first premise. Generally, however, the work was accomplished without violence. Shooting was usually confined to feuds between cattle and sheep men. Cattlemen would shoot up the sheepherder and his camp and destroy his herd by driving them over a high cliff. This feeling between sheep owners posed quite a problem in range management. Sheepmen felt they were entitled to range privileges within the forests which sometimes conflicted with what the cattleman thought was his range. It took a great deal of diplomatic maneuvering to establish peace between these conflicting interests and establish the practice of “common use” on range allotments.

THE FIRST RANGER EXAM

Submitted by John F. Lewis

I dug the following out of my voluminous files on the Sierra National Forest in California, which is a letter from Audie Wofford in his own words and spelling:

"When I started to work April 1, 1905, under Mr. Shinn on the Sierra Reserve, there was quite an additional force added to the personal. 1904—there was Gene Tulley, Jack Noddin, Boot Taylor, Edd Mainwering—do not believe they had job desinations then. Those working on the Reserve before 1905 did not have to take ranger examination. All entering the Service in 1905 had to take and pass the civil service examination. The first civil service held in the West of California was at Ellis Meadow summer camp on Peckinpah Mountain, 7-1/2 miles east of Northfork. As I remember, there was 9 of us takened it on June 15th, 1905—Jim Gettis, George Prettyman, Barber, Bill Gibbons, and myself—cannot think of the other's names. The exam consisted of surveying 40 acres (timber land), each student taking his turn at the compass and chaining, then drawing map, estemating timber in board feet. Riding saddle horse over 200 yds of road, walk horse up to end, trot him back, gallop horse up to end, then run horse back as fast as he could run—us mountain boys could do that OK. Then the examiner would pick out a tree 12 feet to 24 feet in diameter for each contestor, set a stak 60 to 70 feet from the tree in the direction the tree did not lean and tell each contestor to fall the tree with ax (no saw allowed) so tree would drive the stake he had placed. If I remember correctly, three of us knocked our stake down. Some of the boys trees went almost opposite direction, while poor Barbers tree was one of the largest of the bunch—he had only managed to hack a small grove all around it, so Mr. Langel, the examiner from Washington, D. C., told Jim and myself to finish Barbers tree for him wich we did in approx. 20 minutes. The next contest was to put a pack on a horse without kayks. There was a pack saddle, lash rope, 2 sack rolled barley, camp cooking utencils and dishes for two men, 1 bed role and previsions made up of part sack flour, several caned foods, and several other ods and ends, the pack was to stay on in good shape while the examiner turned the horse around twice and led horse 100 feet. Most of us got by OK on this test, but poor Barber, after taking 3 times as long as necessary, when the horse was turned around, part of the pack fell off, and the rest hung under the horses belly. Then the last and toughest part of the exam, at least for me, was two full pages of practaly questions on forestry and other Government regulations, tough for me becaus I had to stop going to school before I had goten through fourth grade."

John F. Lewis added the following information:

With respect to the account of the first ranger examination, the forest inspector was Inspector Langille (for whom a mountain peak has been named in the High Sierra).

Stewart Edward White, the famous author, must have been present at that affair for he describes it almost exactly in his book, *The Rules of the Game*, which tells of the coming of forestry to what is now the Sierra National Forest. White and his wife had a small cabin about three miles from Ellis Meadow, the summer headquarters of the forest supervisor, and knew the Forest Supervisor Charles Howard Shinn (a onetime roommate of Woodrow Wilson at Johns Hopkins) and his wife, Julia Tyler Shinn as friends. In fact, in his book, *The Cabin*, White not only features them but dedicates his book to them.

TALES OF YORE

Howard T. Phelps

In the early days of the century, the site of the Twisp Ranger Station was a favorite fall camping place for Indian women who came to trade for deer skins from which to make moccasins and gloves. During the years we lived at the station, in the twenties, Matilda Wapato came up every fall from the Chelan country, camped on a corner of the station grounds and exchanged her skill at moccasin and glove making for deer hides. The first time she camped there in an old battered lodge, the poles of which were covered with a weathered assortment of canvas and hides, my wife asked if it would be considered proper and in accordance with Indian custom to call upon her. I assured her that it could do no harm and might be welcome. Armed with a homemade cake as a peace offering, we accordingly called upon the old woman and her companion, another old Indian woman of about Matilda's age.

We and the cake received a warm welcome. We found the two old women sitting on opposite sides of a small fire in the center of the lodge. A pot of meat stew was bubbling over the flames. Both were busily scraping the hair and meat from deer skins. The dirt floor of the lodge was well sprinkled with tufts of deer hair and I surmised that the stew was seasoned with it as well. A place was made for us beside the fire, the cake was sampled and we were made to feel very welcome.

Thus began a friendship between my wife and Matilda Wapato which lasted through our years at Twisp. Every fall the old woman, usually with a companion, would appear and pitch her lodge for a stay of two or three weeks. Every year she would welcome my wife and her peace offering. Bea's homemade bread was especially relished. Matilda was blind in one eye and her eyes were weeping and inflamed from countless smoky lodge fires. She traveled behind a gaunt Indian pony hitched to a battered buckboard, usually with another pony trailing behind. We never found out how she lived the rest of the year but we surmised that she had resources well beyond the pony and the buckboard and the lodge and that her fall camping trips were, in a way, a vacation and a reversion to tribal custom. She had a son who was a college graduate and a reporter on the *Wenatchee World*, so we were told. She always carried a letter from her son in the bosom of her dress which she would ask my wife to read to her over and over.

One fall Matilda and her companion were camped as usual in a corner of the station grounds and where my wife could watch them from a window of our home adjacent to the station. The usual amenities had been accomplished and the two old women had settled down to treating deer hides and making moccasins and gloves. Our daughter learned to walk in Indian moccasins made by the patient fingers of Matilda Wapato and soft, lovely little moccasins they were. The first late fall storm of the season had covered the ground with three or four inches of wet snow. Smoke from the top of the conical lodge trailed down wind in wispy streamers. The day was grey and overcast. Suddenly the two old women emerged from the lodge engaged in violent altercation, voices shrill, arms waving. Soon they attacked the lodge, stripping half the canvas from it and packing the canvas, some bedding and utensils on the back of one of the Indian ponies. Then Matilda's companion hoisted herself on top of the load and headed down the valley, a lonely figure against the wintry landscape. It seems they had decided to split the blanket and the lodge and everything else along with it. Matilda still had her pony and the buckboard and she patiently set about rebuilding the lodge, removing some poles, making it smaller and adjusting the remaining canvas to fit. Soon Matilda was in business again, snug by her fire in her lodge.

The young lookout from Lookout Mountain, down to Twisp for supplies, dropped into the ranger's office before heading back to his aerie on the mountain. Jack Burghardt, the District Ranger, talking to one of his guards about the day's work, glanced at him somewhat grumpily and returned to his discussion. I grinned secretly. The lookout was a college student and in Jack's book, all college students were wet behind the ears. To compound the felony, he was sparking Jack's daughter, Mary. He was a personable young chap, but obviously embarrassed and ill at ease. I had a hunch about the purpose of his visit and sympathy for his predicament. There was to be a dance in town that evening. The situation was pregnant with unforeseen possibilities and I watched with concealed amusement. I knew that Jack wasn't in a particularly good humor.

The young chap fidgeted but waited patiently until Jack had concluded his discussion with his guard and then said deferentially, "Mr. Burghardt, the fire hazard is very low today. That rain last night was enough to dampen things down pretty good. I'd like awfully to stay over until tomorrow morning. I could get back very early in the morning. I haven't had a day off yet this season. Do you think it's necessary for me to be on lookout today?" Jack grumped, "That's what you hired out for, wasn't it?" He clapped on his hat and left. The screen door banged. And that was that! No day off! No girl! No dance! Those were the days!

My pack trips into the back country of the old Chelan usually coincided with the open season on blue grouse. They were made in late September or early October, after the fire season and it required little ingenuity to fit them together. A twenty-two rifle completed the strategy. A noble bird for the pot!

One fall we made rendezvous at Eight Mile Ranger Station—Frank Burge, the District Ranger; Glenn Mitchell, Assistant Supervisor and myself, and headed our pack string up Boulder Creek. Here we were joined by Fred Wehmeyer, District Ranger of the Conconully District, upon whose district our trip would encroach and whose dry wit and salty humor made him doubly welcome. We traveled north from there, up past the Windy Peak country to Horseshoe Basin. En route we holed up at North Twenty-Mile Lookout to sit out a severe fall snow storm. Glenn was a fabulous camp cook and he rummaged around among our supplies and some odds and ends left by the lookout, intending to bake a pan of biscuits but some way he contrived a cross between biscuits and cake. Like the cross between a mare and a jack, the result was wonderful to behold—a sort of chocolate cupcake that melted in the mouth. Glenn was a sure asset on a pack trip.

We laid over a day at Horseshoe Basin for range inspection in a pleasant camp beside a clear, cold stream. Someone, probably a herder or camp tender, had felled a fair sized fir adjacent to the camp and its needles were tinder dry. We suggested to Frank that it was an unsightly fire hazard and should be burned, and since we were now on his district, he was the only man among us with the authority to set it afire. Frank listened skeptically, making an occasional pointed and ribald remark disparaging our ancestry. Finally, goaded by our insistence, he scanned the clear evening sky, tested the wind and reluctantly and I think against his better judgment, set a match to the tree. It went up with a whoosh. Unfortunately, a freak gust of wind caught it as it was going good and carried the flames toward camp, crowning out a small fir near camp and showering sparks and embers over the camp area. One of the pack horses came jackknifing through camp, scattering dirt and equipment? Frank, beating at his clothes to dislodge sparks while sidestepping the pack horse and trying to rescue his bedroll, swore at us, the pack horse, the fire and the wind with picturesque fluency and vigor, while we, almost helpless with laughter amid the general pandemonium, tried to salvage our own gear. Ah! Me! I haven't had such a muscle tightening belly laugh since. The remainder of the trip was enlivened by caustic invective against know-nothings who desecrated a man's district and darn near burned up the camp.

Here, Fred left us to return home, having assured himself that we committed no vandalism while on the fringes of his district. Then we headed west along Bauerman Ridge, the trail generally at timberline, to Cathedral Lake set so beautifully in its natural amphitheater; over Windy Pass where we passed an uncomfortable night in a snow storm, and so to Spanish Meadows. Here, Frank and Glenn spent a day riding the adjacent range, while I made a sashay west across the Ashnola to Sheep Mountain to look out its possibilities as a fireman lookout point. I found a light tender for the Geodetic Survey staying in a cabin dugout at the base of the mountain, sporting a summer long growth of beard, starved for companionship and fearful that he would be snowed in before being relieved.

The next morning we broke camp and headed down the Andrews Creek Trail and then up the Steep Way Trail to Coleman Ridge. At the north end of the ridge we parted for the day, Glenn and Frank to ride the Coleman Ridge range while I climbed Remmel Peak to see if it would provide detection for the country north and west. We agreed to camp that night in the meadows at the south end of Coleman Ridge, where it breaks down to the Chewack.

It was then, I should guess, about 11 a.m. I climbed Remmel, made a visibility map with come difficulty in the face of a raw wind and returned to my horse somewhere about the middle of the afternoon. I headed south and soon picked up the tracks of my companion's horses and the pack stock. The weather had been threatening all day and soon it began to snow, large wet flakes slanting on a cold wind. The tracks were soon covered with snow, the country was strange to me and so I buttoned up the collar of my mackinaw against the snow and wind and headed for the south end of the ridge and, I hoped camp.

Dusk was falling when I reached the breaks to the Chewack. The ground was covered with three inches of wet snow and it looked like a blustery night ahead. I scanned the openings below me, separated by clumps and fingers of timber, for the glow of a campfire or the sound of horse bells. Nothing but snow and wind and rapidly approaching darkness. I rode the upper fringes of the openings for half an hour, stopping frequently to listen for horse bells, with no luck. I had about decided to pick me a down dry log in the shelter of a clump of timber, build a fire before the light was all gone to sit out the night, when the wind brought me the faintest tinkle of a horse bell. I listened, motionless, my horse restless and uneasy and had about concluded my imagination was playing me tricks, when ever so faintly on the wind came another tinkle. Angling down toward the sound, I soon caught the glow of a camp fire and in a few minutes reached the snug camp Frank and Glenn had made, screened from the wind in a clump of timber. I was cold and stiff, my stomach empty and growling for food and that cheery camp and the faces of my camp mates, who were becoming a bit concerned, looked right good to me. Glenn, always a man to look to the comfort of his horses before his own, helped me with my horse while Frank stirred a savory pot of mulligan bubbling over the fire. Soon I was hunkered down with a heaping plate of mulligan stew and hot black coffee.

That, gentlemen, was without doubt one of the best and most satisfactory meals I ever ate.

What was it?

Blue grouse mulligan, of course.

EARLY DAY REMINISCENCES

A.O. Waha

My first Civil Service appointment as a forest assistant at a salary of $1,000 per year was effective July 1, 1905. Previous to this appointment I had worked for several years as a student assistant in a number of eastern states on timber surveys and related work. So I was delighted when I was assigned to the Gila Forest Reserve in southwestern New Mexico for I was eager to see new country and work on a reserve.

Silver City, a mining and cattle town, was the headquarters of the Gila. R. C. McClure, Forest Supervisor, was a pompous, potbellied gentleman from Kentucky who affected the dress of a Kentucky colonel by wearing a black broad brimmed felt hat, a flowing and black bow tie, a black Prince Albert coat (in those days we called them for some unknown reason "go to hell" coats), and gray checked trousers. His characteristic pose was to hold his cigar in his mouth at an angle of about 45 degrees and keep his thumbs in the armholes of his vest.

The transfer of the reserves from the Department of the Interior to the Department of Agriculture had been made effective February 1, 1905. McClure was without Civil Service standing, having only a political appointment. The day after my arrival on July 18, 1905, I was told by McClure that I was to take charge of all timber sale work. The Gila Reserve at that time comprised about three million acres, a territory about 150 miles long and 75 miles wide. There were five forest rangers. With Rangers Geo. Whidden and Jack Case, two hard-boiled, hard riding rangers who were much older than I and who had been around considerably. Whidden had been in the Army previously while Case had been an itinerant cowpuncher. I shall always have a sneaking suspicion that McClure had either told them to kill me off or that they themselves had planned to do so. The first day we rode 38 miles over the burning plains country. I was quite inexperienced in riding and, as a result, I was so done up after this long ride that I could scarcely crawl out of the saddle. On reaching the ground it was extremely difficult to move, my muscles being so sore. After the first week in the saddle I had become hardened to riding and felt quite at ease in handling my horse.

While working on one of our larger timber sales on March 7, 1906, in scaling and marking timber, I found that the whole "damn" Miller family, as they were called, was engaged as help—the mother, daughter and son were felling trees and bucking them into logs while the "old man" was the bull whacker. All of the members of the family either smoked or chewed tobacco and I was told that their tobacco bill ran even with their flour bill. They had come out from Indiana in a Studebaker wagon and scarcely knew what it was like to live in a house. They never used a tent but instead stretched a wagon sheet over some logs or boards. I was told that while the Miller family was over on the "Blue" (Blue River in western Arizona) the Miller family was camped on one side of the stream while another wagon outfit was camped on the other side. Mrs. Miller yelled over and asked how the kids were and upon being told they were not so well, she asked, "What kind of a wagon do you use?" The other said they used a "Bain." "Well," said Mrs. Miller, "We have found that our kids were always more healthy in a Studebaker."

I had often read stories of the wild West in which cowpunchers would ride their horses into saloons and stores, but here in Alma I actually saw it done. One young puncher who was quite drunk rode right up to our table in the back of the store, where we passed the time of day, and he then turned his horse and rode out. It all added a bit of color to a setting that already was plenty colorful. The saloon remained open all night and there was a hot time in little old Alma.

I always enjoyed stopping over with Ab Alexander and Shack Simmons at the N Bar Ranch. Ab owned cattle and Shack worked for him. They were bachelors and lived very simply. Their log cabin was very comfortable; the meals usually comprising yearling beef cooked in a Dutch oven in deep fat, sourdough biscuits, potatoes and thick canned milk gravy and coffee were just what an outdoor man required and liked; and there was always plenty of horse feed, so I never passed up the opportunity of stopping with them. Besides they were good company. Ab was quiet and reserved while Shack was inclined to be quite outspoken. His language was most picturesque.

A forest guard had stayed with them at the ranch for awhile whom Shack did not like. It seemed that he bragged too much about being a 32nd degree Mason and in other respects showed that he was not the kind of man to be admired by a man of Shack's temperament. So when Shack was later telling me about this guard whose name has long since been forgotten, he said, "Hell, he hasn't brains enough to grease a gimlet. You can knock the pith out of a horse hair and put his brains in and they would rattle like a peanut in a box car."

Life on the Gila was never dull and when I was transferred to Albuquerque as an inspector in June 1907, but not before I had had a detail in the Washington office, it seemed like giving up home ties. Not only was it hard to leave my friends but also my horses. I had become so greatly attached to my horses that it seemed wicked to sell them and not know how they would be treated.

From June 1907 to December 1908, our life was most hectic. In making inspections of forests in those days, we followed a very detailed inspection outline that had been prepared in the Washington office, which was most inclusive. It was contemplated that every timber sale, grazing allotment, special use permit, claim, etc., should be inspected, together with a review of diaries, to determine if time was being spent advantageously and also a check of office procedures, files, etc. Some of our reports were indeed voluminous—200 to 300 pages.

Certain it is, however, when we completed an inspection, we surely had firsthand knowledge of conditions and personnel.

I well recall my first real inspection trip made of the Jemez, Pecos and Taos Forests, which were later known as the Santa Fe and Carson. This inspection required 30 days in the field and about a week in the office going through the files in securing data in preparation for my trip. I covered an estimated 800 miles on horseback or an average of about 27 miles per day. To go to all of the places necessary and check on the work being done or contemplated required long hours and it was often necessary to ride well into the night and also get on my way very early in the mornings. After having been accustomed to riding my own horses, it was tough to have to depend on rented horses; the more I saw of rented horses, the more I appreciated and loved the horses I had owned previously. My longest ride in any one day on this trip covered 60 miles. It was much too far to go in a day, but in the absence of a suitable place to stop over for the night there seemed to be nothing else to do. To reach Antonito, Colorado, which was my destination, the road traversed a low country that had been pretty well flooded by the heavy rains during the day. I struck this boggy section of the road after dark when my horse was so tired he could scarcely make headway on firm ground. After much floundering about we managed to get through and at about 10 o'clock I reached Antonito, most anxious to get some nourishment since I had had only one sandwich since breakfast which I had eaten about 5 a.m. But, much to my dismay, the stores and restaurant were closed and the small hotel had no dining services. So there was nothing for me to do but go to bed hungry, but I was too dog-tired and sleepy to let this bother me unduly.

I believe it was in November 1907, when the headquarters of the Chiricahua Forest was transferred from Paradise, a small mining camp in the mountains, to Douglas, Arizona, a copper smelting town on the Arizona-Mexico border. A leftover from the Interior Department, by the name of McGlone, was the Forest Supervisor. He was a misfit and an inspection made by me proved conclusively that he was almost a total loss in the position he held. I happened along later in the same month and McGlone and I took the mail stage from Douglas to Paradise.

When we approached the post office and general merchandise store, on the porch of which there were a number of loiterers, we were greeted with a yell that went like this—"Moldy bread and moldy hams. To hell with the Forest Service," followed by the typical cowpuncher yell. This was a bit disconcerting, to say the least, and after we had reached the place where we were to stay overnight, I learned upon inquiry from McGlone the significance of this most unusual greeting. It seems that after he had moved his office to Douglas, he had been interviewed by a newspaper reporter who had been interested in learning the reasons for transferring the headquarters. After stating the principal reasons, he went on with disparaging remarks about living conditions in Paradise, particularly the poor food that one had to put up with and made reference to moldy bread and moldy ham. Being quite naive and not having had experience with reporters, he did not realize that his remarks might appear in bold type in the newspaper. So, naturally, when the residents of Paradise read the paper and saw what McGlone had said (unfortunately the article gave the impression that the compelling reason for the transfer was due to the poor food he was able to get in Paradise, mentioning specifically the moldy bread and ham), they were incensed to a fighting degree. McGlone never lived down this indiscretion.

In March 1908, I had to go to Springerville, Arizona, the headquarters of the Black Mesa (now Apache Forest) for the purpose of assisting Supervisor Martin in a reorganization of his office. I had made a trip to Springerville in October 1907, so I knew what to expect. One got off the Santa Fe train at Holbrook about 4:30 in the morning and then waited for the mail stage, a buckboard driven by two horses. The first day one traveled 35 miles to Hunt, which comprised but one house, and there one stayed with a Mormon family. The next morning at 5:30, another stage outfit with a hairy old Mormon as driver took you 25 miles to St. Johns, a sure enough Mormon settlement with Lombardy poplar trees growing along the irrigation ditches. Here you were transferred to a homemade two-wheel cart drawn by a mule and rode 35 miles over the rocky, hilly roads, cramped up in a manner that was not conducive to a pleasant disposition.

After about a week's work with Supervisor Martin, I started out early one morning for Showlow, the headquarters of the Sitgreaves Forest. While the distance was 50 miles I was able to rent a fine "up and coming" horse, so thought I should have no difficulty in reaching Showlow by dark. But the day turned out to be pretty bad, high winds and snow storms and muddy roads. I was in the saddle 13 hours—no, it was probably 12 hours for I had to walk and lead my horse as he had gone lame in one foot and was about to give out. At 10 p.m. I decided it was impracticable to attempt going further, so I made camp on a gently rolling ridge on which there was a scattering of pine trees. First, I gathered needles and twigs and built a small fire and then rustled around in the darkness for some good size chunks for making a big fire, since the night was cold and windy. All I could find was a rather large log and, not having an axe, I had to exert all of the strength still left in me to drag it in. After unsaddling and feeding my horse, luckily I had brought a feed of oats on my saddle, I tied him up and then started out afoot to reconnoiter. I felt quite positive that I was on the right trail and also that I must have covered about 50 miles and should, therefore, be somewhere in the vicinity of Showlow. However, after scouting about for an hour or so, I returned to my camp, hobbled my horse, laid my saddle blanket on some pine needles which I had scraped together to soften the malapais rocks a bit, and sat down before the fire to think it over.

It was surely a lonesome night and, while I managed to doze a little, I couldn't sheep. A bunch of coyotes growling about would howl heathenishly and make one feel a bit shivery. I had a sandwich in my saddle bags which I had saved out from my lunch but refrained from eating, believing that I would be more hungry in the morning. And besides, if my horse got away from me during the night, I figured it wouldn't be so good to be afoot and perhaps lost and with no food whatever.

Along about 2:30 a.m. I was dreamily thinking of the good bed I was missing and many other things when I realized that I could not hear my horse. Immediately I aroused myself and started out to find him. Much to my surprise, for I had thought he was about worn out, I found him about three quarters of a mile from camp and making hurried tracks to Springerville. I brought him back to camp, tied him up and let him feed on "post" hay for the remainder of the night. He was really good company for me and I was amused in watching him for he seemed to have me sized up for somewhat of an idiot and, by his looks, I imagined that he was rather sneering at me. I couldn't blame him a bit.

Daylight came none too soon to suit me. I prepared a hurried breakfast, which consisted of a toasted tongue sandwich, then I saddled up and started. After riding two miles I found myself in Showlow. I had been on the hill above Showlow when I was out reconnoitering and could have seen the town if there had been any lights in any of the houses.

RECOUNTING THE PAST—EARLY STOCKMEN

Grover C. Blake

I have been thinking about that summer of 1906, when I was camped up on Old Baldy on what is now the Ochoco National Forest, with a band of sheep. Old Baldy has since had her name changed to Spanish Peak, but that change in name did not change her general appearance very much.

These Blue Mountain timberlands had recently become a part of the national forests. Many stockmen had long summered their flocks and herds upon the rich forage of the Blue Mountain ranges in Oregon, and there was keen competition for the choice grazing areas in those days. Range wars were common and many sheep and a number of men had been killed. Creation of the national forests, followed by restriction of the grazing privileges, was looked upon by most stockmen as a deliberate attempt on the part of the government to deprive them of their free and unlimited use of the land for grazing. They were bitter and antagonistic and forest officers had few friends among them.

A few men had been appointed that year as forest guards to carry on field administration of the new national forests. One of these men was J. D. "Bert" Fine. He came to my camp one day during the summer of 1906, and made the camp his headquarters for some three weeks. He was wearing the first Forest Service badge I ever saw. It was not quite as large as a saucer but several times the size of the largest now in use. I was to wear a like badge later.

Bert and I talked a lot about the new Forest Service and studied the Use Book together. It sounded like "good medicine" to me and I became much interested and in sympathy with the forestry program. The following year I took the Civil Service examination for forest ranger and served thirty-three years.

The range was first allotted to stockmen on an acreage basis. This did not work since some areas were rich in forage while others were largely scab lands of little forage value. Then the plan of allotting the ranges on a carrying capacity basis was adopted and enlarged upon.

There were many complaints and the stockmen blamed the forest supervisor for everything they did not like, including adverse weather conditions. For the magnificent salary of $1,200 per year, the first supervisor of the old Blue Mountains National Forest, which included all of the Blue Mountains west of the Burns-Canyon wagon road, no doubt took more abuse and carried a greater burden on his shoulders than any who were to follow him.

Much progress was made as the Forest Service went through stage after stage of advancement. Those early stockmen have passed on, as far as I know, to the last man and their descendants accepted the Forest Service policies and liked them.

EARLY DAYS IN CALIFORNIA

C. J. Buck

On April 1, 1906, I was assigned as technical assistant to the Klamath National Forest. The forest had been newly created and the people in the remote forest areas were largely

Log slide to river through second growth timber, Kaniksu NF, Idaho – c1905.

antagonistic. The gold mining companies, largely placer, were found in trespass for indiscriminate cutting of government timber. Timber was needed largely for flumes. Some flumes were 15 to 20 miles in length. Government interference was highly resented. To make it all worse, the ardent conservationists had secured an Act of Congress prohibiting the sale of any national forest timber in California except by advertisement for 30 days and bid, and so I had to search out the timber needed by each mine in the future, cruise, map and advertise it. I was frankly informed by one large mine company manager that I would be destroyed, preferably by being shoved into the raging Salmon River except that the government would keep sending others to take my place.

The sending of drunken Indians after me and the treating of me with the contempt to which I was entitled were mild irritations. To cap the climax and give proper cause for their feelings against me as the representative of the new order, they found out from something the supervisor let slip that I came from Yale, and that was the most unkind cut of all. "He came from Yale." That surely was something to live down. I degraded my grammar to the lowest depths I knew. I tried to live it down but the disgrace foisted on me was abysmal.

Meanwhile, I had to furnish a saddle horse to travel about. Fifty cents for myself a meal and the same for the horse at these mining camps.

In working with local rangers there were other jobs to do. Such jobs as going out to some saloon and forcefully closing it down and boarding up the windows. I felt Mr. Pinchot must be a prohibitionist. Or going out on Forest Service business to interview an isolated recluse prospector who had advertised that he would shoot the first forest ranger who came around.

Six-shooters were often kept loosened and handy. Half-breed Indians or Chinamen were far from law-abiding. We found we had appointed one half-breed forest guard who had committed most of the crimes on the calendar, including murder. The sheriff would send a deputy in to look up serious crimes but the wanted man would disappear as much as 15 minutes before the deputy reached his hangout. Bush telegraph—no telephone lines.

The San Francisco earthquake of April 1906, was learned about some 10 days later, although we knew something had rattled the dishes one night at the Blue Lead Mine. I had tramped 25 miles the day before, leading a pack horse loaded with grub, with Bill Durbin, years later the Supervisor of the Plumas. Tobacco chewing Bill with his dry humor and his expressed resentment of the way the local people treated us. In fact he was highly indignant. One day a man asked Bill, though in less elegant language, "Say, Durbin, does a man have to get a permit to go to the toilet around here?" Then Bill really blew his top!

I suppose we represented law and order in a back country where there was no law and no order, except in the mining of placer gold. And when, after the spring cleanup of the sluice boxes, the bag of gold was dumped on the bank floor at Etna Mills, it might take two bank clerks to lift it.

SAVED BY MISTAKEN IDENTITY

Shirley Buck

About the time heat generated by the Homestead Act of June 11, 1906, reached its highest point, I was sent to help Supervisor Cryder, Paulina Forest, check up on his June 11 situation. Our work was going well and without much incident until one day word was received that a squatter in trespass at Davis Lake would be pleased to kill the supervisor if he would just take time out to come over. Supervisor Cryder was not of the timid sort and, besides, here was a case of "saving face" which meant much to him because he had lived in the Orient. After a brief consideration Mr. Cryder said he would go at once and did I want to go along. Unnecessary question! Of course I did not want to go, but here also was a case of "saving face"—my own to be specific. So, speaking in a casual manner which I by no means felt, I said, "Oh, yes, I would like to go. Seeing a supervisor killed will be a new experience."

Well, unarmed, we started. I wondered en route whether the killer, if it turned out that way, would care to have a surviving witness. Our light wagon, which made plenty of noise, announced us a long ways ahead. Rounding a short turn in the road near the lake we saw a man standing in front of a cabin, gun barrel resting on his forearm. He was about 40 yards distant. We drove on up to him and spoke to him. He made a muttering reply not understood by us. He made no move to shoot and I'm sure we made no effort to provoke him. We then drove on about 60 yards and made camp for the night. It would not be the truth to say we did not sleep any, but it would be a great untruth to say we slept much. We saw nothing more of that man and after an early breakfast we were on our way.

It was reported afterward that the man did not shoot, he said, because he did not want to kill someone he did not know and that someone was me.

EARLY DAYS ON THE IMNAHA

J. Fred McClain

I will try to relate some early experiences which happened on my district. The early rangers had large areas or districts to supervise and, in order to cover the entire district, much of which was isolated and with a lack of roads, trails, or even ways, it was necessary to cover the greater part by saddle horse. I would start out on a saddle horse leading my pack animal carrying full camp outfit consisting of groceries, cooking utensils and bed, and armed with the Use Book and a copy of Teddy Roosevelt's Proclamation to bring the public to understand what the intentions were: These intentions were to conserve the timber, forage, wildlife and all commodities on the forests and to get the cooperation of the people in bringing about these ends as at this time range wars were just starting.

Just across the Snake River from my district one sheepherder had been killed and numerous sheep and cattle shot, the sheep by the cattlemen and the cattle by the sheepmen. This made it imperative to bring about as much harmony as possible among the two factions. The early divisions were made in the supervisor's office by the representatives of the stockmen and the forest officers. Maps of the allotments were set up and descriptions of each allotment or unit were written up. Of course, these first tentative maps and descriptions had to be checked in the field. When complaints came in the ranger took the people concerned and rode over the range and settled the differences right on the ground. The men were consulted and most always agreed, but when they did not agree, the ranger settled it in their presence and made the change, if any, on the map. I remember one case where three sheepmen were concerned over the division of some winter range. They consisted of two Frenchmen and one American. It was in February and in zero weather. We rode all day, during which time the two Frenchmen dismounted five times to fight but never struck one blow. This trip proved to be very satisfactory, however, as the lines we established prevailed through the years and we never had any more trouble from these men.

Once it became necessary for me to stop overnight with one of the Frenchmen. We arrived at his headquarters at 9:00 p.m., cold and hungry. They ushered us through the kitchen where the whole meal was in a large iron kettle, steaming hot. I guess I made some kind of facial expression as I smelled the contents of that kettle which consisted of meat, cabbage, corn, rice, beans and garlic; the garlic in the largest proportion. As I emerged from the washroom a camptender motioned for me to follow him into the cellar. He produced a quart bottle of Old Red Top, contents of which I lowered about four fingers. I was then able to stow away a large quantity of the contents of the kettle without even tasting the obnoxious garlic.

There was a squatter settlement on the Imnaha River in Oregon, known by other settlers as the "Squatters," among whom one outstanding man was designated as the "king" of the squatter settlement. These squatters caused quite a lot of trouble as they were always complaining about the cattle which were allotted on adjoining range. One day I finished counting the cattle allotted to this adjoining range and started down the river. Suddenly a cowboy overtook me and said a man, whom the stockman called the "King of the Squatter Kingdom," had come down the load and forbid their driving the cattle up the road to their range. I told him to tell the men to start the cattle up the road. Sure enough there was the man sitting by the side of the road. He bade me good morning and suggested I dismount and talk, which I did. Presently the cattle were driven by us, by their owners. The "King" never said a word, didn't even mention the cattle or men who drove them by. However, he went to great detail in describing how our government should be run and ended up by asking why the Forest Service was taking charge of this country. He said he had never been notified that there was a Forest Reserve. I then took out of my saddle bags a copy of Teddy Roosevelt's Proclamation and had him read it. After doing so, he handed it back to me with the comment, "Freak legislation." I mounted my horse and rode back down the river.

The June 11, 1906, Homestead Act also caused the forest personnel a lot of extra work. An old horse trader who lived near the forest got the bright idea of locating homesteaders on the forest under the June 11 act. He secured some blank township plats and inserted numerous numbers which he told the would-be homeseekers they would be listed if approved by him as he had a contract with the government to that effect. His method was to have the applicant apply for a June 11 application blank and bring it to him; he would fill in the land numbers and have the applicant sign and mail for report by the Forest Service. For his part in the deal he charged the homeseeker $150. On the first one of these applications, when reported on and rejected for listing by the Forest Service, the homeseeker came to me to inquire why his application had been turned down. After hearing his story I told the applicant he had been swindled, whereby he went to the old horse trader and demanded his money back. The old horse trader went to town for legal advice with the result that he came back and paid back the money to each applicant he had deceived.

I should not leave out the old prospector. Generally speaking, he was a very agreeable fellow to meet and sometimes hard to get away from, as his many days of seclusion and seeing so few people made him loath having you go. I well remember old Jim Farland who said he only had six more feet of tunnel to drive until he could strike a very rich vein of free milling ore surpassing any Cripple Creek, Colorado, ore. Two old prospectors I met on the Imnaha River at the Forks, Sam Davidson and J. Roy Harvey by name, put in four years looking for a ledge that had pieces of float but could not locate the ledge. They said this float contained a metal which would be in great demand in the near future. It was in 1907, and they said they were quitting and said if I happened to be on Boner Flat, a high plateau between the middle and south forks of the Imnaha River, to go to their cabin where I would find one quart of old Scotch whiskey with the seal unbroken and one-half quart with the seal broken. Later that summer I crossed Boner Flat looking over some sheep range. I went to this old cabin and, sure enough, there was the old Scotch.

One more little instance which also happened in the Lake Basin country. On one of my trips up into the Lake Basin in early summer I stopped at one of the beautiful lakes which was still covered with snow and ice except for a small opening at the foot of the lake. My curiosity caused me to walk around this opening to see if I could locate any fish. Did I ever get a wet ducking when I stepped on an area of a large snowdrift which proved to be covered with a coating of ice and slid into the lake over my head. I scrambled out and went to my horse for matches out of my saddle pockets, and a hatchet, and built a fire. I took off my clothes to dry by the fire, but not for long. I had to scramble to put my clothes back on again in a hurry as the mosquitoes came by the millions, lit all over my body and began boring their bills into me. Needless to say, I did not wish to be host to their feast. I mounted my horse and rode back down the trail, thankful for a wind which helped dry my clothes by the time I reached the settlement.

THE EXAM

Rudo L. Fromme

Publicity in Spokane papers brought around twenty applicants to the Civil Service examination for forest ranger in the spring of 1907, about half of whom had never been employed on the Priest River Forest. Supervisor Benjamin F. McConnell asked me to conduct the written, or first day's test, and said that he and Dave Coolin would assist on the second day of field tests, particularly the horse riding and packing. We used the combination lodge and dance hall above the general merchandise store for the written test and Ben and Dave helped distribute the first papers. However, they were "out on the town" as soon as the pen work got underway. About midmorning I heard them coming up the long stairway, giggling like a couple of young school kids, which certainly did not add anything to the instructed quietude of this occasion. In they came, each carrying a tin dipper in the right hand and a large galvanized bucket of sloshing beer in the other. I started to throw a block but Ben countered with "Now look here Fromme, these boys are bound to be terribly dry, puzzling over those heavy questions. Just look how wilted Jim looks there—and Pat—and Joe—and—." More cheers from the McConnell contingent. It left the examiner nothing else to do than to quickly announce, "Everyone remain in his place and the beer will be passed around."

Soon after Ben and Dave had departed with the empty buckets, Ranger McEwen arrived to spell me for an early lunch, after which I slipped over to the railway station to ask Bill, the agent, to connive with me in a scheme which I hoped would throw a scare into Ben. I wanted him to sober up for his agreed part in the diamond hitch part of the next day's field tests. Bill fell in with the idea at once. He had previously tried to council Ben against excess drinking. Not long after I had returned to the exam room, McConnell came puffing up the stairway, beckoned me aside and whispered in quite serious vein that Bill, the railway agent, had sent for him to tell him something that looked rather suspicious. He had delivered a telegram that morning to a stranger at the Beardsley Hotel. He didn't know what was in it 'cause it was in government code but he felt that his "good friend Ben, ought to know about it and lay low just in case." "Gosh," I injected (or maybe something more emphatic), "Maybe the guy's a Civil Service inspector." "Yeh," slowly nodded Ben, as he wiped the beads of cold sweat off his forehead, "That fellow Bill, is a pretty grouchy sort, so I don't believe he's kidding. He said he didn't notice the guy 'specially, except that he was rather tall and dark and was wearing a black overcoat."

"Well," said I, "The best thing you can do is get right to bed. If the guy comes up here, I'll tell him you're down in bed with a bad cold." "Fine," he responded, "Me for the hay," and, although still "green around the gills," he hurried away apparently much relieved. I also, felt relieved, in fact I spent much of that evening congratulating myself on my cleverness, when, behold—the plan jammed and boomeranged. I was just about to retire for the night when one of the prospective new rangers came with the news that Ben and Dave had been "on an awful toot" all evening, scaring the hotel and downtown section with pistols in hand, looking for a tall, dark stranger in a black overcoat. Fortunately, things had quieted down when we walked back. We found Ben leaning against MacIntosh's bar with "a final nightcap" in his hand, looking lonesome. There was no Dave. Dave's wife had arrived to put a clinching damper on their spree by pouring Dave into the family buckboard and driving off into the night toward Priest Lake. Ben promised me with bleary eyes, interrupting hiccups and boozy pats on the shoulder, that he would now go straight (?) to bed and be on the job "tomorrow." When I stopped in the office next morning for compasses, tape, etc., there was the supervisor sound asleep, in his badly mussed uniform. I bedded him down with less clothing, then with the help of McEwen to grade the horse mounting, dismounting, riding and packing, got the field tests completed and the papers off to Washington—with McConnell's shaky signature as "Examiner in Charge."

CITY SLICKERS VERSUS HILLBILLY RANGERS

Loran Cooper

W. B. Milbury and Walter Jones, Siskiyou Rangers, attended the short term forestry course at the University of Washington in 1907, along with my uncle Elmer Bribble, Gilbert Brown and a number of other old-timers. They could tell some great stories about their experiences on the campus and especially about when the college boys undertook to haze the "hillbilly rangers." When the would-be hazers crashed the door to the foresters' quarters, they found themselves facing Milbury and Jones, each one with a Bowie knife in one hand and a pistol in the other. A few shots into the ceiling over their heads and one look at Milbury and Jones with their long hair and walrus type mustaches and the boys lost all desire to have fun at the expense of the hillmen. My uncle said that in the ensuing mad scramble to get out of the place some of the students lost their hats, which no one ever subsequently claimed.

MY FIRST FOREST SERVICE JOB

Melvin M. Lewis

I well remember the first job that I worked on, April 1, 1907. M. J. Anderson, then supervisor of the Siskiyou National Forest, informed me that I had been selected for a job with the Forest Service and was being assigned to boundary survey work with Fred Merrill and George Woolridge and that we should get together our camp equipment and food supplies and arrange for transportation out to where our work was to begin. We hired a livery team for $3.50 and arranged for an early start the next morning. No expense accounts in those days! The service did not even own an ax or a shovel; we supplied our own for the little old Use Book stated that those should be a part of our equipment.

On arrival at the end of the road on Slate Creek, we moved into an old prospector's shack near where our work was to begin. It was nip and tuck for a few nights to see whether we stayed or turned over the place to a colony of skunks that were denned up beneath the floor. We soon left this camp to the skunks and moved our camp three or four miles out along the line. Back packing was the way we moved camp.

In those days from 8:00 a.m. until 5:00 p.m. on the job constituted a day's work and the work week was six days. Sunday was reserved for shaving, washing clothes and taking a bath in the creek—if you felt tough enough to withstand the shock of the ice cold water.

We spent nearly two months on the job and during this time met up with an old miner by the name of Wilson who had a mining location adjacent to the forest boundary and who seriously objected to our running a line across his claim and threatened to feed us a dose of goose shot unless we cleared out. We informed the old-timer that we did not crave the dose of shot and, since it was nearly time to quit work for the day, we returned to our camp and formulated plans for dealing with the situation.

It was agreed that next morning Woolridge, who was somewhat of a daredevil, would buckle on his old six gun and go back ahead of Merrill and me and, if possible, separate the old man from his shotguns. When we arrived some minutes later, we saw George and the miner some distance from the cabin looking in some bedrock where he had done some mining. The old miner did not discover us until we had nearly completed running the line along the property. When he saw us he started for the house. It was then that George saved the day by telling the old fellow that if there was to be any gun play that he would be in on it. The old fellow cooled down and finally gave us permission to camp at his place. We hired his old horse to pack our camp into the place where we stayed for nearly a week. The old man had supper with us every night and seemed real sorry when it came time for us to move on. We loaned him our Use Book to read while we were there and told him that we did not know much more than he did about the Forest Service but thought that it was a step in the right direction. We told him that the timber locators would be shut out, which he allowed was a "dad-burned good thing since those polecats had caused him a lot of trouble by trying to file a claim on his mining property." Before we left the old fellow was very friendly to us and offered his belief that the Service was probably a good thing for the country.

FOUR SHOES AND ONE TALE

Hugh V. Anderson

This was one of the few times, that I recall, that my father, Milton J. Anderson, Supervisor of the Siskiyou National Forest about 1908, ever wore his Forest Service uniform, other than to ranger meetings.

We came out of the woods at the Southern Pacific Railroad station between Roseburg and Grants Pass, Oregon, called West Fork. It was from here that the mail was being carried by pack train through the mountains to Illahee on the Rogue River and from there on down to the town of Brookings at the mouth of the river on the Pacific Coast. Our purpose was to catch the next train south to Grants Pass which was the Siskiyou headquarters.

Our uniforms had been left at the small hotel, as we had gone into the forest, so we washed up and dressed up for the return trip. We had about forty-five minutes to spare before train time so we walked over to the railroad station where the natives usually gathered. On our way over we noticed a circle of men struggling with a horse which they had securely tied and wrapped with rope and which they had thrown to the ground. It developed that they were trying to put shoes on one of the animals of the mail pack train.

My father was known as a practical joker and so when he remarked that if he only had thirty minutes he would show them how to shoe the horse. There followed some joking in which it was pointed out that he had the time. In the end several bets were made that he could not accomplish the job. Dad removed his stiff Stetson hat and his Forest Service coat and cautiously approached the animal which he had ordered released from the many entangling ropes. At least ten minutes were spent in getting a hand on the horse's neck where he petted him gently. The frightened animal calmed down almost immediately, perhaps due in part to the fact that he had previously practically exhausted himself. In any event, my father proceeded to pick up that horse's hoofs, one by one, measure the horse shoes to the foot, cold spring the shoes, and tack them on as only an expert could do. He gleefully collected his bets and in his much bedraggled condition, with horse hair and dirt clinging to him, we barely caught the train.

The men at West Fork did not know that my father had served an early apprenticeship as a fancy carriage ironer at the Union Iron Works in San Francisco. Nor did they know he had been a blacksmith in eastern Oregon after returning from the Alaska gold fields. He had become more proficient in the shoeing of horses than in ironing fancy carriages.

The feat he had performed in shoeing the horse was told far and wide in the community. It demonstrated his ability of a kind but it gained their respect of him as a man among them. He and the type of Forest Service men who served in the early period served it well.

THE BEGINNINGS OF RESEARCH IN DISTRICT 6

Thornton T. Munger

Chief Forester Gifford Pinchot had decided to decentralize the Forest Service and set up six district offices with a miniature Washington office in each. I was told to report to Portland on December 1, 1908.

In a letter written in July 1909, I wrote: "One sees Pinchot's name more than that of almost any other public man except Taft, in these western papers and not entirely in a friendly connection. In Portland, *The Oregonian* is trying to keep alive the Pinchot-Ballinger disagreement and sides wholly against Pinchot. But there is a great deal of friendly feeling toward the Forest Service from all sides and most of the complaints come from disgruntled grafters, whom the Forest Service has very rightly opposed."

Pinchot was dismissed as Chief Forester in January 1910, because he appealed to the public to do something to stop the patenting of illegal coal claims in Alaska by Secretary of the Interior Ballinger, since Secretary of Agriculture Wilson and President Taft would do nothing about it. This was considered "insubordination."

But Pinchot's influence continued to be felt through the warp and woof of the Forest Service led by men trained by him and very loyal to him. Policies and procedures he so wisely founded continued to vitalize the Forest Service; the form and language to be used in correspondence, the policy of decentralized administration, the spirit of public service that pervades the personnel, the principle of conservation through use, the concept that a research unit is a necessity for progressive administration.

With some 25 or 30 people from Washington and others recruited locally, the District VI (name changed to region in 1930) was established in 37 rooms in the Beck Building (northwest corner of Oak and 7th, now Broadway). E. T. Allen was District Forester, Fred Ames, Chief of Silviculture (name changed to forest management in 1920 and to timber management in 1935), and under him Julius Kummel and I had the one-man sections of planting and silvics, respectively. Most of the crew were in their twenties.

It is an evidence of the vision, progressiveness and scientific spirit of the Forest Service that even under the pressure to take over the administration of a tremendous acreage of almost unknown and undeveloped public forests with a ridiculously small crew of very young men, research was not neglected. Coincident with the pressing problems of manning, developing and protecting the national forests, studies were carried on of growth and yield, silvical characteristics of the important species and methods of reforestation.

My first work was as a sort of technical roustabout. I organized and catalogued a library as a necessary foundation for any technical work. I looked into the causes of dying timber reported here and there. I answered questions the supervisors and the public asked on technical matters; wrote instructions for the field on various subjects.

In a swing through eastern and southern Oregon in the spring of 1909 (400 miles in horse stages and riding eight different saddle horses), I put in two pairs of plots on small timber sales to test "no slash disposal vs. piling and burning," for that question was then in my mind, though not settled until the exhaustive study 22 years later. I also started a tiny nursery of 1,700 wildlings of yellow pine at the Rosland Ranger Station, thinking that this pumice country would need some planting; also tried some direct seeding there. In this circuit I talked up forestry at seven supervisors' offices.

Believing that the most important thing to know about the Douglas fir region was the rate of growth of its forests, on July 1, 1909, six of us went into the field. We measured trees and stands up to about 125 years old wherever we could get to them by logging trains, horse stages or walking. We moved often and boarded at farmhouses, logging camps and country hotels, sometimes putting up our tent when there was not room inside. A walk of three or four miles to work was thought nothing of.

The study was resumed in 1911. The results gave the first convincing evidence that Douglas fir forests grew at an astonishing rate, but the industry took little cognizance thereof for many years.

In the spring of 1910, I established three permanent sample plots on the south side of the Middle Fork Willamette River in a beautiful 54 year old stand. These were, I believe, the first growth plots of this nature to be established in the West. They have been measured periodically ever since (several times by myself) and yielded most definitive information on the productivity of such Douglas fir second growth.

In 1910, the section of silvics turned its attention to the growth of western yellow pine (now ponderosa pine), and two three-man crews were in the field all summer under Dean Hugo

Winkenwerder in the Klamath region and George Bright in the Blue Mountains. I worked with both crews. In the course of the summer, when in the Blue Mountains with Bright's party, I wanted to visit Winkenwerder at Fort Klamath. Rather than take the trains from Austin to Baker to Portland to Weed to Klamath Falls and thence by boat up Klamath Lake, I decided it quicker and cheaper to go by stage and private conveyance across the state. By horse stage I got as far as Crescent, except for one 50 mile gap, I hired a team and driver and one short auto lift. But there was no stage from Crescent south. I thought to hitchhike but in the one and a half days I waited at Crescent only one automobile went through and he wouldn't take me. So I ended up by borrowing a horse and saddle and riding to Klamath. Such were the problems of transportation in those days.

There being no reliable volume tables for even the major species, a feature of the studies in Douglas fir and pine was the preparation of regional volume tables. Later, with the help of forest assistants on the national forests, we made standard tables of other species hoping to replace the many "local" volume tables made by cruisers for a special job based on just a few trees.

When I was exploring for areas where cedar that we might measure was being cut, I asked a logging foreman about the character of his cedar; was it large or small? He said the trees were about 14 feet in diameter! I thought he was kidding. But when I went up into the felling there were several trees 12 to 14 feet across the stump but they were no good for stem analysis, being hollow and too odd shaped for volume table preparation.

A four weeks circuit in the spring of 1910, started with a voyage on the steamer Ramona from Portland to Coos Bay. Near Marshfield (now Coos Bay), I went over the logging operations of the C. A. Smith Lumber Company with their forester John Lafon, who was trying to install some pioneering good forest practices.

On the sand dunes between Coos Bay and Florence I put in some planting experiments with maritime pine, Scotch broom, willows, etc., to test the practicability of revegetating at least the interdunal areas. Transportation between these points was then by horse stage on the beach at the lowest tide or, as I did a time or two, on foot with shoes off, walking on the wet sand at the water's edge.

From the sand dunes I went over some of the backwoods of the Siuslaw National Forest looking at the results of the recent direct seeding in the deforested fern patches. The ranger at Florence then had a boat, for the only way of getting into the interior was by the river.

Though forest protection always had first priority, the organization was so meager in guards, lookouts and equipment, and often the "attack" so casual that it is a wonder more of the country was not burned up. I, like all forest officers then and now, had a taste of unexpected firefighting now and then. As an example: When in central Oregon in 1908, I ran across a fire two miles from Crescent that had been burning for sometime. The next afternoon I got a shovel and, as my diary states, "fixed all the dangerous places by removing the logs and needles and throwing sand on the vigorous fires and in three or four hours had it in such shape that it will cause no more trouble, I think. The nearest ranger is 21 miles from here, so I felt authorized to take time off from silvical work."

In 1909, when on the Douglas fir growth study, I discovered a 160 acre fire on the Umpqua National Forest on a bad brushy mountain side. I walked 15 miles (round trip) to notify the ranger who was not at home. Next day our crew of three went to work on the fire. The ranger rode 20 miles to get to it and together we surrounded it with a line. That evening I rode horseback several miles to inform the state fire warden of this fire that was on railroad land under his supervision. Then to sleep in the hay in a barn to the tune of cow bells, goat bells, sheep bells and horses munching their hay. After these two strenuous days I wrote in a letter that I wished Senator Heyburn could be with me even for part of a day to disprove his statement that forest officers sit around hotel verandahs in their gay green uniforms all summer.

A DRY FOREST

William G. Weigle

The new Milwaukee and St. Paul Railroad was under construction through the St. Joe Forest in 1909. There were seven tunnels on the forest, one of which was over two miles long. This gives some idea of the number of men employed and also the number of saloons and other unlawful dives, all on the national forest without permit.

The forester sent me an order to clean out these saloons. I sent three rangers to these places to notify the owners that we were requested to close their places and that they would have to go along to Wallace to appear before the district attorney. When the ranger informed me that all but two agreed to comply, I told him that I would be over in the morning. It was 22 miles from Wallace to Grand Forks where the men were located and by trail the entire way through heavy timber. I first went to the sheriff and got two pairs of handcuffs and a deputy sheriff's commission. I had a good saddle horse, so I started about 10 p.m. and reached Grand Forks shortly after daylight. After breakfast we called on the nine men who were supposed to return to Wallace and the two who previously had said they would not go but who now agreed to go. With this understanding I started back to Wallace over the trail. The rangers and the men had to walk three miles to the Northern Pacific Railroad at Taft to catch the train to Wallace. Learning the Northern Pacific had an accident down the line, the rangers had to hold the men at Taft three hours before the train finally left. I reached Wallace before they did. The men appeared before the district attorney and instead of him finding them guilty he turned them all loose. I re-arrested them and took them to Moscow where the court was in session. They were all found guilty and fined for running a saloon on the national forest without a license and were ordered to close up.

THE FIRST FORESTRY CONFERENCE

Grover C. Blake

The District 6 first real forestry conference was held in 1909 (November 22 through 26), when the personnel of the Deschutes, Malheur, Whitman and Umatilla Forests gathered at Mt. Vernon Hot Springs for a five day discussion of the current problems and general policies. These four forests had boundaries which differed greatly from those forests of the same names of the present day. We were three and one half days by horseback from the Prineville area to the conference and another three and one half days returning, so we were seven days going and coming and five days in conference including two Sundays and Thanksgiving Day. The supervisors present were Cy J. Bingham of the Malheur, Thos. E. Chidsey of the Umatilla, Henry Ireland of the Whitman and A. S. Ireland of the Deschutes. Henry and A. S. Ireland were not related.

REMINISCENCES OF A HOOSIER

Harry White

That summer of 1909, a couple of former Hoosier schoolmasters were batching in the outskirts of Salida, Colorado, and often saw a forest ranger riding past their cabin. My buddy remarked that he seemed to have a nice job—nothing to do but ride around. It seems that some of the early employees didn't have a much better idea of what a forest ranger was for. E. T. Allen once said that when he received his first appointment and appeared before D. B. Sheller, old-time supervisor of the Washington Forest, and asked about his duties, D. B. scratched his head awhile and finally said, "You're a ranger aintcha? Well, get out and range."

So it was that on a day in June 1910, I got off the train at Wenatchee, Washington, and took the boat up the Columbia River to Chelan Falls, forty miles and seven hours away. From that point a four horse stage coach took me another five miles to the little town of Chelan at the foot of the lake. Little did I suspect that beginning the next day I would be on the Forest Service payroll for nearly thirty-nine years.

It must be difficult for men of a later generation to visualize conditions as they were in the national forests during the first two decades of the current century. In our little forest they were crude indeed during the five years I was there. We had few roads, good trails, telephone lines and buildings. The annual allotment for improvements, including maintenance, was generally less than $700, plus a very small amount from the ten percent road fund. A windfall appropriation in 1908, had permitted the building of a fifty mile telephone line up the lake, but otherwise communication was over a few farmers' lines. Until 1914, not a single lookout was established. The few guards rode the trails or cross country and looked for smokes from vantage points along the way. If a fire was seen at a distance, they could judge its location only by using a small crude compass and by their knowledge of the country. Then it could be reported only by riding a long distance. As I look back over the later years, when large fires burned on both sides of the lake, it is a wonder to me that the whole country didn't burn over in the earlier years.

The salary situation in the early years was discouraging. We had what was called the statutory roll, which meant that positions were listed by grade and salary in the Appropriation Act. Each district had a certain number of positions in each salary grade and promotions depended on vacancies in higher grades. When I entered the Service, Supervisor Milham and many other supervisors were getting $1,800 and nearly all rangers $1,100. Forest clerks got $1,100 and $1,200, with two or three positions a little higher. After the United States got into World War I, the salary situation became almost intolerable. Albert Cousins got tired of handling fiscal control for the statutory salary of $2,120 and went to Ogden as property clerk at $3,000. His successor, H. I. Loving, got so discouraged that he quit for awhile. With the increased activity in Portland and other cities, stenographers and clerks began leaving us for private industry or to enter newly created bureaus, which were not hampered by the statutory roll and we couldn't fill their places because we couldn't pay more than the entrance salary of $900. When my successor on the Rainier resigned to enter private industry there was only one name on the forest clerk register and I had no hope that the man would accept appointment, but he did. His name was George Griffith. The increased cost of living due to the war soon caused the statutory system of fixing salaries to be abandoned.

In 1916, W. B. "Bush" Osborne went from the Mt. Hood to operation to work on fire control activities. Recognizing that lookouts must have a more accurate instrument for locating fires than a simple compass, he had invented the firefinder that bears his name and he continued to develop it into a fine precision instrument. He also designed nested mess outfits, which were copied in other parts of the country and finally were covered by standard specifications, as were the firefinder and the adz hoe, an excellent digging tool designed by him as a modification of the hazel hoe. It was about this time also, that he put together a lightweight, three day balanced ration for smokechasers, which was the basis for continuing development in the field. A few years later he got the idea of taking panoramic pictures from lookouts to aid in spotting fires and borrowed a couple of cameras from the Geological Survey to experiment with. When these had to be returned he did an outstanding job of drawing plans and specifications for a precision instrument which he named a photo recording transit. With this instrument, pictures were taken from all lookouts and used not only for the original purpose but for mapping detection coverage. In 1918, just before leaving for an officers' training camp, Bush turned out the copy for the Fire Fighting and Lookout Manuals. I remember that very well, for the proofs came from the printer after Bush had gone and I had to read proof on them, thereby learning much about smokechasing problems and firefighting technique.

While working with the rubber companies, Bush developed the lightweight, single-cotton jacketed, rubber-lined hose that would stand better than 600 pounds hydrostatic pressure per square inch and still was very much lighter than the standard mill hose in common use around logging camps. In the later years the engineering laboratory under Ted Flynn, played an important part in equipment development. The laboratory men undoubtedly accelerated the development and improvement of chain saws, not so much perhaps, by their own ingenuity as by the influence their ideas and work had on the chain saw manufacturers.

There was no district office fire plan in those days, no fire dispatcher on the telephone day and night and no fire cache. The district chiefs were listed in priority order to receive night telegrams and phone calls. Long distance telephone service was poor. There were no busses and this continued into the thirties so that firefighters, hired through private employment agencies in Portland, had to be transported by train. In the war years and for sometime after the war ended it was almost impossible to hire good men. I remember signing up 50 men for the Santiam in 1919, of which Supervisor Hall was able to use only 17. Later, when calls kept coming for men on the Portland Creek Fire on the Cascade, we hired 50 more of the hoboes and sent them to Eugene by train in charge of Julius Kummell. Julius wasn't able to get any of them to the fire; the long hike that had to be made was too much. It was on that fire that Chief of Operation A. O. Waha and several other men had to immerse themselves in the creek to keep from being burned. The labor shortage was still so bad in 1920, that Supervisor Fred Brundage had only about 60 men on a 7,000 acre fire on upper Lewis River because additional men worth sending to that isolated area were not available.

The first portable fire pumper I heard of was a heavy "one-lunger," called by the manufacturer the Wonder pump. In 1915, G. F. Allen and his boys were fighting an early snag fire in the Cowlitz and Charles Flory called me from Portland, wanting to know if he should send up one of these Wonders. I relayed the message to G. F., but he turned the offer down, much to Flory's disgust. This heavy, low-powered pumper, however, was the forerunner of many different types of portable pumpers to come—light, medium and heavy; two-cycle and four-cycle; one-cylinder, two-cylinder and four-cylinder. It was interesting to watch and have some part in the development of water and other fire control equipment.

ESCALATOR SERVICE BY MULE'S TAIL

Rudo L. Fromme

Soon after my assignment as supervisor of the Klamath Forest, July 1909, I took the four horse drawn, leather hung stage coach (the same type, maybe the same one, in which Mark Twain made west in "Roughing It") to Scott Valley. District Ranger Robert Finley was not at home but his brother at a neighboring ranch equipped me with a saddle mule, which had just returned from a packing job to a fire almost up against the snow line in the Trinity Mountains at the south end of Bob's district. He told me that Bob had been on this fire for two weeks and that the mule had been used to take in several packs, always returning alone. He was sure that said mule could take me right to Bob, but he added, "Bessie's a bit spoiled when it comes to packing human freight up any right smart pitches. If she stops and looks around at you, all hell can't budge her, but, she won't kick at pulling you up by her tail if you're quick enough to get a good hold on it while you're slidin' off her back."

I didn't particularly like the sound of the word "kick" in the above advice, but I found out on my trip next morning that the assurance I had received of Bessie's peaceful, cooperative nature in this respect was 100 percent. I also learned that if I attempted to thwart her willingness to carry me an her back beyond the point where she had stopped and looked back at me, by admonishing her in the flanks with my heels, she disdainfully turned, in spite of anything I could do, and started back for home. In which case, I had to hastily slide off her back and head her forward, fireward. In which case also, I would be too excited to catch her tail and would have to struggle up the hill under my own power. She might stop now and then to nibble bunch grass just ahead of me but I would never catch her till she reached the top of the rise, where she would patiently wait for me to mount again. After that, I respected her wishes when she turned to look at me on steep grades, and as I gained more faith in her inherent quality of mercy, I was able to adroitly grab her tail on my descent over her rump and enjoy the privilege of being materially assisted on such climbs.

I chanced to meet Ranger Finley en route at a small mining camp store where he was picking up a few extra blankets. He suggested that we swing around by a larger mining camp on the chance of securing additional firefighters. As we rode up in front of the broad verandah of the headquarters building I observed an aged Indian lounging in languid repose. As I sat on my mule waiting for the ranger to return from inside the building, the head man came out and introduced himself. Then, thinking to have a little fun at the old Indian's expense, he called, "Jake, you better get ready to go to the big fire." "Oh! No," yelled Jake, "Me too old. Finley say so." "Is that so?" continued the mine foreman, "Well, this man says you've got to go, and he's the boss." "No! No! wailed the old Indian with a rolling shake of the head, "Him not the boss. Finley ride 'em in front!"

WORK AS A FOREST GUARD

C. C. McGuire

On May 1, 1909, I was appointed as a forest guard on the Mount Baker National Forest. The forest supervisor fitted me out with a badge, the "Use Book" and a marking hatchet. He told me to go to Finney Creek and establish headquarters at the Finney Guard Station.

That fall I was directed to go to Ruby Creek and build a cabin. With Fred Scarlett, a forest guard for an assistant, we built a 16 foot by 18 foot log cabin. There was no timber near suitable for shakes so we repaired an old miner's flume which extended three miles above Ruby Mountain, went up the mountain to the old sawmill, packed one inch by twelve inch boards down to the flume, floated them down the flume and then packed them on our backs one half mile to the cabin. These boards served for roof and floor. By the time the cabin was finished, winter had set in with the usual snow and ice. We had a pack horse with us and when we got to Devils Elbow, the drip from overhead had completely blocked the half tunnel with ice. We chopped our way through, but it was very dangerous for the horse to get through, for one slip would send him over the cliff into the river 50 feet below. So, we tight lined him across. To those who do not know what is meant by a tight line, the following is in order. A rope was fastened to the horse's neck and I carried one end across. Another rope was tied around the horse's tail and the lose end with a couple turns around a tree. I took a turn around a tree with the lead rope and as my partner let out a few inches, I would take up the slack, so at all times the horse was in the center of the tight line. Though the horse fell several times, we inched him across.

That winter the powers that be decided to rotate the ranger to a three month short course in forestry at the University of Washington, Seattle. Tommy Thompson and Ralph Hilligoss got the first nod and I was sent to Texas Pond to replace Thompson scaling shingle bolts on the Sauk Shingle Company sale. There was about seven feet of snow on the sale area that winter and a great part of the time I was snowed in. Occasionally the bolt cutters would go out and fall a few trees. About all I had to do was see that they would dig down in the snow around the tree so the next spring I would not be criticized for the high stumps. Even in those days, our minds were directed towards proper utilization.

A cat hunter named Bill Quall from Sauk used to visit me, staying a week or more at a time. We would get out in the deep snow and hunt wildcats. Bill had a good cat hound and I can't remember a day's hunt when we did not get one or more cats. We could not hunt every day because the dog would play out, particularly from sore feet. Then we would spend the evening soaking his feet in a solution of stewed hemlock bark to toughen them. There was a five dollar bounty on cats, so Bill did pretty well. Mine was just contributed time. Incidentally, Bill could not stand prosperity. It was the practice when skinning a cat to leave the bone of the right front leg attached to the hide, then when it was presented for the bounty, the bone would be clipped off. The next summer Bill gathered up old bones and sewed them to the hides on which bounty had been paid. He got away with this for some time, but one county auditor who paid the bounties, got suspicious, with the result that Bill got six months to two years in the pen.

One day while sitting in my cabin with the windows partially covered with snow, a rap was heard at the door. There stood an old Indian and his kloochman. The old Indian introduced himself as follows: "You know me. Me Cap Moses, Chief on Suiattle Creek. We come to visit." I invited them in and served them coffee and cake. Mrs. Thompson had baked a twelve by twenty-four inch cake for me and I was glad of help in eating it before it became stale. About one half of the cake was left and this I presented to the kloochman wrapped in a tea towel. A few days later she returned the tea towel carefully washed and ironed, though she had to climb up 1,500 feet to my abode. This was the first of many visits by the Suiattle Indians who were all camped on the Sauk River at the mouth of the Suiattle. Being familiar with the Chinook language and knowing a few words of their native tongue, I found them to be friendly and good neighbors. I was always well treated in their camp as we sat around the council fire near their teepees.

In contrast to this it is appropriate to relate Guard Maier's experience. Maier, a typical drugstore cowboy, was appointed as a fire guard on the Suiattle. He was a swashbuckling type who always packed a gun, but who had a wholesome fear of the Indians and of anything else he did not understand, and there were many things he did not understand.

He was hired as a forest guard and sent to the Suiattle with instructions to check on the Indians' shingle bolt cutting on their allotments and see that they were not trespassing by cutting Government timber.

Maier, with a six shooter strapped on his belt, began by issuing orders and soon created the impression that the Forest Service was going to run all the Indians off their lands. Naturally, the Indians resented this and took their troubles to Chief Moses, head of the tribe. Chief Moses called for a conference of all the tribe and wanted Maier to attend. Accordingly, three Indians were sent after Maier and invited him to the conference. They brought along an extra saddle horse for Maier to ride. Since the Indians could speak very little English and Maier could not understand their language or intentions, he thought he was being kidnapped. They finally got Maier on the spare horse and set out for the Indian council. Maier was sure the Indians were going to do away with him. The Indians, sensing this, would let out an occasional war whoop and whip up their ponies. Things were getting pretty grim. Finally they arrived at the camp where a hundred or more Indians were assembled. Then the powwow started and Maier was sure they were going to throw him in the river. The meeting lasted all night with feasting and high jinks. In other words, the Indians were showing him a welcome with no sinister intentions. Maier debated with himself about drawing his gun and trying to shoot his way out. But, since a number of the Indians were armed, he realized he would be fighting a losing battle, so he bided his time.

The next morning, the Indians all shook hands with him, mounted him on a horse and bade him farewell. Maier returned to Bellingham and handed in his badge declaring he would never be induced to go through such an ordeal again. A reporter on a local paper interviewed him and it is interesting to note the caption on his terrible experience. Headline: "Bellingham man, with hand on pistol and nerve steeled for immediate action, spends uncomfortable night surrounded by redskins who discuss advisability of throwing him in the river."

HELLO, RANGER, HELLO!

A. J. Radigan

Back in the fall of 1907, a telephone line was built on the Colville National Forest from Republic to Anglin. This line was partly through timber and partly over open ground. I worked on it. The wire used was No. 12 galvanized and the insulators were No. 16 glass on brackets.

This line was a grounded pole system the entire length unless a tree of suitable size stood directly in the right of way. A suitable tree was one from 10 to 14 inches in diameter. This tree was then topped about 20 feet from the ground and the limbs removed. Bark was peeled off from the top down about two feet and also at the butt. A bracket was then nailed to this stub and the line was attached in the same manner as on regular poles with a Western Union tie. The poles were unseasoned Douglas fir and western larch and were installed without stubs or treatment. A few lasted as long as five years.

Almost constantly from the beginning this line gave trouble. Trees fell across it and the No. 12 wire snapped like string. Brackets were torn loose and split. After a year or two, always traveling by horseback, with the usual tools and extra wire, the repair work which not only was arduous but short lived, led to the conclusion that something must be done. E. W. Wheeler was one of the rangers who had a lot of this repair work to do. He had visions of a line which would stay together after a tree hit it, so he strung the wire through small porcelain insulators and hung the tie wire to the side of trees. This was about 1909. Two forest men from the district office came to view the line and immediately claimed it was not the thing. They said all the slack would creep to the long spans and a lot of other things now forgotten would happen to it. However, three trees were purposely felled across the line within a quarter of a mile of each other and nothing happened except the wire did not break and communication was still uninterrupted.

From this beginning was born our present tree line with all loose ties and oval brown split insulators. When Ranger Wallace Wheeler, formerly of the Wenatchee, took care of his tree telephone maintenance or construction, he derived much pleasure from the fact that, "This was Dad's idea."

INTIMIDATION AND COOPERATION

Grover C. Blake

On May 6, 1909, Forest Guard Douglas C. Ingram and I were sent to Mill Creek, northeast of Prineville, to survey and mark the forest boundary. We made camp and hobbled our four horses. They all disappeared during the night and strayed away, hobbles and all, and it took us two days to find them. During the next several years, hunting for straying saddle and pack horses required a large percentage of our time. Eventually, we were able to build enclosures here and there for holding horses, but before this was accomplished, keeping our saddle and pack stock with us was a major problem. Doug Ingram was born in Scotland and educated in England and had been in the States but a short time. He had had no experience as a woodsman and our straying horses caused him much more worry than they did me.

We did boundary work until June 9th, when we got orders to post the main stock driveway used by stockmen entering the forest enroute to their allotments. We succeeded in marking several miles of driveway before June 15, when sheep were permitted to enter. We were then located at the Trout Creek Counting Corral, where we counted in some 50,000 sheep during the next two weeks.

Ingram and I were each assigned to districts with no definite boundaries, and on July 15, 1909, I began to post and mark out the boundaries of sheep allotments. This called for much surveying and, being a lone wolf in a large district, with some 15 sheep allotments and four cattle ranges, the first year or two I would only hit the high spots where need was greatest.

Trespassing was common on the part of the cattle owners. Little or no effort was made to hold the cattle and horses on the range allotted to them. Most of the cattlemen were not only antagonistic to government administration, but some of them were defiant. One large owner on the Crooked River section had several hundred head of cattle on forest range and I could find up to 200 of his stock on sheep range any day. Warning letters to him were ignored and he openly boasted that he would continue to range his cattle where he had always ranged them and no one was going to stop him from doing so. He did a lot of talking to his associates and some of his threats eventually reached me. It was generally understood that he had been an active member of sheep shooters and would not yield to anyone, not even the federal government. His line of thinking was no doubt similar to that of one Fred Light who carried his case through the courts to the U. S. Supreme Court and lost all the way. He made many threats and told people that my mysterious disappearance would surprise no one, that he intended to maintain his rights and "had money to fight the Government."

I had three trespass cases pending against him before final action was taken on any. The wheels of justice ground slowly in those days and a lot of trespassing could be accomplished before a case could be brought to court. When he was eventually summoned to defend himself in the Federal Court in Portland he yielded without a fight and paid the damages in full, both actual and punitive.

Among sheep men, some herders took pride in committing trespass or stealing grass from someone else. They felt that they were doing their employer a favor if they succeeded in grazing some range that was "over the line." Sometimes they were successful, but often succeeded in getting their employer into trouble. Then again, range employees were constantly making reports to the ranger on trespass on the part of someone else. Upon investigation these reports often proved to be exaggerated or unfounded.

During the season of 1910, renewed efforts were made to get the cattle owners to make reasonable efforts to keep their stock on the allotted range. Salting plans were made and were halfheartedly put into effect by some permittees. In a few instances, line riders were employed but little success was attained until the cattle owners were organized into associations and drift fences constructed. It was several years before this was accomplished, however.

All permittees were furnished with blueprints and written descriptions of their allotments so their employees could locate their own allotment boundaries and get along until the ranger could reach them. If they ran into difficulties they could notify the ranger and get help. Constant cattle trespass called for many written notices to owners promising legal action, and trespass reports to the supervisor's office.

On July 1, 1911, I was again at the Trout Creek entrance counting sheep into the forest. I counted in from two to four bands per day for 15 days and then followed up by rushing from allotment to allotment assisting the men in charge getting established and to get their allotment boundaries located. Forest Guard C. M. Irvine was assigned to assist me during the 1911 field season. It was this year that we began organizing the cattle and horse permittees into stock associations and the first one for my district was the White Butte Cattle and Horse Association, organized on November 11th. In this way we were able to deal with all the users of one allotment as a unit through their advisory board, thus greatly simplifying administration

By the end of the 1911 season, allotment lines were so definitely established and the men in charge of the stock were so familiar with them, that I was able to devote more time to building horse pastures and trails. Oh, how we did need pastures!

Most of the forest users had become reconciled to government administration, and grazing men were beginning to realize that they were being materially benefited, rather than damaged, by the regulations of the Forest Service. As a result, a greater spirit of cooperation on the part of the public was to be noted, and the supervisor and rangers were beginning to have friends.

PINCHOT'S DISMISSAL

Thornton T. Munger

During the height, and heat, of the Pinchot-Ballinger controversy, I had a memorable experience.

The controversy started early in President Taft's administration when Pinchot called attention to the several anticonservation acts of Secretary of the Interior Ballinger, most notable, his recommending for patent the illegal Cunningham coal claims in Alaska. Efforts of Pinchot to get Secretary of Agriculture Wilson to persuade the administration to check the Interior Department having failed, Pinchot appealed to the public. Specifically, he wrote a letter to Senator Dolliver (requested by the latter) explaining the situation. This was considered insubordination, so Pinchot was "fired." Ballinger resigned within a year.

On a Christmas vacation trip east I was visiting in Washington with my sister and her husband, Philip P. Wells, a Yale classmate and close friend of Gifford Pinchot.

On January 7, 1910, Chief Forester Pinchot was invited to a family dinner at their house. He arrived a little late and, now quoting from my diary, "He greeted the company very graciously and cordially took a seat by Eleanor (Mrs. Wells) on the lounge. He had two envelopes, one opened, the other not, which had been given him by a messenger on leaving his house. Soon after seating himself he said very cheerfully—'This tells me I've been bounced. Excuse me if I try to find the reason.'

"After reading himself the two page letter, he said—'possibly you may care to hear what it says,' and proceeded to read the letter from President Taft. After reading this letter he opened that from the Secretary of Agriculture dismissing him and read it aloud at the first reading. He showed great pleasure at having Potter named his successor. During dinner all thoughts were on the matter and Mr. Pinchot referred to it but always in the most cheerful vein. He said his only regret was for the others whose positions his dismissal might affect.

"When asked why he wrote the letter to Senator Dolliver he stated that they were twofold—

1. To clear Price and Shaw;

2. To enable him to have a voice while he was needed, as he expected that if he kept quiet it would have been only a matter of a few months that he could stay."

Still quoting from my diary, written late that memorable evening—"During dinner he went to the phone repeatedly, calling Amos Pinchot, Hill, Shaw and was going to call Price but Phil (Wells) suggested waiting and G. P. said, 'Much better let him get a good night's sleep first.'

"G. P. was much affected by what little Elizabeth whispered to her mothers who repeated it, 'Lewis is crying about Mr. Pinchot.' When Lewis (a 12 year old) came in later, G. P. put his arm about him and said very cheerfully, "I'm all right. I'm just as happy as a clam."

The next morning at the old Atlantic Building, Mr. Pinchot assembled all of us Forest Service people and bid us goodbye, saying, "Stick to the Forest Service."

WHICH WAY IS UP?

Rudo L. Fromme

This story relates to one of Supervisor G. F. Allen's amusing encounters with a somewhat irregular sign posting out of Tacoma when he was supervisor of the old Rainier National Forest.

Mr. Allen was driving to Seattle some years ago over one of the old combination horse and buggy and Model T roads, when he spied an apparently new, but homemade looking roadside sign posted on a tree to his left. Stopping to read it, he noted that the sign appeared to read all right as to distances between the two sites, but the arrow for Seattle was pointing toward Tacoma and the Tacoma arrow toward Seattle. Since the sign's location was directly across the road from a small farm side cottage, Mr. Allen assumed that the resident there might have had something to do with its erection.

Sure enough, a rustic looking woman answered the door knock and said, "Yes, my son made that. Don't you think it's purty?" "Oh, the sign looks well made," replied Mr. Allen, "And it's quite artistic, but the arrows are pointing in the wrong direction." "Oh, we know that," she said, "You see, my son made the sign in the woodshed here and was thinking our side of the road, but when he took it out there, w'y that big tree across from us was the best place to nail it up. Anyway, we like to look at it." "Yes, I see," pursued Mr. Allen, "But it could mislead someone coming in from the side road." "Well, my goodness!" she exclaimed, "Folks that don't know their way around ain't got no business leavin' home."

A FIRST TIMBER SALE

M. L. Merritt

In the spring of 1910, a Mr. Reynolds, who had a small sawmill southeast of Prairie City, applied for the timber on half a section of land to the east of the Blue Mountain Hot Springs. I was sent to cruise and report on the area. It was my first Whitman job. I took the narrow gauge Sumpter Valley Railway to Dixie Summit, thence the stage to Prairie City. I climbed to the top of the old Concord Stage and sat beside the driver as he headed his four horses down the dirt road. There was still much snow at higher elevations. The wheels broke through the packed snow frequently, so that the stage pitched and rolled badly until we were out of the snow zone. I saw Supervisor Cy J. Bingham at Prairie City. He was an outstanding character. Before coming to the Malheur, Cy had been ranger along the divide between the upper Deschutes and the upper Umpqua rivers. For years afterwards, men of that area talked of the trails that Cy had slashed through the lodgepole thickets. These ran straight for his objective. Also, he always carried a timber scribe and left his name and the date expertly and prominently carved at section corners, on lake shores, etc. Such landmarks were characteristics of his territory. Cy was also a graduate of the school of "hard knocks" who won many friends for the Forest Service.

The Forest Service soon had an application from Frank Gardenier, under the name of the Baker White Pine Lumber Company to purchase timber on several sections northeast of Austin. This cruising I did alone, running compass, pacing distances, mapping and estimating volume of timber, staying in a little log cabin on Crawford Meadows. Yellow (ponderosa) pine was new to me and I had much to learn. I made my own volume tables. The ten percent cruise later cut out slightly in excess of the estimate. After the cruise there was much arguing and negotiating in preparing the sale contract. Mr. W. T. Andrews, Forest Service lumberman from Portland, did this negotiating. It was the first large sale on the Whitman and, I believe, the first large yellow pine sale in District 6. So both sides to the negotiation were operating in a pioneer field. A minimum price and conditions of utilization were agreed on. We actually cut about 80 percent of the stand. Too much by later standards, but a considerable volume was required to warrant railroad spur building. All brush was to be piled for burning. Final cutting

Forest ranger, headquarters camp, West Walker River, Stanuslaus NF, California – c1905.

Ranger John D. Maxfield at Ranger meeting, California Hot Springs, Cleveland Reserve, 1906.

report shows the following volume in board feet cut on this sale: yellow pine 18,363,000; Douglas fir 3,511,000; western larch 3,511,000. Receipts were $46,396.78. Not a large sale at present day prices and standards, but it set a pattern.

So started the logging of ponderosa pine (we called it yellow pine then) in Oregon's national forests. During the negotiations many government foresters participated, all of those handling sales in the Portland District office and several from Washington, D. C. From Portland there were Fred Ames, W. T. Andrews and T. T. Munger. The latter deserves much credit for devising the forestry practices which were adopted. Among the Washington men was Rafael Zon, born in Russia, educated at Cornell and the Service's authority on silviculture. He spent several days on the project, during which time we "batched" at the Blue Mountain Ringer Station cabin. This two room cabin had but one bed. The lots fell against Zon and he had to sleep on the floor with Andrews. One morning Zon protested he hadn't slept a wink, that woodrats (perhaps mice) had kept running over him to his great distress. No one extended much sympathy, but as I recall, Zon lived through it without damage.

FIRES ON THE CRATER NATIONAL FOREST

Carl B. Neal

On June 28, 1910, immediately after graduating from the university, I left Eugene by train and arrived in Medford at 4 p.m. Before I went to bed that night I had reported to the supervisor's office and had purchased two horses, a saddle and a pack outfit. On June 29, Deputy Supervisor Sam S. Swenning started me out and I rode to Trail. I put my horses in the barn, had supper in the hotel and slept in the corral. The next morning I packed up for the first time. My roommate at school had shown me how to throw the single diamond hitch. He demonstrated with two chairs, two pillows and a trunk rope. This morning I had some local critics watching me. In making the hitch I threaded the rope through instead of attempting any short cuts. When I finished, one critic said, "You'll do," one of two compliments I received that season.

I went up Elk Creek and in the afternoon reached Ranger Andy Poole's train camp at Dickenson's Cabin, which was on a ridge between Sugar Pine Creek and Bitter Lick Creek.

Sometime in July, Supervisor M. L. Erickson spent two nights with us.

I had great respect for Erickson and Poole. The previous winter the chief forester's office had authorized Erickson to designate three rangers to take a "short horn" ranger's course at the University of Washington on salary and at government expense. Andy Poole, George West and John Holst were designated. After the rangers had enrolled and started their course the solicitor ruled the procedure as illegal. Erickson advanced the money out of his own pocket for the rangers to complete their course.

The only advance training I received for fire suppression was Andy Poole's talks around the campfire at night and they were good. He covered how to hire men and to make out time slips, what tools and supplies were needed and how to secure them. And he covered firefighting technique. Toward the end of July we became fire conscious. The atmosphere was smoky; we heard rumors of going fires elsewhere. Nearly every day one of us went up on Butlers Butte to look out for fire. We had no telephone.

On August 3, I received written instructions from the deputy supervisor to report to Lodgepole Ranger Station immediately. I was instructed that I would be responsible for the discovery and suppression of all fires in T. 33 and 34 S, R. 3 and 4 E. Rustler Peak would be my lookout. I received an old USGS map which did not show townships and ranges so I just assumed that Lodgepole and Rustler were in the center of my district. Nothing was said about organization or my relations to other officers. I assumed I was directly responsible to the supervisor's office and acted accordingly. En route to Lodgepole I met Guard Ed Kirby at Prospect.

On August 19, I went to Prospect for men and supplies. I learned that Ranger West had several fires and also had all the available manpower. That afternoon all the fires on the Crater Forest blew up and crowned. From Prospect I saw and heard what undoubtedly was the Geppart Fire crown, and going northeast, crossed the Prospect Lodgepole Trail. In the early evening I left Prospect with four men, picked up somewhere, and started for Lodgepole. We ran into what proved to be a spot fire and worked on it all night. The next morning I proceeded alone to Lodgepole. I went through the newly burned over area. My only guide was the trail and in many places it was crisscrossed with down trees. I had to detour many times and sometimes I had to retrace by horse tracks to find my way back to the trail. From Lodgepole I went back to Gepparts, then to Butte Falls for men and supplies and to phone the supervisor's office.

In Butte Falls I picked up 25 soldiers and 10 local men. The soldiers were self-sufficient; they had their own transportation, overhead, equipment and supplies. I got the necessary equipment and supplies for the civilians and hired a team and wagon. I then went on ahead to check over mentally what I had done against what Andy Poole had taught us around campfire. While my horse was drinking in Ginger Creek, I realized I had no cooking equipment. I didn't have the courage to go back and admit my mistake so went slowly ahead. Then, lo and behold, beside the road near the Army headquarters but out of sight from the Army, was a barrel of new cooking equipment. I waited for our wagon and then told a couple of the men to load the barrel on the wagon.

When we arrived at Gepparts, the Army had already set up camp. Their captain had come in with them. He came to see me. He explained that while they were expected to be self-sufficient they had only emergency rations and that this was insufficient for firefighters. I told him the soldiers could have the same food we ate and I sent our wagon back for more provisions. That winter when forest officers got together there was a great difference of opinion as to the effectiveness of the soldiers. Those I came in contact with were very effective and very cooperative.

Some soldiers went out every day deer hunting for camp meat but without luck. Finally, they killed one. At that time there were only four of us in the civilian camp as against 25 soldiers, but they gave us one half of the deer. I asked the soldier how he got his deer. He said, "Well, I saw the deer and cattle running together and the cattle had a cowbell, so I just hung a cowbell around my neck."

Our job now was firefighting. I was on my own without any supervision or any help from any other forest personnel.

In the civilian crew was a peg leg. He was pugnacious and dirty. He would not wash and one morning before breakfast the men threw him in the creek. Another time, on the fireline, a tree started to fall; the men called "Timber-r-r-r" and all we saw of peg leg was one live leg and one wooden leg fanning the atmosphere above the brush.

My second compliment came one day after the fire was on a patrol basis. I went to Butte Falls. On the road I met Major Martin with his orderly. We met at a creek where our horses drank. He introduced himself and I did the same. He was very stiff and formal. I don't remember what he said but I felt like he had said, "Well done, thou good and faithful servant, enter now into the Kingdom of Heaven."

At the end of the fire season I was told we had burned over 50,000 acres and the suppression costs were $50,000. This was at that time a record for district six. Suppression cost of a dollar an acre was also a record.

APPOINTMENT AS DISTRICT RANGER

C. C. McGuire

In the spring of 1910, I passed the ranger's examination and was assigned to the Upper Skagit with headquarters at Reflector Bar so called from the small reflector shelter built there. A shake cabin of three rooms was built there the fall before and that was my home for the season. My district included the upper drainages of the Skagit River, Ruby Creek, Canyon Creek, Slate Creek through to Harts Pass, Thunder Creek to Park Creek Pass and all drainages into those streams. I left Marblemount with one pack horse loaded with a new Kalamazoo cook stove and a No. 3 Oliver, horse feed, my own grub and blankets. It was a toss up who had the bigger load, I or the horse. Anyway, I delivered my own pack safely but the horse broke through a snow bridge over Gorge Creek and rolled down the stream bed to the Skagit River—result, a smashed up cook stove, a badly skinned horse and two hours building a way to get the horse back on Goat Trail.

There was a need for some kind of crossing on the Skagit River and I was instructed to go see what could be done. One day, which I thought was Sunday, I got a cedar log and chopped out a crude canoe. (It later turned out that it was Monday and here I was doing an unauthorized job on a regular work day.) My first trip in the canoe was to explore the canyon where Diablo Dam is now located. I found one spot in the canyon where the river was standing on edge. It was only thirteen feet wide, but 150 feet above me which I had tentatively selected for a bridge site. It was about 100 feet wide.

In an earlier day the miners had a cable bridge across the river near Deer Park. They had seven-eighths galvanized cables to which they bolted 12 inch by 12 inch stringers and decked on top of them. It soon broke down. One end of the cable was still anchored and the bridge swung to the shore. I spent a couple of days chopping the timbers loose and, with a trusty Spanish windlass, hauled the cables out. So, I had a good start for the bridge. After salvaging the cables, Carl Bell, J. A. Monahan and Henry Soll were sent in to help me build the bridge. We dragged the cables down to the bridge site with manpower and set about preparing anchorage. It was necessary to drill holes two inches in diameter in the rock. We collected some drill steel and I ordered some one and one half inch steel to make the big holes. Of course, we had to shoot away and level off a spot for our towers.

If a ranger was not a jack-of-all-trades, he did not accomplish much. Fortunately, I had some experience in blacksmithing, so with fir bark in my forge, I sharpened steel and we started drilling. We got the anchor bolts in, towers rained and a cable strung. In those days engineering did not furnish us with blueprints and specifications, which in some ways was fortunate, since the first thing they would have done would have been to condemn my cables as inadequate. We were not bothered by factors of safety but were on our own. Anyway, we got a swinging bridge which did support a pack train. The bridge stood for about fifteen years and dynamite was used to dislodge it when the City of Seattle built the Diablo Dam.

In the spring of 1911, I was transferred to Darrington and located on the old Blue Bird Ranger Station. No development had been done on the station but there was an old shack adjacent thereto in which I made my headquarters. My first job was to start clearing the tract so ranger buildings could be built. When the fire season started I was kept on the jump with small fires. The Northern Pacific Railway ran through ten miles of my protective area and hardly a day passed that I did not have one or more fires along the right-of-way. One fire near Fortson got away and covered about 200 acres of cutover land. This was my first large fire. Though we finally whipped it, if ten percent of all the misdirected energy expended on that one had been used for training, I would probably have licked it before it started. But, in those days who was to train you! We had to find out by experience and try not to make the same mistakes twice.

Tommy Thompson, who was stationed at Texas Pond, and I had a sort of joint responsibility for looking out for fires on the Suiattle. To this day I don't know just to whom the job belonged. Anyway, we had a one hundred and sixty acre fire south of the river near Buck Creek. Tommy was first on the fire and was handling it alone until I arrived a couple of days later to help. It never occurred to us to hire a crew, establish a fire camp and spend a lot of money. But hell! What were two rangers for if they could not handle that one. We spent about two weeks working around the fire, putting in scratch lines and going around that fire from daylight till long after dark holding it. In the meantime we were living on corn meal and an occasional salmon. We finally won out and started home. When we reached the Sauk River, there was no way to cross as the boat was on the opposite shore. The Sauk Shingle Company was floating shingle bolts on the river so we selected a bolt, got astride it and, with a homemade paddle, worked our way over. Though we drifted down the swift current for a couple hundred yards, we landed safely, wet, cold and hungry but on the grub side of the river.

I believe it was in 1913 that we made our first fence post sales. The prairie states needed post to string their barbed wire.

The specifications called for five and one half foot posts with a circumference of fifteen inches, split cedar. A rather flimsy post, but that's what they wanted. The first sale was made to a local rancher who wanted to get out a carload just to see if he could make wages on the job. Evidently he did, for later I met him on the street in Glacier, he shook hands with me and left a bill deposited in my palm. How large the bill was I never knew, for I handed it back and explained that selling him the timber was just a part of my job for which I had already been paid.

The Miller Pole Company of Seattle also applied for down cedar for poles and we sold them several carloads. All the post had to be stamped and counted, which took considerable time, as well as wood supervision to see that cutters did not abandon a log just because it was tough splitting.

I think it was also in this year that we got a call from the Madison laboratory wanting eight cords of amabilis fir for experimental purposes. Of course, there was no money for the job. This was just a chore for the ranger to do in his spare time. I know I went back into the mountain, cut and split the eight cords and packed it out on my saddle horse and shipped it to Wisconsin. I waited in vain for an acknowledgment but to this day can only surmise what was done with the wood.

CHARGE IT!

Grover C. Blake

It will be remembered by old-timers that our monthly service report once had spaces for classifying the different activities such as: grazing, timber sales, free use, claims, etc. Activities which could not be classified under the headings given were charged to M.E.D. (Miscellaneous Executive Duties). All employees were instructed to keep their diaries up to date and charge each day's work to the proper classification in the diary. Ranger W. A. Donnelly, whose district adjoined mine, told a short-term guard how to write his diary and make the charge. Said he, "In your case you will charge everything to M.E.D." The guard boarded with the ranger while working at his station. Time went on and on and the guard did not offer to pay anything on his board bill and finally Ranger Donnelly called his attention to the matter, explaining that he just had to have the money to buy some more groceries. The guard gave the ranger a surprised look and said, "I thought you said to charge everything to M.E.D.!"

TRAVELS WITH FOUR DONKEYS

C. P. Cronk

This installment deals with a unique feature of the Siuslaw National Forest in 1910-1911—the use of donkeys for packing. Supervisor Cahoon promoted their use in those days of almost no roads, few trails and little pasturage.

On a rainy day about November 1, 1910, I made the acquaintance of two of my traveling companions, the mother, May, and the daughter, Fanny. We started out of Eugene in style and ease, I riding May; Fanny trailing with my belongings stowed in panniers. This was my first acquaintance with the species. I found it highly educational; their actions unpredictable. After a few uneventful miles we came to a point where, due to rain, a waterfall was coming over the cliff from which the road was carved and landing in the middle of the roadway. The burros balked. Finally I pulled Fanny through; May refused to budge. There was nothing to tie Fanny to. The moment I dropped her halter rope she trotted back to her mother, paying no attention to the waterfall she had to go under again. Next I tried to pull May through the waterfall. By putting a half hitch around her nose and pulling steadily on it until she had to move to get a breath of air, I managed to move her about 10 feet at a time. As soon as she got through the waterfall, she became her docile self again and daughter Fanny dutifully followed her. Thus was completed the first of many lessons I was to learn regarding burros.

We continued on toward the coast as far as Indian Creek. Here I picked up the rest of the party. Vic, the biggest and most even tempered of our burros, and Bubbles, the jackass, smartest and smallest. These two had been used during the past field season to pack supplies to the trail along the backbone of the Coast Range from near Mapleton on the Siuslaw River to near Waldport on the Alsea River.

All burros seem to have a sense of humor. Unless led by a horse or bell mule, they delight in exploring all side trails to the discomfort of their driver. Bubbles had learned that he would be given candy if he walked up the steps of the store in Florence.

One of the trail crew told a story on Supervisor Cahoon. When approaching the camp riding Bubbles quite properly, they reached a mud puddle when the burro lowered his head to the ground and the supervisor slid off into the puddle, to the merriment of the crew assembled at the end of the day.

From the end of the ridge trail at the Corvallis-Waldport Road we turned east, there being then no road along the coast. From Corvallis we went north to Sheridan and along the road to Willamina to Dolph and Hebo, then up the mountain to the ranger station. This was approximately 225 miles and 22 days from Eugene. Because the night stops were more or less routine, only one stands out. I discovered that in the Willamette Valley, east over the Coast Range from the Siuslaw National Forest, a ranger was still an oddity. A bit north of Corvallis, I inquired at a house if I might put the burros in a fenced pasture I had noticed. There was hesitancy, but finally it was decided that the burros could be taken care of and I could sleep in the barn and get supper and breakfast. After supper the boys of the family brought out copies of "Youths' Companion" to show me a series of articles on the forest ranger and to ask questions about the life of a ranger. A couple of hours later, as I asked the man of the house if he would give me a call as he went out to the barn in the morning, he said, "If you don't mind sleeping with the boys, we can put you up in the house." I had talked myself into a bed.

The burros were used to pack supplies to the planting crew on Mt. Hebo during the winter of 1910-1911, but I had little to do with them as I was busy elsewhere. It is my recollection that burros, because of their thin skin and consequent tendency to get saddle sores in wet weather and their small hoofs which sank in the mud, did not prove as satisfactory as horses or mules in this region of high rainfall. I did use the burros during the summer and my last job on the Siuslaw Forest was to deliver Fanny and May to Ranger Durbin near Waldport. On September 14, my record shows it was "rainy" as I took the two donkeys down to the Hebo store, where, beside my duffel, I loaded on a mile of telephone wire, 25 brackets, 25 insulators and an extension bell. We stopped at Neskowin for the night, 18 miles. On the second day we left one coil of wire on the way up Neskowin, or Slab Creek and the rest in the Salmon River Valley, where adjoining ranchers were to erect the line. This time we stuck to the road, as on a previous trip to our sorrow, we had taken the shortcut foot trail through second growth hemlock over the ridge between the two streams. There, one of the burros went off the trail, up to her belly in mud from which she couldn't be extricated until I removed the entire load. Even then, it took much persuading and pulling to get her to exert herself to get onto solid ground again where she could be reloaded.

From Rose Lodge on Salmon River, we went up over the ridge and down the Schooner Creek Trail to Taft. It was on this trail, up from Rose Lodge, that I first met the 12 year old and 10 year old Pylkanen sisters (generally known by the American name they chose—Adams). The older girl had a 50 pound sack of flour on her back to pack up the seven miles to the family homestead. Another time when I passed that way, the two young ladies were "manning" a crosscut saw to replenish the fuel supply.

Having delivered all the telephone equipment, we had only my pack from now on. I got someone to row us across Schooner Creek trailing two unwilling burros behind. At Drift Creek and the Siletz River the performance was repeated. The record for this day shows "Sunday, September 17, 6 a.m. to 3 p.m., 19 miles and weather CLEAR."

Putting up at Newport, in the morning we ferried across Yaquina Bay. At Beaver Creek, I phoned Ranger Durbin. He took the Wald-Newport stage along the beach, and upon his arrival took a picture of me, arms around the necks of Fanny and May. I turned the donkeys over to him, thus completing my last act as assistant ranger on the Siuslaw. I took his place on the stage en route to Newport and Eugene.

THE FIRST DISTRICT 6 EXPERIMENT STATION

Thornton T. Munger

Three of the other districts had experiment stations by 1911, so D-6 followed suit. It was placed at Wind River on the Columbia (now Gifford Pinchot) National Forest beside the nursery, largely because it was then thought by some that most of the problems that an experiment station might settle were in connection with artificial reforestation. An office-residence was needed; the statutory limit on any building was then $650 and that was reached before the plumbing fixtures were acquired. So Julius Kummel paid for the toilet and I for the bathtub. A small greenhouse was built at about the same time, mostly by contributed time from a few of us. It was intended mostly for germination tests.

Odds and ends of exotic trees were planted in the stump land adjoining the nursery in 1912, and this was the beginning of the Wind River Arboretum, which was considered to be a desirable adjunct of a research agency. To this arboretum I devoted considerable time up to the time of my retirement.

To get quickly an inventory of the national forests, the "extensive reconnaissance" project was started in 1909. On every forest, the forest assistant or somebody else was assigned to the job of type mapping and ocularly estimating timber volumes. On some forests the task was done in a year; on others, several seasons were spent.

One trip in 1914, illustrates how things had to be done in those days. In late May, project leader Robert Craig finished an intensive reconnaissance project on Salt Creek (Cascade, now Willamette National Forest) near Oakridge, where I joined the party. His next project was in the "panhandle" of the Crater (now Rogue River National Forest) east of Crater Lake. In the move, the crew of five or six walked, their equipment on four pack horses, across the Cascades to Crescent in two days. I hired a horse at Flat Creek so rode ahead and engaged on a four horse rig to take the outfit most of the way. In two more days, the last part through almost trackless pine forest, we reached a camping spot where Cabin Creek sinks into the pumice soil. Craig and I then spent several days on horseback scouting out this project in beautiful pine timber which was soon included in a sale to the Pelican Bay Lumber Company.

In the early '20s, a Model T Ford was purchased to facilitate transporting a reconnaissance crew. This was the first passenger car to be purchased by the forest in the district. There being no way to license it, I designed and had made at a sign shop, two identification plates. Soon thereafter, the government bought many official cars.

In 1915, Julius Kummel and I bought for ourselves a Hupmobile which was much used to take official visitors about at five cents per mile.

Another "first" that I had a part in was the purchase in 1909 or 1910 of the first adding machine in the district office, a Dalton, for use mostly by silviculture's computing clerk, Erma Bell. Up to that time, she had been using mental addition to check all the scale books used through the district.

For several years, five or so of us foresters batched together in three different houses. The group changed with transfers of personnel, but Julius Kummel and I were the mainstay throughout until our marriages in 1916. This house was quite a headquarters for visiting forest officers, both for dinner and for the night. The Society of American Foresters met there for a time. Until after 1915, it was six full days of work, no Saturday half-holidays except in summer. In the field it was very common to work on Sundays too. My field diary frequently states "breakfast at six" and "back to camp at 7 p.m." But we loved it.

With the completion of the new post office at Glisan and Broadway in 1916, we moved into much expanded quarters.

The district office was very small in the 1910s by present day standards; there were relatively few transactions, little money to spend, a minimum of paper work. Several of the present day divisions were not in existence then. There was no separate office of personnel but perhaps the very best kind of personnel management was when one of us from the Portland office rode horseback with a ranger all day, stayed overnight in his log cabin ranger station, played with his kids and really got acquainted with him. Such contacts made friendships that were close and lasting.

In 1910, when I was on one of my many trips to Wind River, T. P. Mackenzie, Chief of the Office of Grazing, asked me to bring his saddle horse to Portland. I rode Rondo to the boat landing near Carson, a cut bank sandbar, and when the sternwheeler came alongside the captain said, "We're late; no time to load horses. Pick you up tomorrow." So rather than wait a day on the Columbia River's bank, I rode on a then partly grass covered wagon track into Portland in a day and a half, spending the night at a house where I asked for accommodations for myself and the horse which I got without cost.

I had many 12 hour horse stage trips and that meant going only 50 or 60 miles. Once I horse staged from Shaniko to Klamath tills via Prineville, Bend, Silver Lake and Lakeview. One stretch of 100 miles in 22 hours, broken only to change horses and get a meal, was from Silver Lake to Lakeview. My diary says, "reach Summer Lake at midnight and have fine lunch there . . . arrive at Paisley about 6 a.m. and have bum breakfast."

In the horse country, long days were spent in the saddle, even running section lines horseback. One season I counted that I had ridden 18 rented horses and sat in as many different saddles. West of the Cascades it was shank's mare for transportation mostly, and hikes of 20 or 25 miles a day were componplace.

A three weeks trip in 1912, to the Okanogan and Colville Forests stands out in my memory. Arriving in Wenatchee by train, I got aboard a Columbia River steamer in the late evening and at dawn we pushed off up the river and bucked the current with the sternwheeler all day. At one or two rapids the deck hand took a line ashore and with the captain the boat pulled itself through to quieter waters. After a night in Brewster a horse stage took me in five and a half hours to Okanogan.

Supervisor Hale and I inspected the work of the reconnaissance crew and worked with them for a few days, traveling on horseback and on foot.

On Buck's Peak, the lookout demonstrated heliographing to other lookouts, this being an innovation. Vernon Harpham picked us up in his auto for a ride back to town, another novelty.

From Okanogan to Republic was a 24 hour trip on a mixed freight train. With Fred Cleator, I visited the reconnaissance party and inspected railroad crosstie and cedar pole sales, then very active on the Colville National Forest. We traveled back and forth on the Great Northern Railroad tracks by Forest Service speeder.

Train, boat, stage, shank's mare were used to get to the Olympic country. A trip to inspect the reconnaissance crew on the Olympic National Forest in 1914, was very pleasant but rather time consuming. I took the train from Portland to Seattle, boarded the steamer Bellingham at midnight for Clallam Bay, where I arrived at 8:15 p.m. the next day and left at once by auto stage for Forks, arriving after midnight. The next day I walked 15 miles to Ed Manzlik's camp, fording the Bogachiel River a couple of times.

When the people of Grants Pass clamored for the Forest Service to do something about caring for the Oregon Caves, District Forester Cecil sent me down to look at them since I had visited several other well-known caves. With Melvin Lewis in May 1917, I rode horseback over to the caves from Page Creek Ranger Station, the last part of the way in old snow. We tied our horses to trees, made camp and after supper with Dick Rowley, a chief explorer of these "Marble Halls," went through the caves. I reported so favorably to the district forester that the Forest Service soon started protection and development of what has since become a great tourist attraction and taken over by the National Park Service

Ranger Lewis related to me how these remarkable caves narrowly missed being lost to the public. Early in 1907, Ranger Lewis got wind of the fact that a couple of parties contemplated filing limestone mining claims on the area over the caves. He at once told Supervisor M. J. Anderson who had the area withdrawn as a national monument, but still a part of the national forest.

In 1924, Congress made possible a regional forest experiment station in the Pacific Northwest to replace the local Wind River Experiment Station. Much to my surprise, District Forester Cecil asked me if I would like to be its director. I was happy in silviculture (now timber management) under Fred Ames, but I accepted this pioneering opportunity at a salary of $3,900. So the Pacific Northwest Forest Experiment Station was born on July 1, 1924.

AN EARLY AUDIT TRIP

Albert H. Cousins

One of the first audit trips I made was in 1911. It was memorable to me, for as you will see, it entailed a lengthy journey across Oregon and transportation by stages and railroad.

I left Portland one evening on the O.W.R.&N., arriving at Baker City in the morning and left Baker City by the Sumpter Valley Railroad, a narrow gauge road and mixed train for Sumpter, Oregon. Sumpter, even then, was a town where gold mining activities had practically ceased. The headquarters of the Whitman NF were located at Sumpter. Henry Ireland was the supervisor and Otto A. Zimmerli the forest clerk.

After inspection of the Whitman, my next point was the Malheur at John Day. To reach John Day, I left Sumpter in the morning on the Sumpter Valley Railroad and went to Austin (the end of the line) where Mrs. Austin had an "Eating House." At Austin, the passengers all piled out for dinner and, as I remember, we had a good meal as she was noted in that section as a good cook. After dinner the next move was by stage, a regular "Buffalo Bill" rig, four horses, two seats inside facing each other and the body hung on straps.

There were eight passengers, including the driver, mail bags, lots of packages and bundles for the trip. I could not figure where it all could be stowed. However, the driver did not seem to be worried. The packages were stowed on the back seat up to the roof. Three passengers were put on the other seat; two passengers occupied the driver's box with him and the other passenger and I climbed up on the roof and sat on the mail bags. It certainly seemed far off the ground and a rather precarious seat. To make things more uncertain, we started up grade. On my side I could look down into a canyon that seemed about a mile deep. However, after a mile or two I got my sense of balance and "risked" looking around at the scenery and even let go of one hand grip.

We jogged along until we reached Prairie City where the stage unloaded and we had dinner and then resumed travel to John Day, arriving there as it was getting dark. Passengers who were going through to Burns got "comfortable" on the stage for the night trip. John Day was a "wide-place in the road" but there was a hotel of sorts. The next morning I went to the Malheur office. Cy Bingham was supervisor and Leigh Pratt, the clerk. Each of them weighed 200 pounds or over, so a good sized office was needed to accommodate them. The office was a one story stone building with thick walls, a relic of the days when the Indians roamed that country. It had iron shutters, a coping around the roof with dirt on the roof as a protection against flaming arrows shot by the Indians. A well was in the back of the building. Once inside one was fully protected from Indian raids and could get along if there was food enough.

Finishing the Malheur audit, my next was the Ochoco at Prineville. Bingham made arrangements for me to secure transportation on the stage from Daysville to Mitchell. Leaving John Day by stage at 4:00 a.m. before breakfast, we arrived at Daysville at noon. After dinner, I left on the stage, a one seat open rig for Mitchell. En route the driver had some packages to deliver and each time the rancher stood at his gate awaiting delivery and each time the rancher insisted we stop while he pulled the cork and showed his "Western Hospitality" (being with the driver, I was included in the "Hospitality"). When we reached the last four miles of the down grade into Mitchell, the driver whipped up his team, remarking "he would show me there was still some life left in the team in spite of the distance we had traveled." Downhill we went and the team certainly had "life" in it. I hung on the seat and prayed the road was straight and not too narrow. As we jostled along, the bigness and beauty of the October moon seemed to make my trials and tribulations pale and fade away.

We made it all right and pulled into Mitchell in the evening. I had a room over the post office grocery store. After dinner I met the boy who was to drive me to Prineville, went to bed and left a call for 6:00 a.m.

In the morning the boy was not around and the storekeeper advised me the boy and friends had been playing poker that evening. The sheriff had run them out of town and they were camped outside the town limits. We passed them on the way out of town and the boy said he was sorry he could not make the trip with me. The storekeeper had hired as a substitute, an old man who was somewhat the worse for wear from the night before but said he knew the way and could make it. We started out and went over Blue Mountains but took the wrong road and had to climb the mountains again. That put extra mileage on the team and by the time we were in sight of Prineville one of the horses went down on its knees, about all in. We finally pulled into Prineville about 9:00 p.m. The livery keeper gave the old man a bawling out for misusing the team and would not let it return to Mitchell for a couple of days.

The supervisor of the Ochoco was Addison S. Ireland. Prineville was a sleepy looking town, dirt streets and plenty of vacant lots.

After the Ochoco audit my last point was Portland, via stage to Shaniko, a sixty-five mile trip.

Supervisor Ireland and I left by stage at noon Saturday and reached Shaniko early Sunday morning. En route we and others walked a bit up Cow Creek Canyon to stretch our legs and take out the kinks. At Shaniko, there were about ten thousand sheep awaiting shipment and almost as many sheep dogs it seemed to me, yapping and barking. We stretched out on the hard station chairs until seven when we boarded the Sunday local of the O.W.R.&N. Co. for Portland.

Thus ended one of my first audit trips which took about fifteen days. Three audits were made and a lot of ground covered.

TRANSPORTATION ON THE SIUSLAW

Duffy Lewis

In the spring of 1912, having passed the ranger's exam, I was assigned to the Gardiner District on the Siuslaw Forest as acting ranger. The district was about forty miles long and twenty wide, along the Pacific Ocean from Coos Bay north to a point just south of Florence. There was no railroad at that time and the only road was a dirt one from Gardiner to Florence. The land was cut up by many creeks and wherever the bottom land was wide enough, there were beautiful dairy ranches operated by the most hospitable folks I have ever known. The only means of transporting the ranch products was by boat on rivers and lakes.

I soon found that my boat, scow and horse were nearly useless. The boat (when I could get the engine to run) would be at the wrong end of the river and the scow and horse on the wrong side. I finally got rid of them and used the mail boats and hiked the rest of the time. I seldom tell people how far I used to hike because it is pretty hard for me to believe it myself! It was on one of these hikes along the Florence-Gardener Road (at night) that I had quite an experience. It was so dark that when I held up my hand I could barely see the outline. Suddenly there was a "BOO-BOO-BOO" and a violent stamping of feet. I nearly had the "blind staggers!" I said, "What the h____?" A voice said "God, I thought you were a bear!" We lit a match and looked at each other. We had never seen each other before nor since.

Our supervisor at that time was very frugal, to say the least. When I had to go to Coos Bay the fare was around $7, but with my $75 a month salary as an acting ranger, I could hardly stand much of that. Passengers for the Coos Bay stage would ride the sternwheeler "Eva" from Gardiner to the mouth of the

Umpqua. When I got ashore I would start walking down the beach. When the stage came up behind me I glanced back to see which side the curtain was on (it was always on the windy side). Then, as it passed me on the blind side, I would hop on the trunk rack. This gave me a one mile lift to a convenient location just before the stage reached Ten Mile Creek. I had to leave it there lest they discovered how I had been able to keep up with them for the last ten miles. From there I still had to walk the ten miles or so to Coos River.

A DISASTER NAMED PAT

Grover C. Blake

From August 5 to August 24, 1912, inclusive, I spent with Deputy Supervisor Allan E. Hodgson, in doing extensive range reconnaissance work for my district. We would work one area which could be reached from a central camp then move camp to a convenient spot for another area until the entire district was covered. Mr. Hodgson brought with him a newly appointed forest guard to serve as packer and cook. Pat was his name and we soon learned that Pat knew nothing about either cooking or packing. However, we gave him a break by letting him try. He was probably the worst misfit either of us ever encountered and his presence with us caused many anxious moments. In one instance he put a pack on a horse and tied it down. Before the horse had moved out of his tracks the pack turned under the horses belly and caused him to stampede, scattering the contents of the pack over several acres of ground!

Allan and I had our wives (at that time very young girls) along. Pat's blunders caused a number of trying experiences while we were in the field. On one occasion he came dashing into camp on one of the horses at full gallop and ran over Mrs. Hodgson's pet dog and killed it. This caused much weeping on the part of the owner.

We relieved Pat of most of the cooking duties but on one occasion he was permitted to make biscuits. He mistook air-slaked lime for baking powder and the result cannot be described—only imagined! On another occasion we had prepared dinner and had set the victuals on a cloth which had been spread upon the ground. As we began to gather around to partake of the food, Pat headed for his place but got his feet tangled up and started stumbling, and continued to stumble over our carefully placed pans and kettles, upsetting the entire menu.

On August 14, 1912, Hodgson and I left our wives at my homestead which still served as district headquarters and worked westerly about 12 miles. We planned to make camp at the forks of Bear Creek which was our objective for the day. We decided to pack up the horses and send Pat (although we had never before trusted him alone) with the pack string over the easiest route we could select which was a road traversing through the settlements. We were afraid to trust him alone on forest trails but felt reasonably certain that he could follow a road after we had given him specific instructions about the route, and told him how to recognize the camping place and where to make camp. He should have arrived at the designated spot shortly after noon. Hodgson and I worked along, mapping in the various types of range lands and as we approached the Bear Creek forks, darkness was near and a storm was coming up rapidly. We hurried in order to get to camp and the shelter of a tent ahead of the storm. When we arrived at the camping place there was no sign of Pat, the camp, or the horses. We spent some time riding up and down the two creek forks, calling loudly, but soon became convinced that Pat had never reached the place. By this time it was dark. Rain was coming down in torrents and lightning flashes furnished the only light.

I knew of a ranch some two miles away and along the road over which Pat was supposed to travel, so we headed for that, drenched to the skin and guided by lightning flashes. We knew the ranch would provide shelter and food for us but our concern was for our outfit, as we could imagine all kinds of possible disasters which might have befallen our packer and pack string. Someone at the ranch remembered that he had seen some strange horses on a hillside a mile or so beyond and not far from the road. The horses answered the descriptions of our pack animals. Without hesitating we worked our way through the downpour and blackness of the night (except for the help of the lightning) and found Pat and the outfit on a dry hillside (dry as far as drinking water was concerned) and within 200 yards of level ground and a nice stream of water. Just why he decided to camp here instead of continuing on some three miles or so to where he was supposed to go has never been explained.

Wet to the skin, we made our bed and piled in. The thunder and lightning became so intense that we decided to move our bed from beneath the large, lone pine tree on the hill to lower ground. Hodgson and I started to carry our bed down the hillside through the pouring rain, he at one end and I at the other. It would have been a comical sight had we been visible as we stumbled over the sagebrush down the hill. However, it was far from amusing to us at the time and we were very unhappy over the ordeal, but during the years to come we enjoyed many a hearty laugh as we recalled these events. We had only ourselves to blame for the most part because we knew Pat's limitations and should never have sent him out alone.

I will say this, as I look back to 1912, that Pat was always willing to try to do what he was told. He was a good worker as long as someone was near to tell him what to do and how to do it. No doubt he did the best he knew how and the supervisor felt it would be best to keep him on the payroll until the end of the season and do our best to find jobs for him which he could handle.

Sometime later Pat was alone at Ochoco Ranger Station and was the nearest Forest Service employees to a fire which was reported about four miles from there. I got Pat on the phone and instructed him to go to the fire. I also started for the fire but had 15 miles farther to go. After traveling the 15 miles, I found Pat still at Ochoco Ranger Station. I asked him what the trouble was and he replied that his horse was locked in the pasture and he had lost his key. True, the gate was locked but the fence was made of four barbed wires and could have been taken down anywhere, or Pat could have walked to the fire.

THE TEENS

POLITICS!

Rudo L. Fromme

Am I really one of those persons of questionable morals, a lobbyist? This is what I asked myself on one memorable occasion. Mr. George Joy of the Western Washington Forest Protection Association, and Messrs. Pope and Goodyear, of the State Forester's office, assured me, however, that my action was not actually lobbying. Even if it could be so construed, it was definitely for the welfare of both forestry and the State of Washington or the public.

These men had called me to the old capitol building one morning while the legislature was in session, to advise that they had been "tipped off" that a certain King County Senator was all primed to protest an important forestry bill. This bill was to subject all cut over, privately owned forest land to a State tax for fire protection or require that the owner join the association just mentioned. They felt sure that this senator, known to be a rather special dance friend of mine, would listen to my advocacy of this progressive legislation. Personally, I wasn't so sure. They knew that I carried a card which admitted me to the floor of the house or senate at any time.

The clerk was reading the conditions of this particular bill as I hurried past the "doorman" to the said King County Senator's desk. It happened that he was the really "big man" of the Washington State Senate during this session, and I rather reluctantly whispered my mission. "What do you care?" was his whispered and slightly irritated response. "This doesn't concern the U. S. Forest Service." "Oh, but it does," I responded, "We are certainly most definitely concerned in keeping down fire and smoke wherever it occurs," and I started to whisper the reasons, but I didn't get very far.

"Look here," he interrupted, "I've got a stack of written protests from important and influential timberland owners, and I also have the majority of this senate pledged to back me up in defeating this bill, but," and he looked at me with his jaw tightly set, "If you say that this bill is a good thing for the general public and the United States Government, then these go into the waste basket." He paused, with his hands clutching his notes and the protesting letters, and looking me straight in the eye, whispered tensely, "Is that what you say?" I quivered a bit inside as I answered, "Yes." Into the waste basket promptly dropped the notes and protesting correspondence.

"Does the Senator from King County wish to be heard on this bill?" asked the President of the Senate (the lieutenant governor). "No," came the terse reply, as I was sneaking silently but swiftly toward the nearest exit, hot sweat under my collar and sensing many mystified eyes following me.

Congratulations from George Joy and the State Forestry men, who were watching from the gallery helped to calm me, but I wondered if I weren't using the pressure of "social chairman" somewhat recklessly. However, I was sure that what I did was for the public good, as it resulted in a highly important advancement in concerted forest fire protection.

I don't recall, perhaps I never knew, why it was so important that a certain house member be on the floor ready to speak for a certain bill. It was near the end of a session, and this particular representative went "out with the boys," following a prolonged committee meeting.

It was around midnight, so I was informed, that he began to feel unusually congenial and sufficiently courageous to call at the home of one of the young lady clerks who served as his secretary. A couple of his friends went along but stood in the background while he pushed the doorbell of the darkened house. Eventually a hall light came on and the door was opened by the girl's belligerent looking mother, hurriedly shrouded in a wrapper, or housecoat.

He asked if the certain young lady was in. "She's gone to bed," snapped the mother. "What do you want of her?" "Oh, some of us are goin' out on a little party and I thought she might like to join us." "Is that so!" said mamma. "What you need is to sober up." With that she handed him a "haymaker" or maybe it was a straight jolt from the shoulder. At any rate, it landed squarely on one of his eyes, then the door almost got his nose as it slammed shut.

I was called at my home about 7:00 a.m. by one of these friends, who chanced to know of my interest in stage makeup. He told me that a certain Washington State representative had met with an accident late the night before and had suffered a black eye. He was badly needed for pushing through certain important legislation that morning but refused to show up unless the black eye could be made less conspicuous. I responded, did what seemed to be a sufficiently satisfactory paint job for the moment, then on to the forest office for less likely excitement. I didn't learn the causative details of the damaged eye until a day or so later.

AN OREGON MULE STORY

Ira E. Jones

After I was married in 1912, I asked Supervisor Henry Ireland to give me a ranger district (both the North Powder and the Sumpter Districts were vacant). He said, "Yes, how would you like the Sumpter District?" I really wanted the North Powder District as Sumpter was the supervisor's headquarters and I was too handy and was sent out on jobs, but I said OK.

In 1913, Supervisor Ireland, who had been master of the Sumpter Masonic Lodge, was asked to conduct a funeral at Audry, about 30 miles south. We had a team of young and partly broken mules. We thought this would be a good trip for them so we hand a two-seated hack, hitched them up and started off. The front seat was set high up and the brake was worked from it. I drove and Henry Ireland worked the brake. R.

M. Evans and Harry Wilson (a local jeweler) rode in the back seat.

We made it all right until we reached the top of the Whitney Hill and started down. Henry shoved the brake handle forward but it jabbed the mule in his rump and away they went. It was about two miles to the bottom of the hill. Every attempt to use the brake only made them go faster. The road was narrow and crooked, but fortunately, it was early morning and we met no one. We made the bottom and after half a mile got them slowed down but we had lost our hats.

We made the rest of the trip OK. On the way back we gave a ride to a man who had been fishing in Camp Creek. When we stopped to let him off at the Whitney Mill, one of the single trees dropped off. The mule gave a jump, the tongue dropped down, and away they went again. After a short distance the tongue ran into the ground and broke. We all jumped out. I got mixed up with the lines and was dragged 30 or 40 feet before I got loose. Aside from ruining a suit and losing some skin, I came out all right. The team broke loose, ran into the slab pile at Whitney and stopped.

THE FOREST BECOMES MOTORIZED

M. L. Merritt

Henry Ford put out his Model T. Many were sold in Bend, Oregon. I think it was in 1913 that Ranger Earl Austin at Crescent bought a Ford. He did much chauffeuring for the supervisor. One day, going south from Crescent on a fair dirt road, his speedometer registered the dangerously alarming and very unusual speed of 17 miles per hour.

By the end of 1914, however, all rangers and the supervisor had cars—all Fords. This brought to light the poor condition or lack of roads. Most roads were mere ways through the timber with roots and rocks in the tracks.

At Crescent, Perry South, who followed Earl Austin as ranger, organized the local men into volunteer road crews and attacked the roots and rocks vigorously. Soon roads to Crescent and Odell Lakes and south to the Klamath Indian reservation were passably good. This continued until there were usable ways to most lower elevation objectives in this area. Other rangers acted similarly. Never were so many miles of road built with so little. None was good, but you could drive a Ford over them and by the time I left in 1916, one could drive to every ranger and to most guard stations on the forest.

Some of these roads later became important travel routes. Ranger Harriman's "China Hat Road," from Cabin Lake Ranger Station north to the Bend-Burns Road and Ranger John Curl's road from Paulina to East Lake are now main roads. I think I was the first person to drive a car to East Lake. Wasn't sure that I'd get back, but did.

JUSTICE ON THE RANGE

Harold E. Smith

When I took over the Pine Mountain Ranger District, Deschutes National Forest, I was told that I might expect some trouble from the lawless elements operating in that area. Antelope Spring seemed to be the operating center for at least one gang. The so-called spring was merely a water hole scooped out in the center of a sand flat. Pine Mountain, just north of the spring, separated the Antelope Spring area from Millican Valley where George Millican operated his cattle and horse ranch. Pine Mountain was prime grazing land. Thus the Millican stock grazed the north side of the mountain while the Katzman boys, John and Charley, ran horses on the south slope adjacent to Antelope Spring. Both outfits paid the usual grazing fees.

The first evidence that I had that the Katzmans were not to be trusted was when a side of bacon disappeared from my camp. I first suspected the dog but when I found that he showed no sign of having eaten anything more than his normal ration, I decided that Katzmans had taken the bacon but I remained silent. Next I lost a fountain pen but again lack of evidence kept me from making a fuss.

As the 1913 fire season approached, Supervisor M. L. Merritt showed me a letter from my brother Chester at Grants Pass, asking for the job of forest guard on my district. Although Chester lacked two months of being 21 years old, he was hired on my recommendation. I had a feeling that Merritt was somewhat disappointed when he met Chester because of the boy's youthful appearance.

Things went smoothly on the Pine Mountain District until mid-August when an order came from the Portland office detailing me to the Siuslaw Forest. Merritt and I sent in a joint protest. George Cecil replied that they had duly considered that angle and that I was to proceed as instructed.

So I rolled my bundle and hit for the coast, leaving Chester in charge at Pine Mountain under a sort of remote control of John Curl, the ranger at LaPine.

It was generally known that Katzmans were secretly hostile to the Forest Service, especially since we were virtually located in their dooryard. It appears that they reasoned that with me out of the way they could intimidate the guard and that he would voluntarily remove himself from the area. That line of reasoning proved to be entirely wrong. I had barely left the scene when a rash of fires broke out. Evidence indicated that they had been set by the Katzmans. On one occasion, a string of 13 fires was set along the north slope of the Paulinas. Chester discovered the smoke from his Pine Mountain Lookout and immediately set out to bring the situation under control. By the time he had subdued eleven of the fires, he realized that the last two were beyond his control so he got word to Bend that he needed help. Merritt sent out a crew to control the fires. He also sent an investigator. The suppression crew soon quieted the fires and the investigator pinned the blame for the fires squarely on John Katzman. First, he found horse tracks leading from Katzman's corral direct to the line of fires, then returning by another route. Entry back into the Katzman pasture was made by opening a gap in the wire where the fence joined the rimrock. Apparently this was done so he would not be seen returning the horse through the corral gate. Positive identification of John's horse was made by a peculiar track. John's horse had an injured foot which required a special shoe with a crossbar connecting the heel. This shoe print showed plainly in the tracks. The route was carefully mapped, including location of the fires, and the horse tracks were photographed. This evidence was carefully filed in the office at Bend. Merritt wrote me of these developments and finished by saying, "I must say your brother used rare judgment in handling this matter."

It was a typical September day on the high desert when an elderly man pulled into Antelope Springs driving a span of donkeys hitched to a cut down buckboard. Except for his clothing he might have been mistaken for one of the "Mountain Men" of the 1850 period. Instead of the fringed buckskin, his wearing apparel consisted of a pair of faded blue jeans, a jumper of the same material and a slouch hat pulled down over a balding head. He was of slightly less than average size, lean as a piece of jerked venison, slightly bowlegged and his frame tilted slightly forward from the waist up. His equipment consisted of the tools customarily carried by the typical prospector. After hobbling his burros and putting them out to graze, Wells set up a small tent and busied himself tidying up his camp.

That evening when Katzman came to the spring for water, Wells made himself known as a dyed-in-the-wool prospector. However, at this particular time he had turned his attention to searching for uranium instead of gold and the other common metals. He also made it clear that he had no use for the Forest Service, having been engaged in legal battles with them aver mineral claims, timber, etc. Katzman had no use for the Forest Service either because of their restrictions on grazing. So this seemed to be the beginning of a real buddy-buddy relationship with the Forest Service as a common enemy.

Later that evening, under cover of darkness, Wells delivered a letter to Chester. The letter was from Supervisor Merritt stating that Wells was a private detective hired for the sole purpose of spying on the lawless elements operating in the district, especially the Katzmans. It asked Chester to cooperate fully with Wells but warned against becoming too "friendly" lest the Katzmans might get wise to the conspiracy.

Next morning Wells climbed the side of the mountain up to the rimrock and there began to excavate. That night he brought down a sample from his "mine" and showed it to the Katzmans who knew nothing whatever about minerals. Actually, the only trace of mineral in the rock was a rather ample showing of mica. Wells played that up as a sure indication of uranium and the range riders went wild with enthusiasm. Immediately they wanted to stake mineral rights around the Wells discovery. But the old "prospector" counseled against any hasty action. "We are doing all right," he said, "And we don't want a swarm of claim jumpers in here, which is what they were sure to have if the news of his strike ever leaked out to the public." "Wait," he said, "Until we are sure of our values and then we will file on every foot of that mountain that the law will permit."

Wells, it was soon learned, was an excellent storyteller and his camp gradually took on the role of a social center. Every night after work the Katzmans and the Forest Service representatives would gather around the Well's campfire and listen to the old "prospector" spin yarns about his exploits in the various boom camps throughout the West. My brother told me that though he knew that there was scarcely a grain of truth in the yarns, it was still entertaining and the range boys lapped it up with a high degree of relish.

As the prospect hole became deeper, more samples were displayed to the Katzmans. A sample was sent out for analysis. Presumably it was going to an assay office. Instead it went to Supervisor Merritt. Merritt, with some outside help, concocted a fake assay report and sent it back to Wells. Merritt's report was highly encouraging and Wells had a hard time convincing his associates that they were not yet ready to stake the ground. "We need to know more about it," he said, "In order to forestall the possibility of a law suit. In the meantime, silence is by far the better strategy."

By this time the ties of friendship between the prospector and his range riding acquaintances had cemented to a point where Wells was invited to share the Katzman house as a hedge against the approaching fall weather. Wells readily accepted and moved in. Not only did they share the house and housekeeping facilities, they also shared their trade secrets with the old man. Within a short time he knew of their horse rustling; he also knew of their arsonist activities, erroneously designed to remove the Forest Service influence from that part of Deschutes County.

Food supplies became a matter of concern. These hearty men eating three meals a day was making a dent in their rations. When John brought up the subject of going to Bend for more supplies, Wells saw no reason for such extravagance. He said he knew of a homesteaders' cabin nearby, well stocked with food; the owner was away engaged in logging for Brooks-Scanlon in Bend. Actually, the food had been placed in the cabin by the Forest Service. Labels had been removed, the cans branded and labels carefully replaced. A night trip was planned. Wells pleaded fatigue as the reason for not going, saying he needed a good night's rest in order that he might continue his mining activities the next day.

Armed with a description of the homestead cabin location and the best means of access, the Katzman boys lifted a substantial supply of food and returned to their headquarters before daylight. Thus, they had sprung the trap that had been set. Wells passed the word to Chester and he relayed the message to Merritt. Merritt notified the sheriff and he proceeded to the Katzman hangout, armed with a search warrant. There he seized the stolen goods, removed a few labels revealing the identification marks placed there by the Forest Service before the goods were planted in the homestead cabin. Both Katzmans were arrested and taken to Bend and charged with larceny from a dwelling. The wheels of justice must have been turning faster than they do now. Within a week the boys were tried, found guilty and sentenced to a year in the correctional institution.

SHAW'S SOLILOQUY ON HER SOJOURN IN LANDS

Helen D. Shaw

Arthur Salinger was primarily responsible for inflicting me on the Forest Service. However, he can't be blamed too severely for this, because the poor guy expected me to be around for only about five months—he didn't realize that I'd finagle it into a 23 and one half year stay! He interviewed and hired me on a Friday afternoon and, since in those days we worked Saturday mornings too, he had me come to work the next day, giving me an extra couple of days on salary and I certainly needed them.

On Saturday morning he led me into recreation and lands and introduced me to Jack Horton, then chief of the division. My name was still Daugherty at that time, and when I was introduced, Mr. Horton's only greeting was, "My God, another Irishman!" Since he was married to an Irish gal, that wasn't too much against me.

The next member of the division was Laura Orser, an auburn haired, twinkly eyed little gal smart as a whip. She and I immediately became friends. When I was introduced to Laura she was rapidly converting the worst mess of hieroglyphics I'd ever seen into a neatly typed copy. My eyes bugged out at this awesome sight but she assured me that I'd soon be able to read it as easily as she. It certainly didn't seem probable but she proved to be right. The above mentioned hieroglyphics were the brain children of one Fred Cleator, a gum chewing, Will Rogerish type person who truly loved the forests, and described them in flowery prose that is instantly recognized as his by old-timers. I did most of his work, so it was well that I learned to translate his chirography. He had one habit that made this especially important—he'd wake up in the middle of the night with an inspiration and scribble it on a bedside pad without turning on the light! The next morning, likely as not, he'd be unable to decipher it and would be forced to bring it to the office for me to read. Believe it or not, I never failed him! One thing that endeared him to me was his willingness to let me bully him! He let me change his dictation and longhand notes any way I wanted to, in fact he expected me to do it. One time, when they took me off of his work and assigned another girl to do his work, he almost wrecked the office. She was a perfect lady (something nobody ever called me), and typed his work the way he gave it to her. It made him very unhappy.

When I landed in lands, Althea M. Wheeler was chief clerk. She had practically grown up with the Forest Service and carried an unbelievable amount of Forest Service knowledge in her clever brain. She knew the law pertaining to land acquisition better than the lawyers and they frequently consulted with her. She didn't care a hoot about exercising her supervisory privileges and as long as things ran smoothly, she attended to her work and left us to ours.

For the first several years I was in the Forest Service, second in command of our division was red haired Tom Sherrard, ex Supervisor of the Mt. Hood Forest, complete with the temper of a buzz saw and a heart as big as a cow and he frequently displayed both characteristics. He and Fred Cleater occupied the room next to the division "harem," and one day when I had opened my brand new oiled silk umbrella and parked it in their room, he kicked it clear across the office. It was through no good management on his part that it wasn't split from stem to stern. Another time, believe it or not, he actually spanked me! His reason for this action was that I persisted in pawing through his basket searching for a file he wanted. He insisted it wasn't there and threatened to spank me, but I kept on. Even after he whacked me, I kept on and found the file! Having Bud Burgess as a witness to the whole affair roused his red headed ire. He paid off, though. Returning late from lunch with a package under his arm, he stumped into his office. A couple of minutes later, he bellowed, "Why in h___ don't you ever clean out this basket?" When I answered on the double, I saw what appeared to be a ream of paper with the file fastened to it with a rubber band. Close inspection revealed a three pound box of chocolates! On the other side of his ledger was the way he drove far out of his way to pick me up every morning and take me home every night after I had sprained my ankle and pulled the ligaments so badly that I was on crutches for six weeks. What a man!

I hadn't been in the division too long when Jenckes Mason came in from the field. He and I didn't click very well—he wouldn't let me bully him! He'd dictate words of which I didn't approve and wouldn't let me change them. It was practically the case of the irresistible force meeting an immovable object, only I don't know which was which. Finally, he gave up and migrated to timber management. When he came back into our division as its chief a few years later, things were different. One or both of us had mellowed and when he graciously suggested that I change his dictation if I didn't like it, I didn't care to. By that time he dictated beautiful letters and memorandums. We got along wonderfully.

LETTER TO THE CHIEF

Submitted by Carl Wilson

Editor's Note: Carl Wilson didn't exactly "remember" the following early day letter written at a time when regions were known as districts. He dug it out of his files. We think it is a gem, a diamond because of the many facets we see in it. Carl thinks the author displays a special kind of fearlessness. –E.J.

The Forester, May 18, 1916.
Washington, D. C.

Dear Sir:

Your letter of May 10 is received. While this article was being prepared we realized fully that it was not strictly correct to attribute entire forest growth to fire protection. As you were well aware, the preparation of this article was undertaken at Mr. Graves' request. This particular point was discussed with him in a preliminary way and he felt that for the purposes of

Angora goats feeding on alligator juniper, Datil NF, New Mexico - c1906.

Riverdrivers pulling a "plug" on Douglas Creek, caused by running too heavily at low water. Cheyenne NF, Wyoming - c1908.

this article it would be best to attribute all growth to fire protection in view of the difficulty in making a segregation.

In reference to the criticism of the growth figures, which appears on page 7 of your letter, general rules for estimating growth were sent to the various forests for application at the time that the articles were requested from each. The growth figures given by each forest are assumed to be the best available. There did not appear to be enough material at hand here to justify any criticism of the figures submitted. The discrepancy between the Tahoe and Eldorado growth figures is attributable to the large amount of second growth on the Tahoe and the small amount on the Eldorado.

We have no one available here who can undertake the revision of this article before next winter, and even at that time I do not believe that we could handle it as well as you can with your available editorial staff. I suggest, therefore, that you shape the material up for publication in any way that is satisfactory to you. This will probably involve but little more work than would be involved in the preparation of criticisms of our attempts. Very truly yours,

COERT DUBOIS, District Forester.

A SPECIAL USE PROBLEM

Rudo L. Fromme

I was sent to Alaska in the early summer of 1917, by former District Forester George Cecil to familiarize myself with administrative problems on the Tongass and Chugach forests, on the possibility of later being called on to try to temporarily fill the large launch lurching and muskey mushing shoes of veteran Supervisor Bill Weigle. He had applied for a forestry regiment major's commission in the first world war. Unfortunately for Bill, but otherwise for Alaska, neither plan materialized. However, I achieved a most enjoyable two months sight seeing trip throughout the coast areas of these forests with the congenial cooperation of Bill, his Assistant Supervisor George Drake and various rangers.

As Bill was giving me an itinerary for the final Anchorage corner of the Chugach, he added, with a slightly skullduggery grin, "You might want to look at a housing tract (or words to that effect), which happens to be just outside the city limits of Anchorage but just within the boundary of the national forest. It looks definitely like a special use case, but I have not felt authorized to issue permits for that type of occupancy." He further mentioned that the Forester was highly incensed and felt that we should insist that the City of Anchorage accept the problem as strictly its own, and either annex the land and buildings by elimination from the forest, or wipe it off the landscape. Maybe you've guessed it by now—a red light district.

The City of Anchorage seems to have had more of a moral movement than other Alaska towns at that time. At any rate, the law enforcement officers were persuaded to do something about "cleaning up the town;" with the result as indicated. The gals of easy virtue were persuaded to move out, which they did, about two miles from the city center to where the city limits butted against the Chugach Forest.

I was, naturally (perhaps the adjective should have been "improperly") inquisitive to get at least a cursory birds-eye view of this special housing unit. There being no plane or balloon available at that time, I had to view it on foot, but I had my willing Kodak on easy leash. Glancing at the old prints now, all outside views, I note two double rows of small one-to-three room huts facing each other across a dirt street, one set of about ten each extending for the equivalent of a short city block and the other about one half block. There seems to be about 30 of these batten board buildings laid out in strict geometric lineup, as though established by city or other surveyor. The settings is extremely bleak with unscreened outside "plumbing," a very thin stand of small trees in the immediate background, but in the far distance a splendid range of snow-capped peaks dominated by Mt. McKinley.

On the sidewalk I set up my Kodak on a tripod for a side view shot of the front portion of one of these tiny huts, but primarily to include a protruding bracket sign above the doorway which read "IRISH ROSE." Rose became inquisitive, stretched her neck out the open front window and asked What and Why? I told her what was actually my spur-of-the-moment deep dyed plot. I said that I was contemplating moving to Alaska very shortly from the States, and was now taking a brief look around for the best location. Also, that my wife had particularly requested that I bring back photos of any houses and surroundings which would likely appeal to her. "It looks to me," I continued, "As though this little abode might ring the bell. She's part Irish herself, and this looks like a nice, quiet neighborhood, everything so neat, and—"Phooey," snapped Rose; jerked in her neck and slammed down the window.

THE INITIATION

H. R. Richards

January 1919, was cold on the St. Regis River in western Montana when the "kid" ranger and his bride arrived at the two-room log cabin which served as living quarters and district ranger office. So cold, in fact, that 20 degrees below was far from uncommon and shoes would freeze to the floor overnight.

Spring warmth took away the ice and snow but brought complications familiar to all woods-living people in those days. Bugs of all kinds and their offspring that had had their basic training in lumber camps considered no obstacle too tough to surmount. To get to the bed, all else failing, they would crawl along the rafters and chute from there. One application of calcium cyanide and sulfuric acid usually turned the trick though, and was accomplished as follows: Put calcium cyanide in a paper bag; tie the bag with string; run one string through fence staple in ceiling and out the window; seal all cracks; put the sulfuric acid in an iron pot; get out and through convenient crack, to be plugged up later, let paper sack down into pot. Results? Gratifying and stupendous. Just like a baby A-bomb!

Having no roads to the station, transportation was limited to a hand-operated speeder on the railroad and horses on the trail.

Horses were wintered on the open range and were more than just a little owly when caught and saddled for the first time in the spring. The natives predicted the brute would kill the ranger and this he wholeheartedly tried to do the minute he was boarded.

The speeder was used one dark and stormy night to visit the closest neighbor two miles up the grade to see if he would be the one and only lookout for the summer. He agreed to take on the job and, because his lamp had been broken, the ranger left his lantern. Going back down the track in the dark, however, was something else. Out of the wet and inky darkness a black object loomed ahead. There was just time to holler and push the little woman off on the river side. Down the bank she rolled, straight toward the icy water. The crash came and speeder wheels and ranger flew in all directions. It seems the section crew from a neighboring town had been to a dance out of their section and, to avoid detection because of unauthorized use, had been traveling without a light. Everyone was soon accounted for except the ranger's wife who, after much fearful searching, was found hanging to a willow half in and half out of the river. Quite an experience for the little school teacher from Ohio!

There was soon to be an addition to the family and with no grocery, dairy, or milkman handy, a cow was a necessity. Only range cows were to be had. After a hard, hot day of dragging one of the spooky, half-wild critters up the trail and track, the drive was met near the station by the little woman anxious to view the stock. One look was enough for the cow who had never seen a woman and off through the brush she went, dragging other things besides her tail, namely the ranger. As the dust and leaves slowly settled, the young lady thought the honeymoon must surely be over. The howling and cursing coming through the air were hardly the things one expects to hear, especially on one's first wedding anniversary. Crouched down behind a rock the tears flowed like wine.

As isolated as they were, the great flu epidemic hit them, too. No doctors were available of course, and this was the real McCoy. During the peak, the ranger gave his wife a half glass of castor oil with peach juice as a chaser, and it was many years before she could even smell peaches without gagging.

Fire season opened early that year. It hadn't rained since March, and soon after April, smoke could be seen almost anywhere in that area. No checks had been received from the regional fiscal agent since April and the crews were more than just a little unhappy, and justly so. The ranger realized that something had to be done and done quickly. The sympathetic Milwaukee Railroad agent at St. Regis wired his dispatcher to have the Olympian stop and pick up the chief forester. It stopped. The ranger, to the agent, was just as important as the chief, and he didn't want to know the difference. In Missoula the fiscal agent was indifferent to inquiries about the delinquent checks. Seemed as if perhaps he thought all that money was coming out of his own pocket. As it turned out later, he was caught trying to put about two hundred and fifty thousand dollars in his own pocket. After a big hassle and being told that the ranger wasn't returning without the checks, he finally wrote them out and the ranger was assured then that he wouldn't be mobbed on his return.

The return was just as dramatic as the departure, if not more so. An ancient switch engine without screens on the stack or grates had been ordered out for some reason or another and the engineer, peeved at having to run such a clunk, had roared wide open the length of three ranger districts, setting more than fifty fires before he was finally stopped under arrest in Avery, Idaho. The one lone guard had put out twelve fires by himself, but it was just too much for him and one fire took off and the fight was on. All available tools and men were gathered and the crack westbound Olympian boarded. The conductor said he'd be damned if he was going to stop his train at every old black stump along the right-of-way but, when the fire was reached and the ranger said to stop, he stopped. They were all feeling pretty good about catching those fires so soon, but their elation was short lived. Burning conditions went from bad to worse. Just like a charge of dynamite with a short fuse, it wasn't long before things exploded. Dry lightning storms and the railroads soon had more fires going than the ranger's limited forces could compete with, and it was many crews and some 20 camps later in late October before the snow buried the last ember. Cold

Creek exploded and the winter survey crew found thirty thousand acres blackened,

The Chief in Washington, D.C., wanted to see a big fire and they had some dillys for him to see. The neighboring ranger up the river had troubles. A jim-dandy on his district broke loose and came roaring down the canyon past the ranger station and wound up 17 miles from where it started, burning a large part of the district. The ranger below lost one that joined with our Cold Creek Fire and soon St. Regis and the station was just an island in a sea of fire. Searing flames, choking boiling clouds of smoke and drafting hot embers were everywhere. The sun was a mere orange ball for days on end. It was reported at one time that the Red Cross in Minneapolis was coming out to evacuate the town of St. Regis, but nothing ever came of it.

When Chief Graves arrived, the ranger was appointed a committee of one to show him around. He took him out to the firelines and beyond for three miles across the country to another ranger's fire and camp. They crossed over windfalls from the fire of 1910, that in many places, were higher than their heads. The good Lord protected them that day. Later it was learned where they had traveled, making less time than a mile an hour, the entire basin had been consumed by a cyclone of fire in just a few minutes. So far as we know, the Chief just gave a verbal report stating that it was awful.

The mineral claims expert was assigned to gather evidence and make up a trespass case against the railroads. After the survey and much expense and months later, the case was brought to court. It was promptly dismissed since it was decided that one branch of the federal government could not sue another. The railroads were still being operated by the government, having been requisitioned during World War I.

These are just highlights of the first year as a ranger. Other things come to mind though, such as the time the ranger flagged down the passenger to load a crate of chickens. Besides not having any authority for such acts at this point, the rails were icy and it was one fourth mile past the crossing which was on a curve, before the train stopped with great screeching and groaning. The engineer was all smiles and good natured as he backed up, but the conductor was on the verge of apoplexy and the ranger feared for his life.

The time the ranger had his pants burned off and had to borrow a pair from the camp cook before he could return to the firelines and had three pairs of caulked boots burned out from under him during the season.

The time the little woman had to evacuate the cabin because of fire and packed her most precious possessions—her pajamas, her husband's picture and material for a new dress.

A few lines here too, would be appropriate in tribute to the wives of the early foresters. Many suffered hardships not to be found today and many made homes against difficult odds, also acting as clerk, dispatcher and roustabout in the absence of the ranger which was at least 80 to 90 percent of the time during field season without reimbursement. They are women we can all be proud to have as wives and mothers.

THE TWENTIES

NORTH FORK

H. B. "Dusty" Miller

To the rest of the world North Fork was "No Where." The only claim to distinction of this small Madera County community is the fact that, for nearly forty years in the first half of this century, it was the site of the supervisor's headquarters of the Sierra National Forest.

My North Fork experience spanned a period of about six years from early 1919 through 1925. I was not there as a Forest Service employee, but rather as an eight year old school boy. My family moved there from Ashland, Oregon, immediately after World War I. My father, John M. Miller, was to set up an investigation and control project for the pine bark beetle epidemic in that area. He was actually under the (then) Bureau of Entomology, so our family was not Forest Service per se, but for all practical and social purposes, we were part of the Forest Service community. Dad had begun a Forest Service career on the same Sierra Forest in 1909, and was ranger at Mariposa when his interest in forest insect problems led to a transfer to the Bureau of Entomology. Because of this, he was already well acquainted with the area and Service personnel of the Sierra Forest.

In 1919, the Sierra National Forest (originally the North Sierra Forest Reserve) was a sprawling wilderness reaching from the crest of the Sierra Nevada westward to the foothills along the San Joaquin Valley and from Yosemite National Park south to the Middle Fork of the Kings River. It included a vast treasure of natural resources, the water for San Joaquin Valley agriculture and some of the most breathtaking scenery in these

United States. Road access existed only along the western fringe and interior access was strictly by trail and horseback. The resources were undeveloped and recreational travel was virtually nil. The nerve center of all of this was a small, two story frame building in North Fork.

At the time of my arrival on the North Fork scene, M. A. Benedict was the forest supervisor. Because of the remote location, it was virtually a “must” that the Service furnish living quarters for the employees and their families. In 1919, there were four residences, in addition to the headquarter’s office, warehouse and barn. When our family arrived, there were no quarters available at the compound, so we had to settle for what we could get and this turned out to be Sight Rock Ranger Station. This was a dilapidated old building on the top of Manzanita Ridge, about two miles from the compound by road but only a half mile by shortcut trail down the ridge. The place was liberally supplied with bugs, snakes and the usual varmints, but, as an eight year old, I saw nothing wrong with it. I do recall my parents having lengthy discussions about the merits of the place. Within the year a more modern abode became available and we moved to Douglas Ranger Station, just above South Fork, where we resided for the next three years. Douglas Station was the only house I can recall in which the bathtub was built into the kitchen.

The community of North Fork remained fairly static during my years there. The village and surrounding area (including South Fork) probably did not number over two hundred souls. The village itself was a collection of rather drab buildings scattered along a dusty road on a gentle hillside with black oaks, brush patches and scattered openings. There were two general stores, Franklin’s and McDonald’s, and the latter of these hosted the local post office. There was a large vine-covered building that served as a hotel of sorts and I believe there was a small blacksmith shop. About eight or ten other buildings, mostly residences, completed the village scene. Fresh milk and produce were not available at the local stores since refrigeration was yet to be invented. It wouldn’t have sold anyway because a number of the locals had milk cows and virtually every family raised a garden. Both stores had a single hand-operated gasoline pump, but only the one at McDonald's Store had a hose that could deliver directly into a car. Fresh meat and milk could be obtained at Thornberg's Ranch and Hotel at South Fork, but it was a by-appointment arrangement because of lack of refrigeration. This latter satellite community, about two miles south of North Fork, was about the only place where meals could be obtained and the hostelry where most RO or WO visitors stayed when they were visiting the Sierra. There was no garage or car repair facility available in the area—further evidence that the automobile invasion was only in the beachhead phase.

Paradoxically, for all of its lack of living amenities, the North Fork area did have commercial power. In fact, this was about the only commodity that the area exported at that time. The San Joaquin Light & Power Co. had established a generating system of three powerhouses along the North Fork drainage to serve the City of Fresno. Most of the electrical use was devoted to illumination since virtually none of the modern day kitchen appliances existed at that time.

The lines of communication and supply to the outside world consisted of fifty miles of county road via O'Neals and Friant to reach Fresno. The first thirty miles in Madera County were all unsurfaced, dirt (mud in winter) road. The final twenty miles, after reaching Fresno County, were dust oiled. Depending on the weather and the car you were driving, it was about two and one half hours to Fresno. A grounded telephone line ran down to the valley, but during my time there this was upgraded to a metallic circuit. Any telegraph messages had to be called over the line from Fresno. A stage line did operate from Fresno to North Fork six days a week. This consisted of a Cadillac touring car which left Fresno at 8:30 a.m. and arrived at North Fork about 11:30 with the return trip after lunch. This transportation delivered the mail and sundry items that were to come to North Fork. It was always referred to as "the stage," and I never did hear the term "bus" used. Entertainment in the Forest Service community consisted of occasional potluck gatherings on summer evenings or card parties during the winter. Incidentally, the card game played was not bridge, but rather, a game called

"500." It was similar to bridge in some ways but I have never heard of the game since.

The nearest movie theaters were in Fresno which was out of the question for casual entertainment. We did make special overnight trips when features like "Ten Commandments" or "Covered Wagon" were playing there. The summer of 1924 was special because a motion picture company was on location with their people and actors staying at South Fork. They had with them a motion picture projector and pictures they showed outdoors after dark. All locals were invited. The films were all black and white "shoot-em-up" serials to the delight of our younger generation. Surely the career of "cowboy" was the only worthwhile endeavor in this life.

The routine diversions of the summer were swimming in the local creeks (North and South Forks) and an occasional trip to Bass Lake. The same two creeks also had a scattered fish population that we constantly attacked but never seriously depleted. I believe our favorite diversion was listening to the stories and experiences of the high country when our respective fathers returned from one of their field trips. As the years went by and Dudley Booth and I grew a little older, we received the ultimate reward. We were invited to go along too.

Time did not stand still at North Fork despite its isolation, and I have to review the significant events that happened while I was there. The Southern California Edison project was under construction and its engineering wonders were one of the favorite topics of conversation in the Forest Service community. However, it was on the south side of the San Joaquin River and removed from the North Fork area, so I did not see the construction phase. The San Joaquin & Eastern Railroad, built in connection with the project, was visible on our trips through Auberry when we traveled that route to the valley towns. In 1924, the construction of the Minarets & Western Railway was commenced to serve the logging of Wishon timber around Bass Lake by the Fresno Sugar Pine Lumber Company. This line did not go through North Fork, but it was close enough to hear the train whistles. The project created an influx of new people and employment in the area and put some new faces in the local school. Suddenly, life seemed more exciting. Also, at about this same time, the idea of using aircraft for fire detection was implemented. These were military trainers flown from the Bay Area (I believe) which left them with a rather limited time for patrol. I am uncertain if they ever made an initial detection, but one afternoon the flight circled the headquarters office and dropped a message. There was no radio communication between air and ground, so that was the only way a fire could be reported.

About 1923, we became aware of a new word that was creeping into the American vocabulary; RADIO. As I understood it at the time, it was sort of like a telephone, only you didn't need a wire. It was late 1924, when the first demonstration was set up in the warehouse one evening and everyone gathered not knowing what to expect. There were no commercial programs available, so the event was a bust.

However, commercial radio and programs were on their way. Ben (Supervisor M. A. Benedict) bought a set which I believe was the first privately owned radio set in the area. On January 1, 1925, our family was invited to his house to listen to the Rose Bowl game. It was a landmark experience, and from that point on the isolation of North Fork and its unique life style gradually became only memories.

UP A TREE WITHOUT A SADDLE

Grover C. Blake

On July 3, 1920, on the Ochoco National Forest, I witnessed what I believe to be the most unusual of all the unusual spectacles of my career. I saw it with my own eyes and still I don't believe it, so I will not expect the readers of this tale to believe it either. I found a full grown horse fast in the forks of a tree.

Virgil Allison, foreman for Elliott, Scoggins & Wolfe, road contractors, and his wife were riding with me along the Vowell Trail near the summit of the mountain when we saw this horse in the tree not far from the trail. He was an unbroken range horse about three or four years old and probably weighed about 1,100 pounds. The tree forked about two feet from the ground and the spread at six feet was not more than 15 inches. The hind feet of the horse were on the ground on one side while the head, neck and shoulders were on the opposite side of the tree with the front feet about four feet from the ground. His body was wedged between the forks until he was pinched as tight as it was possible for him to get. His struggles had worn all the hair and most of the skin off his sides where they contacted the tree.

He tried to fight us when we came near. I took the axe off our pack horse and we started to chop off the smaller fork, about 16 inches in diameter. While we were so engaged, another man, Mr. Bill Peterson, came along and assisted. When the horse was finally released he was in a bad way and very wobbly. He was able to keep on his feet, however, and soon wobbled away without saying "thank you." No doubt he had been fast in the tree for at least two or three days. The question that bothered us was, "How did he get there?" The tree stood alone in an opening of considerable size and the only theory I could advance was that a bunch of range horses were standing in the shade of the tree, fighting flies as they would likely be doing at this time of year and started fighting each other and this horse was cornered somehow and jumped at the only opening he could see. It took a tremendous leap to get his body high enough to get between the forks of this tree. However, it may have happened some other way, I do not know. I have always regretted that we did not have a camera on that day of all days as I realize I need proof.

THE HOH GUARD'S HO! HO!

Rudo L. Fromme

As forest supervisor I was invited to show Olympic Forest lantern slides to the Port Angeles Mountain Climbing Club on one occasion. They organized along about 1920, with much the same objectives as the Seattle Mountaineers, but on a much smaller and more local basis. While visiting their Lake Crescent Lodge under special use permit at one time, I was informed that they were planning a climb of Mt. Olympus on a specified date. I declined an invitation to accompany them because of other prearranged plans.

However, while visiting the Upper Hoh Guard Station shortly thereafter, I suggested to the guard that it would be a nice gesture if he could climb up the Hoh Glacier and meet them as they came over Blizzard Pass from the Humes Glacier. He nodded halfheartedly but was noncommittal. I knew him to be of the "wild and woolly" type, not particularly "in step" with the school teachers and office nature lovers, so didn't press the suggestion. It was some weeks later, when visiting the district ranger again that I learned what a twist my proposal begot.

Ranger Morgenroth learned of it from one of the club members, then wormed this "joke out of said practical joker." On the day just preceding the planned Mt. Olympus climb, he went to his saddle horse stable, shoveled up a gunny sack of fresh manure and made the strenuous backpacking climb up the long Blue Glacier, then over onto the Upper Hoh Glacier. Here he proceeded to empty the sack in judicious clumps from the point where Blizzard Pass climbers would most likely hit the Hoh Glacier, then along the left edge, heading the normal exposed crevasses, toward the main peaks. It undoubtedly took him a full day of very hard work, but he was buoyed up with self-congratulations of the appalling wonderment and possible belittlement he would cause these comparative tenderfeet. Of course, it was absolutely impossible to put a horse over this route.

I should also mention that he carried a horseshoe, not so much for luck, but to press onto the hard snow just enough to look mysterious. I wonder how he would have accounted for his time in that planned trip and hour schedules of later years. It was a time of more freedom and inspired initiative.

COMMUNICATIONS ON THE WEST BOUNDARY RANGER STATION

Harold R. Bowerman

The West Boundary Ranger Station on the Cascade National Forest was located on the Middle Fork of the Willamette River at Lawler, just above Eula, and nine miles above Lowell, Oregon. The ranger station has long since drowned in the pool behind the Lookout Point Dam at Lowell.

The Southern Pacific Railroad finally was punched over the Cascades past Oakridge. This then became the main line and carried tremendous tonnage within less than 300 feet of the office. The giant malletts, with five or six helpers per train, roared their long string of cars past the office window. All conversation ceased, including fire reports, during the passing. West Boundary was on the forest boundary and at the head of a long grade. Sanding of the flues was done here because of fire restrictions on the national forest. This additional din added greatly to sound "pollution."

The West Boundary Ranger District maintained 30 miles of metallic telephone line on the Southern Pacific right-of-way. Transportation for line maintenance was a hand powered, three-wheeled velocopide. Heavy rail traffic, compounded by frequent tunnels and signals hidden by a curving track, would have curled the hair of the present day safety officer. A packsack of tools and maintenance supplies on the car's outrigger made the car ride smoothly. The maintenance man soon learned to keep the outrigger free because of the necessity to make an occasional derailment to escape an unexpected mallett engine on the silent downgrade. The square sawn poles presented an additional hazard because of the larking spike knots and the barbed wire fence stapled to the base of the pole.

The single wire telephone lines along the trails and roads were hung loosely from split insulators to permit the wire to come to the ground twice in each one quarter mile. This construction helped prevent the breakage of the wire when caught under falling forest debris. Maintenance and fire crews carried hand telephones equipped with oscillators or buzzers with which to call the ranger station. A loud speaker, or howler, on each line amplified the buzzer signal in the office. The howlers also proved the bain of uninformed field men who were prone to gossip after hours on the telephone. The simple old telephone receiver, hooked between line and ground frequently simultaneously served as transmitter and receiver. They were much lighter to carry.

Trail and telephone maintenance were carried along by the same two or three man crew. The mule string moved the crew between campsites. Moving schedules were dependent upon predetermined schedules when there were no telephone lines in the area. Rations and tempers tended to become short when unexpected maintenance problems prevented the crews from meeting the packer. Power saws were nonexistent. Hand powered crosscut saws (misery whips), heavy mature timber and experience forced the crews to take advantage of labor-saving devices. Cuts in fallen trees were flared apart on the lower side to permit the section of log on the trail to roll free. A pole was laid under the section to be cut so that the cut could be pivoted out of the trail and rolled to the lower side.

INCIDENTS

Ira E. Jones

When I first started work on the Whitman National Forest, there were seven ranger districts. Each ranger had only one guard on duty during the summer months.

In the early 1920s, you could not get a raise without a change in title (due to the old statutory roll). One of the peculiar results of this was that at one time on the Whitman there were four men with the title of Forest Supervisor—R. M. Evans, Johnnie Irwin, Otto Zimmerli and myself.

A great many people said I'd never live to an old age for I used to like to have fun at other people's expense. We had a ranger who was a very good bread cook and he always bragged about it. So, once when a bunch of us were at the old Blue

Mountain Ranger Station (since torn down), this ranger was to make the bread. We emptied the baking powder can and filled it with flour and got quite a kick out of watching him. He sat by the oven door and kept watching the bread. He said, "I don't know what's the matter. I can make better bread, of course." The bread didn't raise but we ate it. He never again mentioned his breadmaking ability.

Another time, Henry Ireland, Frank Mattson, the North Power ranger and I were at Anthony Lake Ranger Station and cooking supper. I was making bread and Mattson was cleaning fish. The ranger spoke up and said, "I don't like fish and don't even want to eat from a pan in which fish have been cooked." I reached over to where Mattson was working, picked out a small fish head, wrapped it in biscuit dough, marked it and saw to it that the ranger got this biscuit. He was quite surprised and thoroughly disgusted when he broke it open.

I recall also stopping at the Blue Mountain Station when some Burnt River ranchers were there overnight. They wanted to get an early start, so I set the alarm clock. When it went off we got up, cooked and ate breakfast and sat around waiting for daylight. Finally someone looked at his watch. It was just twelve o'clock midnight, so we all went back to bed. They never knew why this happened.

K. P. Cecil often paid us a visit. He was kind of a roaming supervisor to coordinate the Umatilla—Wallowa—Malheur and Whitman National Forests. K. P. was hard to wake up in the morning, so one morning at Desolation Creek, I went in and gave him a big dose of sneeze root. He woke up sneezing but only covered up his head, thinking he was catching cold. The men were outside standing around the fire laughing, so pretty soon it dawned on him that he was being "jobbed." He had a Model T pickup and when we went back to Baker, my bedroll was tied on top of other things on the back. I kept watching it. Finally K. P. said, "You don't need to watch that. I can see it through the rear view mirror." But when we got to Sumpter the bed had fallen off and had been drug behind till it was about worn out. I often wondered if he didn't chuckle just a little over this.

Once my plans backfired on me. We were at Baker. Howard Stratford, the chief clerk, had quite a family and towards the end of the month he was frequently short of funds. One of these times was in early winter when he was running his car with his last year's license plate, I went across the street and phoned him and said, "I am Sheriff McKenney and you will have to buy a new license plate or I will have to stop you from running your car." Soon after I went back to the office, Howard came in and said, "Say, I. E., the sheriff has just called me and says I must get a new license plate. Won't you loan me $5 till payday?" I had to come through, but he paid it back.

During early days the rangers were bunched at least once a year for a week or ten days on some construction project, usually a telephone line. We always had a penny ante game with a ten cent limit with a rule that lights had to be out at ten. It was hard to hold the men together longer than about ten days. They began to think of things that needed to be done on their districts. Strange excuses were offered, such as that they had to split wood for the wife. But, we got lots of work done. I think the biggest day's work was a telephone line from Tipton to Whitney. We were stringing a #9 metallic line on the P.T.&T. poles along the Sumpter Valley Railway. We got a push car from the railroad. On one day we strung out and tied six miles of metallic #9 wire on 10 pin cross arms (already on the poles) and walked back to Tipton, then coasted on the push car to Austin.

One spring while on one of these jobs, Supervisor Walt Dutton, from the Malheur National Forest, visited camp and sat in on our penny ante game. He got away with our money. When he returned in a few days someone had fixed a cold deck. This time he brought along one of his rangers. We sure took him to a cleaning. In one hand he held four queens but our man had four aces so we took his money and later had a treat of candy, nuts, etc. Supervisor Kuhns was with us that night. He didn't know about the cold deck. I sat near him and gave him what advice I could and advised him to drop out when he rode along on a pair of kings.

Road building had quite an evolution. We first started with a plow, a wooden V and a slip scraper. Then we advanced to a Martin ditcher and a Fresno scraper. Starting around a hillside, we first opened a trail by hand for one horse to walk in and shoveled out by hand until we could get a team over it. We then used a ditcher until wide enough for a small grader. The first bulldozer was a two ton Holt tractor with a floating blade that was worked on a counter balance beam that extended back on both sides of the tractor. All rock work was done by hand drilling and using only caps and fuse for blasting.

The lookout man on Bald Mountain got quite a shock one summer. The trail from Burnt River to John Day Valley runs right by the lookout. A traveling circus moving from Burnt River to John Day decided to follow this trail, as it was several miles shorter than the road. So, unknown to the lookout man, they walked among other things, a camel and an elephant right past the lookout. The lookout was reading or napping in the cabin when the camel and the elephant walked by and he saw them for the first time. It so startled him that he sure thought he must be dreaming.

SMITH'S SENSATIONAL SECESSION FROM OUR SYLVAN SERVICE

Rudo L. Fromme

It was in the spring of 1920, just a few months after the passage of the national prohibition law. Spirituous beverages had taken "to the bushes," giving way to "Bevo Beer" and other even less sizz-sudsing slush.

Our banquet that spring was held at the Multnomah Hotel in Portland, Oregon, and it shortly took on the appearance of an open defiance of the above legal restrictions. Smith C. Bartrum was the long time popular and valuable supervisor of the Umpqua. It had come to us by the grapevine during our sessions that this was to be Smith's last appearance at a Forest Service gathering. We were, therefore, not too surprised when we had taken seats at several group tables, to see him rise to his feet and step to the end of the head table as if to unceremoniously sound off his sob song before the meeting became involved in its orderly program.

"Fellows," said he, glancing toward the middle of the head table, "With due respect to the presence of Chief Forester Henry S. Graves and District Forester George Cecil, this is a sad occasion for me." Now, turning toward us lesser lights, he continued, as near as I can remember at this long lapse of time as follows: "Circumstances, which I shall not attempt to explain, have engendered my resignation from a Service for which I have given the best years of my life and, I might add, gladly so. Such an unusual occasion, I think you'll agree, justifies an unusual ceremony. I desire to leave the Service in the best of spirits so have ordered the best of spirits for our mutual celebration of the many past years of cordial fellowship."

During this last more buoyant remark, he was waving the "come-on" signal toward the service door where a flock of waiters were nervously waiting with trays held high, to speedily supply our several group tables with highball glasses, "White Rock," cracked ice and dark bottles, bearing conspicuous whiskey labels. Meanwhile, Bartrum finished with, "Mix your own, boys, to suit your varied tastes. The treat is all on me."

But, what about the head table? There, matters appeared to be getting a bit out of hand, to say the least. Even at the very start of Bartrum's remarks, Mr. Graves seemed to be looking at him with some apprehension. As Smith continued, he appeared to grow increasingly nervous, and, when the speaker started his play on "spirits," he turned toward Cecil and seemed to be urging him to start a "backfire." He was apparently trying to get him to comprehend that we were rapidly approaching a hazard. Well, when the waiters started distributing the delectables among the "boys," he could contain himself no longer. He rose to his feet with jaw set and dark eyes flashing

This went unnoticed by most of the "boys," as they were thirstily watching the waiters. Charles Flory, Chief of Operation, and I were, however, watching the Chief Forester quite closely. We were "in on the gag" with Smith, we three constituting the program committee. Just as Mr. Graves was about to take the protest action into his own hands, Cecil was seen to be pulling him back to his chair by the coattails and whispering into his near ear. George had to give away the "stage secret," that the "whiskey" was nothing more than apple cider and quite tame cider at that, as the "boys" soon learned to their apparent sorrow. However, they responded with good-natured laughter, and both Graves and Cecil beamed with peaceful delight.

Cecil told us afterward that Mr. Graves had been quite definitely unnerved and had urged him to stop "this crazy deal at once," that it was strictly against the law and would be bound to get into the morning papers. I was glad that we had finally seen fit to tip Cecil off to this hoax just before the banquet. Bartrum was opposed to this, but Flory and I both knew something about Mr. Graves' serious and conscientious nature from having studied (?) under him at Yale. It could have easily backfired and jumped the fire trail.

MY WELCOME TO THE HAPPY CAMP DISTRICT

James D. McNeill

A lot of water has gone down the Salmon River, under the bridge and by the old Ranger Station at Sawyers Bar since that day long ago (to be exact, May 25, 1921) when I shipped my bedroll, via parcel post, tied the rest of my worldly belongings behind the cantle of my saddle, said good-bye to Ranger Ed Hickey and Protective Assistant E. V. Patterson, mounted my sandy horse and headed out of Sawyers Bar on my way to the wilds of the Klamath River country and the Happy Camp District.

Three days of hard riding after leaving Sawyers Bar brought me to the pine covered flat overlooking the then sleepy little town of Happy Camp. There it lay, my future hometown, spread out below me dozing peacefully in the late afternoon sun. Five minutes later I was introducing myself to Ranger Sutcliffe and his Protective Assistant, Nickolas Effman.

A lot of water has gone down the Klamath River and under the Indian Creek Bridge and many changes have come to the Klamath River country and particularly to the town of Happy

Camp since I walked into the old Ranger Station and said, "I am McNeill." The station then was in a small three room frame building that was formally used as a residence by the resident doctor of the now worked out and defunct Grey Eagle Mine. The building was located directly across the street from the present post office. The building was long ago torn down to make room for the wheels of progress.

In the days that followed I met eleven of the fire protection and improvement personnel who were working on the district. This left one other man, Sam Bratt, who was stationed at the old Elk Creek Guard Station, some sixteen miles south of Happy Camp. He was doing trail work on his unit.

All of those whom I met were friendly, but all seemed to leave me with a feeling that I was a new man on trial and they seemed to be reserving judgment on me until a future date.

For a few days after reporting for duty I was kept busy with various details and district chores, such as logging out the Grider Ridge Trail to the district's only lookout station, Happy Camp Baldy, splitting pickets and peeling fir saplings for use in construction of a new fence around the station. These I packed in from the woods on Jack, the one government owned pack mule assigned to the district.

Then came the day when I was assigned the job of repairing the Elk Creek telephone line from Happy Camp to Whooping Devil Gulch. Somewhere along the trail I was to meet that character of characters, Samuel Bratt.

In those days there were no bridges over the Klamath River from Seiad Valley to Orleans. A few run-down ferries and several uncertain fords afforded the only means of crossing over the river and it was necessary for me to cross from the north to south banks of the river in order to reach the mouth of Elk Creek.

On that early June morning I loaded my telephone repair gear on my back and hitched a ride on an old Maxwell truck which was being used as a candy wagon by the Willets and Burr Construction Company which at the time was engaged in building the new Klamath River Highway from Crawford Creek to Blue Nose Bridge under contract from the old B.P.R. The candy wagon dumped me off along the highway across the river from the mouth of Elk Creek. The boat that was to take me across the river was still on the far shore. There followed a half hour of whooping and yelling at the end of which an old Indian, Capitan by name, came unhurriedly from his cabin and ambled down to his old dugout canoe on the river. Then with my telephone gear again on my back I began my trip up the Elk Creek telephone line.

By noon I had reached a point nearly up to Whooping Devil Gulch. Having no canteen, I decided to hike up the trail to the gulch to get a drink of water and eat my lunch. When I rounded a curve in the trail, there, lying on his stomach in the middle of the trail, his head to the south, fast asleep, lay a mountain of a man. I walked to within a few feet of the man and called out, "hello." The man aroused, came up on all fours and swung around in the trail. He stayed on all fours blinking at me much like a great shaggy bear. After a moment or so he said, "Just who in the h___ are you?" I told him I was the new man down at headquarters. At this he rolled over and squatted on his heels and blinked at me for another moment, then said, "So you are that college so-and-so that they sent down to take Art Blunt's place. Well," he continued, "I am Sam Bratt, Elk Creek Guard, stationed at Bishop's Ranch." This was my introduction to the quaintest character that the forest has ever known.

In those days in the off season months I usually worked as a cowboy and bronc stomper, so during the noon hour I entertained Bratt with tall tales of the ranges I had known and bad horses I had ridden. When the noon hour was over and I was ready to start my tramp back to Happy Camp, Bratt shook my hand with a grip like a bear's and said, "You are a kinda good big so-and-so and I like you. I am going to tell Stub (meaning Ranger Sutcliffe) and have him let you come up to stay a whole week with me."

Then I knew that I had really arrived, had met the man, had been judged, passed a test and at last had been made welcome into the inner sanctum of the employees of the Happy Camp District.

Thus began my pleasant association with the district personnel which was to extend over a period of more than twenty years. This association came to an end only when I laid my pine tree badge on the ranger's desk in a brand new office and hung my shovel and pulaski tool in the tool rack in a brand new warehouse. Both these buildings I had helped to build. The President had sent me his greeting, so I traded badge and fire tools for an 03 rifle with grenade launcher and a T/4 rating in the Army Combat Engineers.

THE FIRELINE

2424 Tice Creek Drive #3
Walnut Creek, Calif. 94595
August 4, 1993.

Bob Olson, Fire Planner
Lassen National Forest
55 S. Sacramento Street
Susanville, Calif. 96130

Dear Bob:

This is in reference to the article in the August 10, 1993 FSX newsletter concerning Lassen National Forest fire history. I have no useful information, just a story my father liked to tell.

During the period of about 1922 to 1923 Show and Kotok were developing fire standards that I am sure you are well acquainted with. One standard had to do with going to a fire on horseback. Five minutes get away time if saddle horse only was used and fifteen minutes if both saddle horse and pack animal were used.

This event probably took place in 1924 the first year the standards were put to use, I believe. The location was somewhere on the Bogard District. Bill Durbin was the Forest Supervisor and my father, Joe Elliott, Sr. was the Deputy Forest Supervisor. The two of them were on an inspection trip on horseback when they sighted smoke from a small fire near the top of a hill; probably started by a lightning strike. As they approached the hill they were aware some kind of activity was taking place, but they were puzzled because there was no metallic clang against rock as would be expected.

Bill Durbin surmised it was the local guard who I will call Fred for lack of knowledge as to his correct name.

Bill Durbin: "Fred is that you up there?"

Fred: "Yes."

Bill Durbin: "What are you doing?"

Fred: "I am scratching a line around this fire with a stick."

Bill Durbin: "Haven't you any tools with you."

Fred: "Tools by God, how would I get away in fifteen minutes and bring any tools."

Sincerely,
/S/ J. E. Elliott, Jr.

THE HOLE

One evening in 1923, Ranger Ellis of the old Core District on the Whitman Forest, instructed me by telephone to take a sack of TNT down from the loft of the Catherine Creek Cabin and bury it out in the meadow. This sack of about sixty pounds of TNT had been hanging in the cabin for a couple of years and none of us except Ellis knew about it. Evidently, due to some safety drive in the Forest Service, it had been suggested to eliminate the keeping of all explosives around cabins and ranger stations. Either they were to be kept in a powder house or else disposed of. So, I proceeded next morning to follow out the instructions. I took the sack down quite carefully, and it weighed nearer one hundred pounds than sixty pounds. I took the sack outdoors and carried it in the meadow a couple of hundred feet and contemplated the job of diging a hole in which to bury it. Then I had a brilliant thought. Why not just leave it there and go back to the cabin and get my trusty 30-30 and shoot the sack and set the powder off? I wondered how much of a boom it would make and took the precaution to stand by the corner of the cabin and sighting around the corner. After careful aim, I pulled the trigger. Well, the boom was awful! The concussion knocked all the dishes off the shelves in the cabin. The stove pipe was knocked off of the roof and I was set back a foot or two. All of the glass in the two windows of the cabin was blown out and there was a hole in the ground where the powder had been big enough to bury a horse in. After making the best of the situation, reputting the stove pipe back on, etc., I still had no way to fix the windows. So, it became quite necessary to report to the ranger that I needed X number of glass panes of such and such measurements and enough putty to place them in the windows. Also, a few dishes. The report was made, not because I wanted to, but because it had to be done. Ranger Ellis came out a few days later with the necessary items and somehow he failed to give me the reprimand that I expected. He stayed a couple of days, helped fix up the windows, went with me to look over the trail work being done and left with final instructions. All the time I was uneasy and I thought I detected at times a look of amusement on his face.

A LIVING LOOKOUT

Grover C. Blake

When I arrived at Crowflat Ranger Station to take over the Burns District in 1924, on the Malheur National Forest, there was no one to show me around or to introduce me to the new territory so I found my way around alone. There was a short-term man located at Calamity Guard Station near Drewsey and a lookout fireman at West Myrtle Butte on the opposite end of the district. When I reached Myrtle Butte on my preliminary rounds I found it to be a butte covered with a heavy stand of mature timber with a commanding view when an opening could be found between trees. An Osborne firefinder was set up on a wobbly table about four feet high, constructed of small, round sticks wired together with emergency telephone wire. When a smoke was sighted, the lookout fireman would proceed to carry the table and firefinder to a spot from which the smoke would be visible between trees. He would set the table down and orient the finder as best he could, as the table wobbled and shook, then take a leading and report.

I sized up the situation and said to the guard, "Don't you think we can rig up a better setup than this?" He thought it might be worth a try. So we felled two fir poles about 75 feet in height and I prevailed on a road maintenance crew not far away to send a team and driver over and drag the poles over to one of the tallest trees. I found some lumber and nails and we made a 50 foot ladder and got it raised to the side of the tree. At the top of the ladder we built a platform. Then we made a 30 foot ladder and pulled it up the side of the tree until it rested on the platform. Now we were up 80 feet and another platform was made. About three feet above the upper platform we cut the top off and set up the firefinder on the stub. We now had a platform which did not wobble and in a permanent location. Three years later, when I left the Malheur for the Umatilla, we were still using the tree lookout. I have been informed that a steel tower later replaced our tree lookout on West Myrtle Butte.

FIRE ON THE OLYMPIC FOREST

W. D. Bryan

I was designated acting district ranger at Quinault until Joe Fulton arrived in July 1924. A few days later the Bailey Mill and Quinault Hotel were destroyed by fire. I was transferred to Quilcene about August lst. Bill Vallad was the district assistant.

Most of my memories of the 1924-30 period tie in with fires. I arrived at Quilcene in time to see a fire starting at Dry Creek on the upper end of the Penny Creek Road and sweep to the top of Green Mountain. Neither Vallad nor I had had any previous fire experience. The fire reached a size of about 1,700 acres. Most of the crew of about 150 men came from the Seattle "skid road," a common practice continuing up to the time of the CCC programs. Many of the crew were "slick shod" misfits, but there were a few experienced firefighters and their advice in fire technique was a great help. In view of the present policy, it is hard to believe that Vallad and myself were the only overhead. August 25 is a firm date. It rained hard that day and resulted in the first real sleep either of us had had for the three weeks.

The first time that a fire crew was supplied by aerial drop was in 1929, when lightning fires hit at the head of Tunnel Creek. A crew of about forty men were working out of a central camp on two fires in mugged country without trails. Included in the crew were Walter Lund and his party of cruisers, which also included Allen R. Cochran, recently retired from R-7. J. R. Bruckart, Assistant Supervisor, arranged for two drops on successive days. Both canned and fresh supplies were packed in burlap sacks together with blankets to offer some protection. The sacks were stacked in a cabin plane with Robert McClay of the supervisor's staff acting as dropper, or perhaps bombardier is a better word. The fire camp was located in a small meadow near the edge of a shallow pond. On the first trip the plane circled the camp to get a bearing. On the next circle the plane door flew open and McClay started kicking out the sacks with his feet while holding on to the inside of the plane with his hands. The first two sacks out were filled with bread and both landed in the pond. Bruckert and I had stationed ourselves on a ridge a short distance from and about 200 feet above the camp to spot the fall of the sacks. One sack landed near us and canned corn flew like shrapnel. These were free drops from an altitude of about 300 feet above camp and to 50 where the plane crossed the Tunnel Creek-Dosewallips Divide. Supplies came through in surprisingly good shape. In all, about 60 sacks were dropped in the two days and all but one was recovered. The second drop included a copy of the *Seattle Times* with a vivid description of men surrounded by fire being supplied from the air.

A GAME OF CHECKERBOARDS

Don Bauer

This story was told me by Gus Boulden after he had been assigned to the regional office as a fire prevention specialist during World War II. He stopped by and visited with me in the San Jacinto Ranger District Office in Idyllwild and gave me most of the information that I am repeating now.

There on the San Jacinto, where he had been ranger following World War I, he loved horses and the outdoors and it was obvious to me that he loved his job and was deeply attached to it. He was a top ranger with deep dedication to the Forest Service. He went on from that job to become Supervisor of the Cleveland Forest and then after the 1924 fire season, where losses were excessive from fire and public reaction required positive action, the San Bernardino Forest was re-established by President Coolidge in 1925.

At that time, Gus Boulden was selected as the new supervisor of the new San Bernardino Forest. Gus is remembered for taking a number of very basic, fundamental and constructive steps. One was the San Jacinto District was largely checkerboarded because of its proximity to the railroad going through San Gorgonio Pass. At that time, he told me, there was strong pressure to get state parks established in different parts of the state. San Jacinto was selected as one and the idea of taking Tahquitz Valley and Round Valley and all the high country of the San Jacinto and putting it into a state park was approached and received quite a bit of public support

Gus knew of the problems of checkerboard ownership and came up with the idea that he could work with the state very effectively if the state would buy all the private land sections in the high country. Then, the state and Forest Service could work out an exchange and consolidate the wild or primitive type lands and enable the state to have fee title to San Jacinto Peak itself, along with Round Valley, Hidden Lake and Long Valley. Then, the railroad lands that the state would have acquired to the south and north would be traded to the national forest to consolidate Tahquitz Valley and Upper Snow Creek on each side of the park under national forest management as a primitive area.

This was done with total public control of the entire high country, which I think experience shows, was a very forward looking step and points up Gus Boulden's foresight and dedication to accomplishing a sound management objective that is paying benefits to all of us today, and that is public control of the entire San Jacinto high country.

YOU ASKED FOR IT

Dorothy Burgess

I had come out from Minnesota to teach school in Bend in 1925, when I met Bud, who was a USFS ranger on the Deschutes Forest, stationed at LaPine. Before long we knew there was something special between us. As we discussed marriage, he said he had a house that was completely furnished and all I needed to do was to pack my clothes and move in.

I always accused him of marrying me for my money. I had recently lost my parents and so had a bit of inheritance. He said he was in debt, but could get "leave" (a new meaning for me) if I would finance a trip to Portland. He must have known that women's lib was in the offing, but I was brought up to believe that the man paid for the honeymoon, and so we settled for the Ochoco Inn in Prineville because Bud had a friend who rented a room there by the month and was not using it that particular weekend.

So after our weekend honeymoon In a rent free room, we returned to the "completely furnished" FS cabin that my groom had provided for me. My heart really sank as I looked at that forlorn Rosland Ranger Station sitting all by itself several miles from civilization. It was two rooms, one a combined living, kitchen and bedroom, measuring about 15 feet by 18 feet, the other was Bud's office with a breezeway in between.

Our living quarters were furnished with a combined heating and cooking stove, a built-in cupboard, a small unpainted drop leaf table and four straight back chairs with rawhide bottoms and an army cot. The walls were decorated with coyote hides. The only convenience was a pump in the breezeway.

Bud said he was a good cook and would get our evening meal. And so as I kept the tears back and unpacked, hanging my clothes that had been bought in some of St. Paul's better stores, in the corner of the kitchen, Bud prepared our dinner. He had tomato soup with hunks of tomato floating around and cocoa. I thought it was a strange menu but my only comment was, as I pushed the hunks of tomato to one side, "It is very good." The next morning when he suggested the same menu for breakfast, I took over.

The dishes were granite and the pots and pans a miscellaneous collection mostly of a size that would accommodate cooking for an army. Now that so many years have elapsed and I know I cannot be thrown in jail, I will admit that I became attached to some of those FS cooking utensils, as I am still using an old, beat up granite roaster, a big granite cup and an antique food grinder.

But I knew I could not go on using granite dishes, so our first morning we went back to Bend, bought a set of dishes and a few other things, including a day bed that looked quite elegant during the day and was more comfortable at night than a single army cot.

We had boarders from almost the first day of our marriage, a never ending succession of foresters from the Bend office. They occupied the bunk beds in the breezeway.

While the men were out cruising or doing whatever they had to do, I kept busy cooking and making that one room more livable. I painted the furniture grey and blue—never my favorite colors but the floor was grey, the house was painted grey and of course grey was the standard FS color at that time, so I guess I was just trying to stay in tune. Bud had built a closet in one corner of the room, so with our clothes out of sight, the colorful day bed and some new scatter rugs and the newly painted furniture, it was quite cheery.

At least I thought so until my family came out one by one from Minnesota to see what I had gotten into. I had no parents but I had older brothers and sisters. One sister was so depressed over the way I was living that she only stayed a couple of days. Then too, I guess she did not like sleeping in a sleeping bag at the end of the day bed. When my oldest brother came to visit I can still see him looking around and saying, "Jabes, (a nickname) to think you would ever come to living like this! I think the only good thing about this deal is the fellow you married." And they all liked Bud so that made me feel good.

I was afraid of everything! First it was the switchboard. Bud had failed to instruct me how to use it and I thought all I had to do when the phone rang was to say "hello." Finally, after pulling plug after plug I got a voice—Mr. Keefer from Crane Prairie, and he instructed me how to operate the switchboard. That was a big mistake. From then on I was operator without pay and during the busy fire season I was often there by myself manning the switchboard.

The wild animals frightened me. Bud said to ignore the coyotes as they were more afraid of me than I was of them. I did not believe that and would make a hasty retreat for the house whenever I saw one. One day I heard this eerie noise when I was out walking and looked up to see a great big owl blinking at me. He really scared me as I thought owls were only found in captivity.

I was afraid to stay alone and made myself sick so Bud couldn't leave me for the first couple of times. He solved that by making arrangements for me to stay with the Clarks who operated the store in LaPine. But I didn't care for that and finally decided to be brave and stay alone. As a result, Bud's bulldog Jack became my good friend. Prior to that I was afraid of him, as he seemed to resent me, but from then on he sensed he had to take care of me.

And the car—Bud said I had to drive as it would help him a lot. Shortly after we were married, he wanted to pick up a horse at Crescent, a distance of 20 miles. The highway was glare ice. I was driving and looked at Bud to see that he had his hat pulled over his eyes. When I asked him why he was doing that, he said if he were going to be killed he didn't want to know about it and that when I was going downhill I should take my foot off the accelerator. I said, "What's that?" With those instructions I learned to drive. When we got to Crescent Ranger Station, Mrs. Floe couldn't believe that I had driven for the first time. She really gave Bud a bad time and said that she had been driving for years but Sanford wouldn't think of letting her drive on that icy highway. Bud said, "Oh, she can do it!" And I did. Bud rode the horse home.

One day I heard a knock on the door. I looked out to see a grizzly haired old man. I quickly latched the screen door but before I could slam the door in his face he said, "Why, you're Bud's new bride and old Jake wouldn't harm a hair on your head. I just stopped by to see if Bud had any coyote hides to sell." I couldn't believe that I could get rid of those unsightly things and get money for them. When Bud got home that evening he was really pleased. He said Jake had given me twice what the hides were worth.

And then came that first baby and I was afraid something would happen to him. I spent three weeks in the hospital in Bend and when we came home a distance of 30 miles, I had a regular drug store with me. I thought if this is what it takes to raise a baby, he'll never live. Fortunately, he was a good baby and I did not have to use any of those drugs but he was brought up according to all of the government bulletins that I had amassed. I worried if I was five minutes late feeding, bathing or napping him. I wouldn't let anyone near him for fear they would contaminate him. Bob's first Thanksgiving we spent with Bud's brother where there were five children. They had looked forward to playing with their new cousin. I am sure I did not endear myself to them, as I kept Bob in the car all of the time we were there. I still marvel at the casual way young mothers care for their babies today.

I was a bit more relaxed with our second son who arrived a year and a half later. With no neighbors, close friends or modern conveniences, it was hard. With Bud's help we did the babies' laundry but the rest of it went weekly on the mail stage to Bend.

The only time we had fresh meat or vegetables would be the first few days after we had been to Bend, and in those days, 30 miles was a long trip. For a few months in the winter we could keep the meat by freezing and Bud did try one summer to have an ice house. But that didn't work and we lost some good meat. Mostly we missed fresh vegetables and to this day, I cannot stand any kind of canned vagetables. We all built up an aversion in later years to ham.

Forest officers ready to leave a tool box to go to a fire, Arapaho NF, Colorado – c1910.

Forest rangers repairing bridge trail, Arapaho NF, Colorado – c1910.

THE BOULDEN HELIOGRAPH CODE

Joe Elliott, Jr.

Gus Boulden, upon becoming the first forest supervisor of the San Bernardino National Forest, was replaced by my father as forest supervisor of the Cleveland in April of 1925.

In April of 1929, Gus Boulden was transferred to the Santa Barbara (now Los Padres) National Forest and my father from the Cleveland to the San Bernardino. Thus Dad followed in Gus Boulden's footsteps twice.

Probably the Cleveland had more need for the Boulden code than most forests because the three districts were so separated that I do not believe there were telephone lines between them. The only telephone lines, as I recall, connected primary lookout stations to their respective district headquarters. Thus, the three district rangers and the forest supervisor (no staff 1925-1929 other than an administrative assistant and two clerks) had little opportunity to use telephones. Heliographs were the principal means of communication with a lookout while traveling by car or horseback, providing the traveler was intervisible with the lookout and the sun.

Lookouts had a large model heliograph that operated off a tripod about three feet high. Portable models (which fit into two small leather cases for carrying in a car or on a horse) were mounted on a tripod probably 18 inches high. Both had the same parts; a shutter, two arms that supported mirrors and sighting devices that were used to aim the sun flash much the same as aiming a rifle. If the sun shone in front of the operator, only the mirror was required. If the sun shone in back of the operator, the second mirror was used to center the flash on the primary mirror focused on the shutter.

To initiate a call, the calling party would open the shutter, look through a small hole in the center of the mirror and adjust a pointer located midway between the mirror and the shutter so that the tip was just under the lookout house. A set screw made it possible to raise the pointer up and down; lateral adjustment was accomplished by pivoting the whole instrument. Once aimed, the shutter was closed and the flash centered on the shutter in such a way that the small dot created by the hole in the mirror rested right on top of the image created on the shutter by the pointer. There were verniers to make this fine adjustment.

The shutter was then opened and if aimed properly, the lookout would see the incoming steady flash and use it for his target. Upon seeing the steady flash from the lookout, the initial sender would flash his shutter rapidly a few times to acknowledge the answer and both would close their shutters preparatory to exchanging signals.

The beauty of the Boulden code was that you did not need to memorize it. You sang along with the sender so to speak. Gus Boulden divided the alphabet into four major units of six letters each and one of two letters. The major units, A, G, M, S and Y, were indicated by one, two, three, four or five long flashes, respectively. Individual letters within the major units (i.e., a, b, c, d, e and f) were indicated by short flashes, respectively, from one to six. If two or more consecutive letters of the message were in the same major unit, it was not necessary to repeat the long flashes; a person sent the first series (from one to six short flashes) and then waited a moment before starting the next.

There were several abbreviations for commonly used messages. There was also a way to indicate numbers.

A typical exchange was as follows: Initial sender—one long flash followed by a short flash and a second series of six short flashes (AF – any fires?). The hoped for answer—three long flashes followed by two short flashes and then three short flashes (No). If there was no further business, the initial sender would send three short flashes followed by a second set of three short flashes (adios). The lookout did the same and that was the end of the exchange.

If the response to the initial inquiry was five long flashes followed by one short one (Y – Yes), communication procedures and immediate travel plans were both likely to become more complicated.

SAGA OF A FOREST FIRE DISPATCHER

Les Colvill

I was the fire dispatcher for the Deschutes National Forest during the period 1926 to 1931, located in the supervisor's office at Bend. During this period the primary lookouts were Walker Mountain and Maiden Peak (Crescent District), Paulina Peak and Mt. Bachelor (Lapine District) and Fox Butte and Pine Mountain (Sisters District). The principle equipment used consisted of the telephone, plotting board (a steel plate) on which was mounted a one half inch to the mile scale forest map with a three-sixteenth inch hole bored at the exact location of each firefinder, a twelve inch brass protractor, wall maps and filing baskets. A wander type telephone was hooked up to a four line plug and jack switchboard. The Crescent line terminated at Lapine. The others came in direct to the switchboard. Ringing was done manually and four rings was the fire call.

On one occasion I was talking to Walker Mountain Lookout approximately 60 miles distant and, due to interference, was having difficulty getting my message across. I must have been shouting pretty loud because the Supervisor, Carl Neal, whose office was two doors down the hall, came busting in thoroughly mad and asked who I was talking to. I replied, "To Walker Mountain." He said, "Why don't you use the telephone?" and stalked out slamming the door. My loud voice was not all bad. On occasions, during a fire bust and with windows open, people would gather on the sidewalk and get fire information first hand. Even now my wife reminds me to lower my voice when talking over the telephone but I counter that it was this voice coming through the open window of the dispatcher's office at Bend that caught her attention and proved irresistible.

The longest known sleeper fire was 22 days and happened to be one of the most unusual fire fighting jobs reported. The fireman originally dispatched found a large lightning struck pine. The only visible fire was high up in a branch crutch. Not feeling equal to the task of cutting down the tree, he decided to build a control line that would place the limb inside if and when it fell. His calculation missed by a couple of feet.

Pine Mountain Lookout was the location of a baffling incident. Gail Simpson of the Portland Radio Laboratory had designed a lightning detector meter and arranged for the Deschutes personnel to give it a field test. Pine Mountain was chosen to make the test. A day or two later he called to tell me the needle was acting up. It was a clear day with only a slight breeze. I called Walker Mountain Lookout to determine if there were any clouds to the south that would indicate an approaching lightning storm. The report was negative. By this time the needle indicated extreme lightning conditions. The only visible weather change was an increase in the wind velocity. Ernie Putnam, the lookout, noticed the flag flapping violently and whether he guessed the probable cause or hauled the flag down to protect it from fraying, I do not know, but with the flag down the needle resumed its rightful position.

I recall the 1926 Fox Butte Fire, which burned an estimated 10,000 acres. The first morning the crew had to fry bacon and eggs in shovels due to failure of the truck driver to load the mess gear. The regional office made it a practice to have an assistant regional forester on large project fires. Fred Ames of timber management was assigned to the Fox Butte Fire and he arrived in Bend late one evening dressed in faded blue jeans, heavy mackinaw and slough hat. He looked unlike the immaculately groomed gentlemen we were accustomed to seeing. He arrived at the fire camp the next morning via a supply truck just as the men were eating breakfast. He walked over to the warming fire and was standing there when spotted by the camp boss. Not knowing who he was and assuming he was a firefighter, the camp boss told him to get something to eat, grab a fire tool and get going. After considerable prodding, Fred broke down and told the camp boss who he was and his mission. The camp boss, an English gentleman, never got over apologizing to me for what he considered an unpardonable mistake.

The Maklaks Mountain Fire produced an incident I will always remember. I got word late one evening a member of the night crew had a cut throat, panicked in the dark and could not be found. The next morning I was completing my notes of the night's activities and, looking up, saw before me an unshaven man whose shirt front was covered with dried blood. When he lifted his head I could see the slit in his throat. I immediately called Dr. Vandervert, who had offices in the same building. He responded promptly and took the man away. The doctor reported later the man had cut his own throat and it was not an accident as we supposed.

The Fremont asked for help to control a fire burning under the Summer Lake rim. We had a small road crew working in the Fort Rock District. There was telephone communication to the camp but none to the crew when out on the job. To meet this situation we had rigged up a call signal, consisting of three sticks of dynamite, each with a different length fuse so that when fired the time interval between shots would be about one minute. Following the Fremont call I instructed the camp cook to fire the shots and prepare for an immediate takeoff when the crew arrived. Everything worked as planned. The crew arrived in camp, had lunch and took off, including the cook. Soon after the crew departed, I received a fire report from Paulina Peak that plotted at the exact location of the camp. The powder had ignited the paper wrapping which in turn ignited some forest debris.

The dispatcher functioned in many capacities. I recall once helping R. C. "Bud" Burgess with a school fire prevention program. Bud was the ranger at Lapine. The school principal was an elderly scholar who enjoyed oratory. When the time came to present Bud, he used a varied assortment of complimentary adjectives and closed his introduction with a dramatic "and now we will hear from our own dear ranger." Bud took off from the back of the room where we were standing and, firm in step, headed for the stage. He mounted the one step, faced the packed house with one hand raised in true oratorical fashion. He got by the opening welcome but no farther. That awful thing overcame him that has happened to so many unaccustomed to speaking, and he closed with a subdued "thank you." I chided Bud the next day that he was a victim of the flowery introduction.

Later in the season Bud was asked to be marshal of the town's Fourth of July celebration and accepted. A big cowboy from Silver Lake was entered in the rodeo. He thought his ride should have received first prize. Bud judged otherwise amid much protesting. Toward evening some of the rougher elements were getting pretty well liquored up. They were having difficulty getting someone to referee the boxing match and finally persuaded Bud to accept the job. One of the contestants was the big Silver Lake cowboy. At the conclusion of the match Bud raised the arm of the cowboy's opponent and as he did so the cowboy landed a right on Bud's jaw. Bud retaliated and the free-for-all was on. Thus ended a perfect day for many but not one becoming a young aspiring forest officer.

EARLY TRAINING

Rupert Asplund

I started to work as a patrolman out of North Fork in 1927. My patrol area consisted of the front country between Auberry and the Merced River. I furnished my own car and received $150 per month. During the day, I would contact stockmen and ranchers; camp in the campgrounds at night and visit with the campers.

In 1928, Supervisor M. A. Benedict of the Sierra NF urged that I take the ranger's exam which was given in the fall. The day of the examination I was relief lookout on Shuteye Peak and Earl Morrow came up to relieve me. I walked two miles down the trail and drove to North Fork in time to take the exam at 9 a.m. Shuteye Peak on the Sierra and Claremont on the Plumas were the first two FS lookouts in the California Region. I did pass the examination and received an appointment in 1929 as assistant ranger under Frank Sweeley on the Pine Ridge District. At that time, planning and development of the recreation area around Huntington Lake was at its peak. The big job was to plan and survey the summer homesites and resort areas. Most of the campground sites had been selected but development, including digging toilet pits, was just getting started.

Power development by the Southern California Edison was also at its peak and I received a good introduction to the relationship of dams, lakes and other facilities to national forest resources.

I received my introduction to timber management and sales by Burt Hurt on the Fresno Sugar Pine Lumber Company sales area at Central Camp above Bass Lake. Burt had worked for the Shaver Lake Lumber Company for many years, so was familiar with logging practices and how to get along with loggers. The company mill was at Pinedale near Fresno and connected with Central Camp by wide gauge railroad. Logging donkeys were 2,000 hp and electrified, which required a high voltage line and also a railroad to each siding. Since white fir was not being logged, large fir trees were used as power poles with the line 100 feet in the air to permit use of large equipment below the line. This required some rather difficult limbing and topping, and only the best climbers stayed on the job. One of Burt's jobs was to lay out the line and mark the trees to be used.

I received my introduction to grazing and range management under Jim Poore, District Ranger on the then High Sierra District. One day we stopped to visit the occupant of a cabin near the headwaters of Mono Creek. His first remarks were, "I told you we would get rid of the sheep! I have written to my congressman in Washington." I looked at Jim and he shook his head as an indication that I was to keep my mouth shut! Down the trail a few miles we had counted 2,000 head of sheep. Jim explained to the occupant that the sheep would not bother him any more. After we left the cabin and headed up the trail, Jim explained to me that he had told the herder to keep the sheep away from the cabin, and whenever he drove them across the trail, to cross in the rocks.

I received my first introduction to the pine beetles before I went to work for the Forest Service. While I was living at Cascadel, above North Fork, my partner Milt Blakely and I had rented the ranch which consisted of a large area of mature pine. The years 1924-25 were low in rainfall, now considered drought, and the pine beetle infestation had reached epidemic stage. Hubert Persons, with the Bureau of Entomology which later joined with the Forest Service, was on the Sierra to make a study of the beetle activity. There were many "bug" killed trees on the ranch and Percy decided to make the study in that area, setting up a tent camp on the ranch near Whiskey Creek. This gave me the opportunity to become familiar with the various species and the control program. Percy and his helper felled several of the dead trees on the ranch to study the beetle's activities. One of the studies was to determine the number of larvae to hatch, bore out and leave from a tree as adults. After marking off one foot square areas of bark every few feet from top to bottom, they would fill the holes with tooth picks (each square foot resembled a porcupine) and count the toothpicks!

FIRE MONEY

Grover C. Blake

It was at 7 a.m. on July 20, 1927, that I left the camp of the telephone line construction crew at Seven Sisters Spring, on the Asotin (now Pomeroy) District of the Umatilla NF to go to the forks of Wenatchee Creek to examine the burned-over area of a fire we had recently extinguished, and to try to determine cause and fix responsibility.

I took with me Forest Guard Roy Madison from the Wenatchee Guard Station. Roy had been raised in the vicinity and had worked, short-term, for the Forest Service for a number of years. He knew his way around and was a lot of help to me in learning about my new district.

It was hot that July day, and the sun was really pouring it on in Wenatchee Canyon. At the forks we came upon a campfire burning merrily and no one near. I surmised that the person responsible was no doubt busy fishing somewhere down the creek. We put the fire out and then scouted around for tracks. We soon found tracks leading up the creek. We followed the creek about two miles before we came upon a man whipping the stream with a line and hook. I told him why we were there and he admitted that he had left his campfire burning "because it could do no harm nohow." I informed him that he was under arrest. After explaining to him how only bad men left fires burning round about, and how tough the laws were with culprits who did such evil deeds. I asked him to sign a statement admitting responsibility for the fire. He agreed after some hesitancy, and I wrote out the statement. He signed and Roy witnessed the signature. I told him that, since he had signed the statement, he could go ahead with his fishing, but I would expect him to meet me at the office of the Justice of the Peace in Asotin, Washington, the next morning at 11 o'clock. He did not like it but agreed. Asotin was the county seat and was some 40 miles or more away. Quite a trip for both of us.

Roy and I returned to the forks of the creek, wet with perspiration from the long hike in the hot sun. We found that the previous fire of several acres had started on a private road leading down Wenatchee Creek, and was used only by a rancher who lived at the mouth of the creek, on the Grand Ronde River. We walked the four miles to the river and the ranch and found the owner busy putting up his hay crop. He admitted that he built a fire under a log across the road to burn it out of the road. He did not expect it to spread and "didn't mean no harm nohow." When I mentioned a trip to Asotin to talk the matter over with the J.P., he objected as he did not want to leave his harvesting. I could not blame him. He was willing to cooperate in any other way. I was sorry that there was no J.P. closer than Asotin, but I could not help that. At this point Ray remembered that a farmer up in the hills five or six miles away had been elected J.P., but had never had occasion to make use of the office. So we (Roy, the rancher and I) proceeded to the home of the country J.P. He admitted that he was a duly elected J.P. but did not know much about the duties of the job. I explained the customary procedure, so he held court with me instructing him as to his duties. Rather an unusual procedures but no one objected. I made the complaint and the rancher pleaded guilty. He was fined $10 and costs. Then the J.P. asked me, "'What will I do with the money?" This caused a round of merriment. Later, the rancher had to pay the cost of putting out the fire a total of $12, in addition to the fine.

Roy and I went back to map the fire area and I became quite ill from heat exhaustion, but we made it back to Wenatchee GS at 7 p.m., very weary, and feeling that our day had been about as busy as we could reasonably have made it and looking forward to a good night's rest before proceeding to Asotin to take care of the law enforcement case connected with the fisherman. But the end was not yet.

At Wenatchee GS we were informed that a fire was burning at Grouse Flat over in the south part of the district. Roy remarked, "It looks like we are going to get in a full day." I called the telephone crew and got them started for the fire. Then Roy and I left and we reached the fire with the crew at midnight. By this time we were moving on stored up energy, I guess. For my part, I was beginning to feel like a wooden man. At 5:30 a.m. on July 21, Roy and I left the fire in charge of the very efficient telephone crew and left for Asotin. We stopped enroute at Wenatchee GS for breakfast and to change cars. We arrived at Asotin at 10:00 a.m. In the meantime, Roy had told me about the unhappy experiences the Forest Service had had in getting the officials at Asotin to cooperate in law enforcement in fire trespass cases. It seemed that the J.P. in Asotin considered himself to be a very important individual and liked to be called, "Judge," and was very officious. He did not believe in penalizing fire trespassers and the common procedure had been to call court to order and promptly dismiss the case when suspects had been brought before him. I had all these things to mull over in my mind as we approached the county seat.

When we reached Asotin we had one hour before our fisherman was to appear in court. I told Roy I wanted to meet the J.P. first of all. As we drove into town we saw a man walking down the street. Roy said, "There's the J.P. now." We stopped and I walked over to him and introduced myself. I told him that I was the new forest ranger; that I had heard that we could not expect law enforcement in his court and all the other things I had heard. I would probably have been a little more lenient and cordial in my approach had it not been for the strain of the past 30 hours. He began to defend himself and assured me that he was much in favor of law enforcement and could be depended upon to do his duty, etc. I told him that we had a case coming before him at 11 o'clock and I hoped that he would be able to hear the case at that time. He could.

Roy said he had never known him to act so humble and his attitude was entirely different from the normal. I thanked him and told him how glad I was to know him. We then left for Clarkston, six miles away, to confer with the district attorney and get a complaint against the fisherman. The D.A. was an elderly man. I told him what I had heard about previous fire trespass cases, and he assured me that he would assist us all he could; that we could depend upon him as D.A., and that he would prosecute the case at hand if the party did not plead guilty. He also advised me to bring future cases to Clarkston if things did not turn out well at Asotin.

We got back to Asotin in time for me to have a talk with Sheriff Wayne Bazona. At 11 o'clock our fisherman showed up and he had a real mad on. He had had 24 hours to think things over. He did not like people who went about spoiling other people's vacations and he had a special dislike for nosey forest rangers. If he hadn't signed the statement, he would get his lawyer from Spokane and he would show us all where to head in. That outburst did not help his case any as it riled the J.P. who informed him that if he did not want to pay a fine for contempt of court, he had better keep silent and respect the dignity of the court. The complaint was read. He pleaded guilty, paid his fine and court was adjourned. Before leaving, he turned to Sheriff Bazona who was sitting by my side, and said he had a request to make and that was that the case would not be reported to the newspapers. The sheriff made no reply, but as the fisherman left he (the sheriff) reached for the telephone and called the newspaper and gave them the story.

LIFE ON THE RANGE (SURVEY)

Lloyd W. Swift

The summer of 1927, I was a member of the Lassen range survey crew. Chief of party, Russ Beeson, a lifelong friend, and I took the train to Davis and from there rode with Russ and Mrs. Beeson to Susanville via Alturas, since the direct routes were not cleared of snow. The first camp was Norvelle Flat on the Bogard District. Perhaps floored tents with zippered screens and sleeping bags with air mattresses were available but not to my knowledge. Anyway, excerpts from letters written many years ago relating to my introduction to life in the Forest Service may be of interest.

May ?, 1927. "There are six chow houses in Susanville, all but one apparently owned by Chinese. I started with the exception and ordered a vegetable plate and, after sampling the array of debris, decided I had better drink a glass of milk and call it a meal, but the milk was warm and tasted like the vegetables smelled.

"I made the rounds of the other eating places. Tonight got me down for sure! I sat at the counter as most working men do and pondered over the menu to the music of a man picking and sucking his teeth. The chap wore out most of the toothpicks and departed only to be replaced by another with the same musical talent. Eventually my order was served, but not in a palatable way. The beef had died years ago; the gravy would make a poor grade of rancid tallow; the potatoes were greasy, fried, and the buns too young, but thank goodness the milk was fresh and cold. I must pick an edible dessert, so after considering the array of pies and other concoctions, I ordered prunes as surely they would be good, but they were wrinkled little chaps, too small and too bitter to leave their sticky syrup."

May 20: "Thursday morning we checked out of Susanville and headed for the woods bucking a nice snow storm until mid-afternoon. Amidst snow, a flat tire, and depleted gas we were separated from our guide. We sought him diligently until 6:30 p.m. turning to Westwood baffled. On phoning we found he was quite comfortable at home. I am thankful for the mishap as it resulted in our staying at a steam heated hotel on one of the coldest nights I've ever experienced.

"The next day we slid out to camp, located on the coldest knoll in the Lassen National Forest. To the east we had the Susan River, to the north and south snow, and on the west a promising frog pond. Last night it was cold. I had not arranged my bed and so slept in a knot with my hips crying for mercy; However, I am actually enjoying the situation and feeling tiptop. Stanislaus Sam, the cook, said, 'It's too damn cold,' and says he plans to leave immediately, but how I don't know. We are about

40 miles from Susanville, isolated, and he has no transportation."

May 29: "Friday morning I was awakened by snow drifting through a hole in my tent upon my face. This downfall of snow flakes continued until Saturday building up a nice white coating—quiet, beautiful and not at all warm. Oh yes, it had rained before snowfall. Last night it cleared and became beastly cold. My bedding was moist so I put in a bum night. My head became so darn cold I had to cover it with a sweat shirt.

We did have some fun with Shirley Bliss when crossing the streams. We were mapping a large meadow through which a number of streams meandered. Our method of crossing was by vaulting as done in pole vaulting at track meets. Bliss had never done any vaulting so furnished us with glee by landing in the middle of the stream."

June 5: "The biological influences here are a damn nuisance. The mosquitoes are endemic and so thick the only escape is to crawl into a tent, have a smudge of lodgepole pine needles and smoke them out. At night the mice combined with mosquitoes to conduct an organized raid on us mortals. At my tent, for instance, the mice struggle to the top and slide down the canvas with a disturbing rasping sound. Then they conduct a track meet along the edge of the door I am sleeping on. Having had their fun, a construction "or destruction" gang is organized to dig, scratch and bang under the door for the remainder of the night. I had revenge yesterday, crowding a mouse into a corner and stabbing it with my pocket knife. It was a well executed feat and showed much technique."

RANGE DIPLOMACY

Philip L. Paine

I was placed in full charge in July 1928, of the Steele Creek Ranger District, Beaverhead National Forest, Region 1, Montana. There were cattle, sheep and horses grazing on the district. The post and pole timber sale activity was heavy.

Nearly all fences, of which there were hundreds of miles in the Big Hole Basin, were of the lodgepole pine jackleg type. There also was one small Douglas fir logging operation. I was a graduate forester—had had some timber survey experience and was now in full charge of a ranger district. Others, no doubt, have had the same experience, but it was rough going to say the least. Even though I was "just college trained," I survived!

The Big Hole Basin Battle Ground Ranger Station (site of Chief Joseph's last stand) was located 30 miles west of my station across Big Hole Valley at the foot of the Rocky Mountains. Ranger Ramsey was an old-timer and a real forester. One of his first words of wisdom was, "Well, kid (age 24) you are hardly dry behind the ears. You won't find any better, kinder people in the whole country, many of them ran stock here before the Forest Reserves were set aside by Teddy Roosevelt—but don't throw the book at them—use good horse sense in your dealing with them." How good his advice proved to be! Late that fall, after the grazing season, I was called to the south end of my district to mark some posts and poles. I found some horses trespassing on my one horse allotment and the brand was not that of the permittee. In those days we had one book which gave all the answers. This book was called the Administrative Regulations. I had studied it in one of my forestry classes and was generally acquainted with its provisions. I was supposed to immediately initiate trespass proceedings. Instead, I remembered Ranger Ramsey's advice. When I got back to my station, a log cabin six miles east of Wisdom, Montana, I "hunt and pecked" a letter out on my trusty Oliver typewriter advising the offender to remove his stock within a week's time. Travel being by saddle horse, I felt the need of a reasonable time limit. A week later I visited the range. The stock was gone and the boundary fence gate was closed. Sometime later I met Ranger Ramsey in the post office at Wisdom. He greeted me with a big smile and said, "You did a good stroke of business in handling the horse trespass on Fox Creek. All the trespasser's friends and neighbors are giving him the big laugh. You did more good and gained more respect than if you had thrown the book at him."

The next spring I was counting one of my permittee's cattle through a range fence. I was on my horse counting the critters and he was pushing them through. His permit was for 60 head, but before I could get the gate closed, two extras got through. He asked me how many head I had counted, and I said 62. (I suspect he had made a count also.) As we rode back towards his ranch he said, "Last spring only one extra critter got through the gate and that young ranger (new like I was) made me go out and get the one extra back into my pasture. The next day I went back and let 40 more head out on the forest range. I won't do that to you." I told him I was allowed a one percent error—six-tenths of a cow—and I was not worried about the balance of one and four-tenths cow. Again, I profited by Ranger Ramsey's advice.

1928—UNUSUAL FIRE CASE ON THE ANGELES

A. M. Longacre

District Ranger N.E. Peterson of the Saugus District had what might have been a perfect case for the machinery of law enforcement had it not been for very unusual developments.

A Los Angeles resident, known nationally as a racing driver, owned a summer home in San Francisquito Canyon and had been clearing his lot at various times that spring. He wanted to burn the material which he had cleared on May 12, and called at the Newhall Ranger Station for a burning permit. Assistant Ranger Longacre issued him the permit but cautioned him very emphatically about what he could burn. The permittee thought, however, that he could safely burn after the hour of 8 a.m. which was the time limit set in the permit, and the result was that the fire escaped from him about 10 a.m. Sunday morning, May 12. Things looked pretty bad to the permittee—as if the whole world was going to burn up, and he probably was afraid that he was due for an encounter with John Law. Inasmuch as it is understood that he had had financial trouble lately, he no doubt thought that his life was no longer worth living, went into his cabin and shot himself.

A Mexican who has done work for the permittee happened on the scene just as the fire was picking up to make a run. He jumped into action with a shovel and kept the fire down until help arrived. About a half acre was burned over, but the fire could have been extinguished by the permittee with very little trouble. Fate was against him, however, in his mind and he took what he thought was the easiest way out.

CUYAMA GRAZING RECONNAISSANCE

Russell Beeson

Editor's Note: Russ Beeson has favored us with a backward look at a range survey job he carried out on the Santa Barbara NF(now Los Padres). There is only a passing glance at survey techniques. Instead we learn that hot, sweaty days were more than compensated for by hospitality and good food at the various "campsites." –E.J.

I went to the district office in the Ferry Building as Jess W. Nelson's assistant in June of 1929. At that time there was a request, several years old, from Nash Boulden, Supervisor of the Santa Barbara National Forest for grazing reconnaissance. The area under consideration was located both north and south of the Cuyama River east of Santa Maria. During the winter of 1929 and 1930, we made plans for the survey. Early in March, I took the Daylight train to Santa Barbara and got a room at the old Santa Barbara Hotel.

For a week, Judell Salmon and I drove over range allotments in the Santa Ynez drainage. We also collected the equipment needed for the job. Of interest to me were two small plane tables. Something I hadn't used before.

During this week, Judell took it upon himself to introduce me to Santa Barbara. Of special interest were the new courthouse, mission and botanical gardens, plus an unusual number of good restaurants. Our sightseeing in the courthouse had me a little nervous. It was a Sunday and Judell was lamenting the fact that it was closed. At that time we noticed there was a door open on a balcony. We entered and spent a couple of hours going from room to room. I was a little hesitant at first but Judell reassured me by saying, "It's all right, I know every law enforcement man in town."

On a Monday morning we drove to Santa Maria and then on to Bernard Permasse's place up in the Cuyama Canyon. He had a gas station and also took care of travelers. His wife did all of the housekeeping and cooking. Bernard told us of early days in the Santa Maria-Guadeloupe country. It was a pleasure to stay there. Mrs. Permasse was a native Spanish-Californian and served many dishes that were new to me.

Shortly after we got there, Cyril Robinson and Fletcher Hayward arrived. Even in March, the annual grasses were mature and the heat on those steep canyonsides was something to remember. Fletch was my partner and he took it upon himself to give me all the assistance possible. In spite of my objections, he insisted on carrying that plane table at all times. This equipment came in handy, the country was so rough and in places brush, that it was impossible to run lines everywhere. One thing I learned in a hurry was not to set that plane table too close to a clump of Spanish bayonet. I surprised myself by locating a few corners by triangulation, but I have to admit they were corners witnessed by blue oaks.

We left Permasse's after about ten days. Fletch had made arrangements for us to stay at the Andrew Jouchin Ranch. When we pulled into the ranch yard, Mr. Jouchin came out of the house to greet us. After introductions, I picked up some of my plunder and started to walk to the bunkhouse. Mr. Jouchin stopped me and said, "I will show you your room in the house." I attempted to enter an objection but after talking for a few seconds I realize that he was waiting for me to stop talking. Then he said, "I asked for this survey; we want you to stay in the house."

Although the ranch buildings were very old, the ranch house was new, big and beautiful. Fletch and I later came to call it the Hollywood ranch house. Our bedroom was very large, beautifully furnished and contained the largest bed we had ever seen. When we turned down the covers on the bed we got a surprise, the sheets were heavy, stiff and obviously linen. We had been introduced to our first kingsized bed.

Mr. and Mrs. Jouchin treated us so hospitably that it made us uncomfortable. After struggling through dusty brush on hot hillsides, we would go back to that luxurious bathroom and almost scrape the skin off ourselves trying to get as clean as possible.

I have a vivid memory of two gracious persons in Mr. and Mrs. Jouchin. One tangible thing we were able to do was to show him the location of quite a few corners on his property, lines that he had never seen before.

INCIDENT AT CAMP TOUGH-IT

Everett Jensen

Our trail crew was camped near the western edge of the Kootenai NF in Montana near the Idaho line. We may have been in Idaho for all I knew; I don't recall seeing a map all that summer of 1929. Our tent was pitched on a small level area high on a slope that angled up and westward to the main divide. We had about everything that a trail crew could want; shelter, tools, grub, a good cook and a fine view. What we lacked was a firm water supply. Slim Basham, the packer, did bring us water on a regular schedule unless he got tied up on a higher priority job.

Ranger Woods had made it clear that he wanted a neat camp. We slept four in a tent and beds were made up each morning. A fly stretched out from the tent over the shiny Kimmel stove. The stove pipe was arranged to protect the canvas and was equipped with a spark arrester. Garbage pit and latrine were located a suitable distance away and were limed regularly to prevent odors and flies. Tools were racked at night. We were a model camp.

Woods must have wondered a little at the crew he had hired. Al Tate, the foreman, was a little, old man who had spent a lot of time in the mountains but seemed lost when he got ten feet off the trail. The most noticeable thing about Al was the cartridge belt, holster and heavy revolver that he packed on his person. Joe Martin knew all the jungle hangouts between the Missouri River and the West Coast; not a tramp, a hobo. The IWW days were in the past, but Joe still carried his card, or so he said. Bill Green, our cook, knew all the words a camp cook was supposed to know, plus some he had learned from mule skinners. He too, was a wanderer and did not outlast our dry camp. I was last and least on the roster, without camp skills or woodcraft, a "student." My presence was the product of my own desire and the annual student lottery conducted by the forestry schools in cooperation with the Forest Service.

One late afternoon our packer arrived with a much needed load of water. Woods was along to inspect our camp and to see if we had been doing what we were hired to do. All was in good shape and, after considerable talk around the stove, the crew retired to the tent and cots. Slim and the ranger rolled up in blankets outside.

Sometime during the night, we were awakened by shouting that sounded a lot like swearing and a rapid fire of pistol shots. We tumbled out in the dark, a blackness soon broken by a flashlight held by Woods. The beam roved here and there, showing Al Tate in a night gown and the rest of us not so thoughtfully attired.

We learned from an excited Tate that he had heard a bear investigating the camp, perhaps drawn by our garbage pit. He had frightened it away by shooting in the air. By now it was far away and one by one we settled back to bed.

We rolled out soon after daylight. The boss was here and we had to show him that we were an alert crew. With the usual preliminaries out of the way, we gravitated toward the stove to find the ranger ahead of us. He turned to the foreman, who was coming forward from nearby.

"Al, you say you heard a bear last night?"

"Sure did, boss."

"And you say you fired up in the air to scare it away?"

"That was the idea. It worked, too, didn't it?"

"Well, Al, if you fired up in the air, how do you account for all those big holes in the stovepipe?"

Picture five men and a boy grouped near the Kimmel stove staring hard at a bullet ridden stovepipe. For a long moment no one dared to look at the tent where three of us had been sleeping.

THE RATTLER

Ira W. Roberts

I had opened the Mt. Lukens LO station on the Arroyo Seco Ranger District, Angeles NF, May 15, the official opening date of the 1929 fire season. It was a wooden structure supported by fairly heavy timber 14 feet above a base of from one to two feet of semiflat stones.

Ensconced comfortably in my swivel chair on a raised platform, map table and alidade before me and nervously anticipating turning in my first smoke report, I suddenly became aware of a large diamondback rattler which apparently had just emerged from that pile of rock beneath the tower and was headed for the brush line near the head of Pickins Canyon, which ended as a steep slope right up to the south side of the lookout and covered with loose gravel. Grabbing my 32 caliber Smith and Wesson, I gave chase and squeezed off a shot when the snake was less than ten feet from safety. At the same time my feet flew out from under me and I just barely missed landing right on top of that old boy and was he mad! I learned later that my first shot had just grazed its back and apparently made it so mad that instead of heading for the brush it headed for me while I was desperately crayfishing back up that gravel covered slope. It would coil and strike repeatedly and was gaining on me each time, until in desperation I squeezed off a quick shot which to my utter amazement severed that snake's head completely.

Just a little piece of skin was holding it on, so after I recovered from the shakes I dragged the carcass up to more level ground and as an experiment stuck a stick in it's mouth. It clamped down on that stick and the venom came oozing out of its fangs.

I take no credit as a marksman due to my lucky shot at the rattler. During prior target practices I was never able to even hit the target at 15 feet. My conclusion is that it was so close that it struck at the motion of the revolver to its own undoing. I don't buy the often stated theory that they strike at the bullet.

THE THIRTIES

THESE CHANGING YEARS

James D. McNeill

As I look back down the years to the late 1920s and the early 1930s, I still feel a little bit of amazement in the amount of work that the district rangers on the Klamath Forest accomplished with the limited amount of manpower and funds that they had to work with.

Very few of the district rangers had the help of assistant rangers. The protective assistant usually filled in in the capacity of the assistant ranger. He was the logical man to have that position because he drew regular funds longer than anyone else as he was paid from May 1 through October 30 out of regular fire funds and was paid primarily to head up the fire control organization. From October 30th, unless we had a tag end fire season, he was paid from whatever money the ranger was able to save during the summer months from his regular activities of trail maintenance, road construction, telephone maintenance and campground maintenance, etc. A few dollars saved here and there paid his salary in post season work until usually around the Christmas holidays. From then on the ranger was alone until the middle of March when the pre-spring money became available. Then he could call in a guard or two who could give him a hand until it was time for the short term forest students to come back on duty again.

That went on all through those years. The Happy Camp District was a little different than most of the districts. Michael Morgan, who worked as Ranger Sutcliff's protective assistant, came to work on May 1 and worked to the first storm in the fall. When the fire season was reasonably on the down grade, he pulled the office door shut behind him and headed back to his ranch and family at Bull Meadows at the end of the Middle Creek Road on Horse Creek. I usually picked up the chips where he laid them down and carried on until the funds were exhausted and then I joined the army of unemployed until March 15th. These conditions existed until the early '30s, when came the Great Depression.

There were no roads, and the district was administered from the hurricane deck of a saddle horse. That ran into quite a little chore.

When the Great Depression struck, those people from the flatlands, from the cities and towns of Southern California, landed along the streams and bars. Every place that was large enough to hold a shack, there was a shack. An old house was cleaned out and made habitable. A tar paper or a shake or a bark shanty of some type sufficient to take care of a family was put up, and of course, that increased our work about a thousand fold. We were paid for a 44 hour week but if we didn't work twice that long, I guess we were disappointed, and mind you now, we were not paid overtime.

The guard station and the ranger station was a source of information to everyone who came inquiring where they could find a place to live; where they could dig gold out of the hills, gold out of the creek, gold out of the bars, how to make rockers, sluice boxes, riffles; anyway to catch the gold so they could buy food and clothing for their families. We were the ones they turned to first. We were busy and it went that way through 1930, and that winter a few extra dollars were made available somehow, someway, and we were able to carry on. Guard Lee Waddell was added to the staff and we worked that winter delivering food, checking on people, seeing that they didn't freeze or starve. We had a sinking fund at Evans' Mercantile Store where every working man along the river donated whatever he could afford. We saw to it that the food was delivered out to the people in the forest. We saw to it that the streams weren't polluted. We saw to the sanitary condition of the place as best we could, and in short, we took care of everything that we could take care of and made those people as comfortable as possible without knowledge of woodcraft and forestry. Still things got away from us. We kept a pack mule on the district that winter, as well as our saddle horses.

I remember that there was an Indian boy in Happy Camp by the name of Raymond Storrs. Ray took it upon himself to visit those people in the outlying country and he wandered those creeks and ridges up and around and about, seeing who was in the worst condition and reporting it to Ranger Sutcliff or the store. Right away we were on our way out there with some kind of relief. Old Joe, the pack mule was making a trip with either Waddell or myself. I think a great many people owe a debt of gratitude to the memory of that boy, Ray Storrs, (he has been dead many years) for I think many of them would have gone to sleep hungry a good many nights, if it hadn't been for his trusty rifle putting meat on their table for dinner at night.

The summer of 1931 we had a rash of forest fires; grub stake fires that just wouldn't quit, but fortunately we didn't have any that really took off. We handled everything that came our way in pretty good shape. As I remember it, we caused only one man to pack up and leave the country. He got a little too brazen, but otherwise we tolerated the breach of forest laws because those people had to eat. That winter it was really tough, but only one death occurred and that was at Classic Hill, and it was unavoidable because we couldn't get in to the man before he was too far gone with pneumonia.

The summer of 1932 was real bad also because the grass, the moss, the crevices and the bedrock were stripped clean of the last vestiges of gold. That spring the administration had decided that all Government employees should make some sort of a donation to the welfare of the country at large, so they had taken a bite of eight percent out of our salaries. That bit back at the people who had been kicking into the sinking fund because we had to live too.

That fall about the middle of November, someone came to the realization that the Klamath Forest had committed the heinous crime of employing one young male veteran of World War I in a full time capacity, where they should have employed two family men in a part time capacity. Therefore, the single man should be separated from his job and be put on leave without pay immediately. That being the case, one Jim McNeill was placed on leave without pay on December lst. I departed from Happy Camp and took a job breaking horses at $15 each, plus my board and smoking tobacco for the winter.

We had been reading a lot about the change of administration and about the CCC. Along about the middle of April, Tom Bigelow came by where I was working and left a note with my boss saying he would like to see my ugly face in Happy Camp not later than May 1st. To make a long story short I made it on time and on the 10th day of May, I was the welcoming committee of one at Camp Clear Creek to meet Company 1235 from Fort Slocum, New York, the boys from the bowery and the sidewalks of New York. I was with that Company for about three weeks as the one and only foreman. I won't take any bouquets for the amount of work I got done, but I at least kept them together and that was saying something. And so began the CCC.

BUGS

Corwin E. "Slim" Hein

Insect enemies of the forest were created for a purpose, no doubt at the same time trees were created. Early in this century, forest managers becoming concerned with the bugs' depredations, researched, studied and began control measures. The old-timers were mostly concerned with the infestations of western pine bark beetles in the ponderosa stands of Eastern Oregon. Because the technical name of <u>dendroctonous brevicomas</u> was too unwieldy for the woodsman of that time, the term "brevi" or just plain "bugs" was used. Early rangers were instructed to be on the lookout for beetle infestations and report their findings.

The story among the old "bug hunters," as the beetle control men were dubbed, was of the early day ranger who glued a brevi to a paper and submitted it to his supervisor with the following bit of doggerel beneath:

This is the beetle dendroctonous,
The killer of the pine.
He lives on pitch,
The son of a ____;
And to kill him will take a long time.

His observations were more true than he realized. In spite of all the control measures taken, the brevi is still very much alive.

Beetle control projects throughout the drought years of the 1930s, provided winter employment for many smokechasers. Autumn of 1929, brought the first bug control projects on the Deschutes Forest in the Fox Butte burn of 1926. The fire weakened timber was heavily infested with brevis. Control procedure was to make a 100 percent cruise of the project area with a team of one compassman and two spotters who marked the trees and located them on an eight inch to one mile map sheet. Treating crews of from two to four men followed later, falling marked trees and peeling the bark on the top two thirds of the log, and with the aid of pitchy wood or dry needles placed under the log, burning the bark sufficiently to kill the overwintering brood of bugs.

Living accommodations in the camp were meager, with six to eight men in a squad tent with a Sibley stove in the center and bedrolls on the ground. Can't remember that I ever slept more soundly, though. Water was hauled in from Cabin Lake Ranger Station in barrels and our bathing facilities consisted of tin basins and Lava soap on a wash bench of poles. There were no R&R trips to town, and after several weeks in camp the BO index was pretty high! Who could tell though, when everyone smelled the same?

Every bug hunter aspired to become a member of the spotting crew, as this was considered less arduous and more interesting than felling trees and chipping bark. It was an unwritten law that to become a spotter one had to eat some brevicomas larvae. Now, this isn't as bad as it sounds. The larvae is pure white, about the size of a grain of rice, and all its life has dined on nothing but good, clean pine bark. The flavor is very much like that of piñon pine seeds. It's really little different from eating shrimp.

With the coming of the CCCs, much of the bugging was done by these boys under the supervision of Forest Service foremen. Transportation facilities to and from the field were far from luxurious in those days. Open stake trucks with benches, too cold to sit on in the often prevailing below zero temperatures, were provided. In spite of the many hardships endured in connection with the bug camps, morale remained high and a fine spirit of camaraderie grew from our close associations at work and long evenings in camp.

A FOREST RANGER? YOU'RE KIDDING

Harold A. Dahl

I was raised on a farm four miles east of Gresham, Oregon, during the days when the nearby lands were forest covered and the streams had fish in them for young boys to catch. There we spent much of our time when we could slip away from the farm chores. My Uncle De often said that I should be a forest ranger because I liked to be in the woods so much. That was a wonderful job. He described the work as hunting, fishing, trapping and all this with a cabin in the forest. I had also been reading magazines of the day in which there were correspondence school advertisements an how to qualify for such jobs. It was obvious that this was my kind of work.

I began my forestry education at Oregon Agricultural College in the fall of 1929. That winter I applied for a lookout job on the Umpqua National Forest. Later a notice arrived stating that I would be a lookout on Elephant Mountain that coming summer. Fire guard school would be held for one week the middle of June at the Tiller Ranger Station, Tiller, Oregon. All of us were to meet at Roseburg, Oregon, the day before school was to start.

Forest Service employees were unknown to me. I had seen pictures of such people in magazines and papers. With these in mind, I bought a pair of choke bore pants, a pair of high top leather boots, a khaki jacket and a tan hat with a brim far too wide for a 130 pound kid. Thus attired, I boarded the bus in Portland, arrived in Roseburg in time and reported to the district ranger thusly: "Harold A. Dahl reporting for duty on the Elephant Mountain Lookout." The ranger kept a straight face, but it took me some time to realize the Forest Service deportment was somewhat different than the ROTC.

Before leaving the farm for Roseburg, mother and others, provided considerable help and advice. Mom was especially concerned about scorpions. She said they lived in damp places such as, rotten wood, bark, under rocks or boards, etc. They made a painful wound if they stung a person, so don't sit down in the woods unless it was on the bare ground or some similar place. She also provided a cardboard box containing about 100 three by five inch cards with recipes on both sides.

While at the school I became acquainted with many fellows. One of these was Henry. I was using some worldly knowledge on him one day—I guess you could say we were sharing our skimpy knowledge with each other (bragging)—including the information about scorpions. He was unbelieving. I offered to show him. We walked out to the bottom of the nearby ridge where there were some rocks. I turned one over and lo and behold there were two under it. He was surely impressed. I must admit I was too because these were the first I had ever seen. As a matter of fact I have never seen another one in the woods, so it really was quite an event.

From the school we went by stake truck to the Big Camas Ranger Station, my first trip in that part of the country. One of the first jobs was to order groceries for the summer for three whole months. I just loved Campbells canned soup, tapioca for pudding, canned peaches and a few other specials. I remember ordering four cases of soup and at least 10 pounds of tapioca. The rest is kind of hazy except for matches which I forgot to order. For years afterwards, I couldn't stand the sight of Campbells soup and tapioca.

The Elephant Mountain lookout was about three or four miles by trail from the Diamond Lake-Big Camas Road. The road was at an elevation of about 3,000 feet where the trail met it. The lookout was close to 2,500 feet higher. The lookout was in two parts. The lookout tower was just that. Four poles about 25 feet long erected vertically with bracing to space them at the top for a platform about six feet square. In the center there was a stand about 14 inches square and three and one half feet high on which the Osborne firefinder was to be mounted. There was a wood railing around the outside of the platform and on both sides of the stairs. For lightning protection the railing was lined with No. 6 copper wire to conduct the current to the ground. It was strictly open air; no chairs and hardly room enough to sit on the platform. When lightning storms came over, it was strictly down to the cabin and it didn't take much urging, The snap and pop of electricity flashes between the plates of the firefinder and parts of the protection system were incentives for a rapid exodus.

One day in the summer of 1931, I had retreated to the cabin as a lightning storm was going over. The hail was coming down in 25 cent bags; that was a lot of hail in those days, much of which was coming through the shakes of the cabin roof on everything inside. Then there was a paralyzing flash of light; an unbearable volume of sound; vibrations of the cabin and grounds, then back to the rattle of hail on the roof and sharp little shocks of the hail hitting my bare flesh. Sometime later, after regaining my senses and confidence, I got up and went outside to see where the bolt of lightning had come to earth. It had to be close. It was. Right back of the cabin about 75 feet away a narrow strip of the bark on a large noble fir had been torn away leaving the raw scar of the exposed wood.

The lookout cabin was about 10 feet by 12 feet, built of logs cut from nearby trees with a roof of noble fir shakes and the outside walls covered with them also. There was a space between the shakes on the log walls and the logs themselves. In this area pack rats made their home. During the day they were very quiet but at night they made a tremendous amount of noise or so it seemed to a kid alone in the woods for the first time.

I tried to scare them away but without results. They would be quiet for a while and then you could hear that thump, thump, thump that sounded like a trip hammer in slow motion. Then they would run around the building, push things into corners, shove canned goods off the shelves, or out of the boxes or make other almost inconceivable sounds. Alone at night in the woods for the first time with packrats is an experience not soon forgotten. I finally gave up trying to live with them and decided it was them or me. I got up, found a piece of cardboard, stuck it into the Kimmel Stove for a light, pulled the torch out and there was a packrat looking me right in the eye on top of the cache box. Instinctively I shoved the faggot under his chin. There was a flare of flame and the packrat disappeared between the logs and shakes. I was in trouble. A fire would surely start, so I stayed up all night looking and smelling but there was no fire.

This incident really got me. I decided the job had to be done in the daytime. After daylight, not much later, peering in the openings between the logs, I finally spotted a pack rat that looked singed. I took the Krag 30-40 rifle and found that he could be hit by shooting between the logs. The crack was wide enough. I didn't consider the fact that there could be some problems by shooting from inside the cabin but I was to find this out shortly. The floor was dirt. When I pulled the trigger the concussion raised every loose particle on the floor and thereabouts and deposited it in many places including over myself, the bed, cache box, store, etc. There were two good things about it. The pack rat had gone speedily in several parcels to his reward and the floor was as clean as a dirt floor can be. The experience resulted in development of other pack rat control methods.

The summer finally ended with a soaking rain in early September. The rush was on to get the lookouts and firemen out of their stations and to other jobs, on the way home, or to school. Mine was to be more college. John called up this fateful day and said a white mule complete with pack outfit and halter would be left at the end of the trail by the road. They were leaving an extra lash rope with him. The mule's name was Whiskey and he was known to bite and kick people. I was to get the mule, pack my stuff on him and bring him back to the road, where hopefully, they would pick us up in a truck. I had never put a pack on an animal.

Down I went; the mule was there; he seemed gentle enough during a cautious get acquainted period, so I untied him and started out. His manners were exemplary. I decided to ride him up the trail remembering it was steep and a long way to the lookout. I led him up to a log beside the trail, tied the other and of the halter rope to the off side of the halter and climbed on. Whiskey came unhinged, piled me up and waited for me to think that over. I got on again and the same thing happened, The point being well made, Whiskey permitted me to lead him up the trail.

With his reputation, caution was essential and some real good ideas how to get the job done safely were in order. Right outside the door of the cabin there were two trees growing just the right distance apart and another about five feet beyond. I got Whiskey to put his head and neck between the two trees and tied him to the third. The two were close enough to just fit his neck and too narrow for his shoulders to go through. Then I took another piece of rope and pulled up one hind leg and tied it close to his stomach. Now he couldn't go ahead, back, sideways or kick. Neither could he bite. Things looked pretty good.

I started piling things on. The first layer wasn't so bad, The first lash rope did the job with a little to spare. But there was more, so a second layer was added with the help of a couple of blocks of wood for me to stand on. The load was getting pretty high. The second lash rope together with the remainder of the first just barely tied things down, That is, I surely hoped they did. Due to the height of the pack and the lack of tying places, I had to crawl under Whiskey's tummy. Some tying places were added with some No. 9 telephone wire and a pair of pliers. Then things could be tied here and there like clothes in a closet, All through the operation, Whiskey hardly moved a muscle. When his foot was released, he showed no emotion. He was most docile as we went down the trail without incident,

Ed, the Forest Service packer and Clay, the stock truck driver, were waiting with the truck when we arrived. Their first question was how in the world did that rope get wound all around that mule. I told them. They expressed considerable surprise and alarm. Ed offered some packing lessons for me soon. I expressed sincere thanks,

Then back to school after a wonderful experience.

POACHERS ON THE MODOC

Alvin Parker

Exerting authority under the Forest Service badge was an everyday affair. However, one incident that really remains in my memory occurred on July 31, 1931, under authority of the deputy game warden badge. It happened while I was stationed at Hackamore, a Pickering Lumber Company portable sawmill camp on the Modoc National Forest. This portable camp was located adjacent to and along the south side of the old dirt road between Canby and Tule Lake. Prior to the time of this episode I had met with Addie Jordon, the state game warden in Alturas and had established a working relationship in connection with game violations. Suspected game violations were to be reported to Addie, however, if I were to observe a violation I should take direct action.

Both Modoc NF and Modoc County had derived their names from the local Indian tribe, the Modocs; remnants of several tribes were still with us at the time. Since the entire area was scented with Indian lore it seemed only natural for me to wear a feather in my hat and carry a side arm.

So much for background to the drama. I was in my office type cabin at the sawmill camp. It was shortly after lunch when Carol, our patrolman, drove up to my cabin with his report.

"I was in the vicinity of Seven Mile Flat when I stopped my car in the shadows of pine timber," he said. "After a few minutes I heard several gun shots which sounded like they were coming from some distance to the south. Since it is closed season I was curious and drove toward the sound of the gunfire. When I came within sight of the flat, I stopped my car, again in pine shadows. Through my binoculars I could see three men. They were dressing out some deer they had killed and were loading them in the back of the coupe. At this point I decided to leave, undetected I hope, and high tailed the seven miles back to camp and report to you."

I thought this might be a case of "direct action." However, I first put in a call for Addie Jordan at Alturas and was fortunate to reach him at his office. Addie gave instructions. "I'll leave right away and intercept them between Canby and Seven Mile if they come this way. If they come your way (toward Klamath Falls) try to follow them and I'll come along as fast as I can."

About ten minutes later a car came by our camp headed for Klamath Falls. Carol was standing with me at the cabin. "There they go now; that's the coupe," he said. At this point I realized that I was more involved than usual. Once more I put in a call, this time to Malin, Oregon, and Curt Higgins, District Ranger on the adjoining district. I was lucky again; Curt was in his office. "You follow them and I'll come to meet you," Curt said.

Five minutes after the coupe had passed our camp I was in pursuit in my car. The road was intermingled lava rock and dust, so travel was slow. However, in a matter of five to ten minutes I began to catch up with their dust trail. Another five minutes and I caught glimpses of the car through dust clouds. Visibility improved and when we passed from the pine timber into juniper and sagebrush, I remained well behind with no effort to catch up. The road and Southern Pacific railroad were parallel here and I had just passed the Perey Railroad Station when I met Curt. His place was to turn around, follow the coupe to the next straight away, then pass and block the road.

The plan was well executed. Curt got out of his car and walked back while I stopped about a car length behind and walked forward with side arm in clear view. Curt was unarmed. The coupe driver rolled his window down and Curt told him that the state game warden had instructed us to stop his car and hold them until he could arrive from Alturas. The men were a bit surprised and as Curt was conferring with the driver, I noticed one of the others place his hand on the breach of a rifle clearly visible to me through the rear window. The rifle was pointed at Curt. All this prompted me to display the old 44, pointed upward. The man looked back, saw me and changed his mind.

As tension subsided I heaved a sign of relief for if this man had gone further there would surely have been fireworks. The rest seemed anticlimactic. The driver got out of the coupe and opened the trunk. There were the carcasses of five deer; two does, two fawns and one small buck. Addie arrived. Not only did he make arrests, he seized their guns and confiscated the meat.

I never told Curt or Addie how close we had come to a shoot-out.

Forest officers scaling dead lodgepole pine, Arapaho NF, Colorado – c1910.

Skidding logs, Coconino NF, Arizona – c1910.

THE MATILIJA FIRE OF 1932

Robert W. Cermak

Editor's Note: The Matilija Fire, famous in its day, has been fading from memory thanks to the thinning ranks of firefighters of that era and to more recent large, problem fires. Cermak's letter transmitting the story states that $120,000 was spent on suppression. This was a lot of money in a depressed economic climate, requiring a lengthy special report in justification. Cermak has a comment. "It is interesting to compare the firefighting efforts on this fire with other very large fires of the last twenty years. Despite modern firefighting technology and trained crews, these fires were controlled only after the weather changed, just as with the Matilija Fire. Makes you wonder." –E.J.

Cool, wet spring weather delayed the onset of the 1932 fire season, and by early September, it looked like Region 5 would have one of its most successful fire seasons. Fire control people on the Santa Barbara National Forest had their fingers crossed. It looked like they might finish the season with their best record ever. Them, at 10 a.m. on September 7th, La Cumbre Peak Lookout reported a smoke rising from the upper end of the North Fork of Matilija Canyon, about ten miles northwest of Ojai, California. Within minutes of the report, the Ojai District Ranger, E. L. Baxter, had ordered 66 men and a fire camp. He left a few minutes later with another man for the fire, but it took two hours to reach the fire, traveling 13 miles by car and five miles on horseback. When he arrived, the fire had already covered 300 acres. Evidence at the start suggested a hunter had left a campfire or dropped a cigarette into dry fuels.

While Baxter was on his way to the fire, men from the supervisor's office ordered more crews and overhead, assembled pack stock and began stringing telephone line from the end of the road to the fire area. At 4 p.m., fire control plans were proceeding smoothly. Forest Supervisor, Gus Nash-Boulden, was hiking up Matilija Canyon from the Ortega Place with 50 reinforcements when a messenger from Baxter arrived saying that the fire was expected to reach 600 acres by evening. He hoped to contain the fire by midnight. At this time, the temperature was 98 degrees, the relative humidity was 11 percent and the wind from the northwest was 11 mph.

At about 5:30 p.m., Nash-Boulden and 50 men were hiking along, when, without warning the fire exploded and began advancing toward them from three sides. The intensity of the fire exhausted most of the oxygen from the canyon and the crew was close to suffocation. Nash-Boulden remembered that the veins in his neck swelled to twice normal size. He immediately ordered the crew to backfire the canyon walls. The backfires rushed up the slopes bringing a draft of lifesaving fresh air behind it. Nash-Boulden and the crew escaped down canyon and he at once began organizing for a project fire.

The extreme fire behavior was probably due to 50 mph Santa Ana winds which had trapped gases from the fire in nearby canyons. When the gases ignited, the fire exploded and ran over 20,000 acres in 12 hours. By noon of the second day of the fire, 28,000 acres had burned and the fire was wildly out of control in three directions. The only safe way to control the fire was to cold trail the flanks and wait for the weather to change. In fact, most of the final control line was cold trailed.

On September 8th, crews and overhead began to arrive from eight other Region 5 forests, the regional office, experiment station, division of forestry, Ventura County and Santa Barbara County. Paul Mantz, a famous movie stunt flier of the time, was under contract to the Forest Service. He flew scouting missions and by request, dropped a side of beef free fall to a fire camp on inaccessible Topa Topa Mountain. The camp cooks reported that the beef was well tenderized by the drop.

Bob Deering from the regional office, visited the fireline with Nash-Boulden near Wheeler Gorge on September 10th. While they watched the fire surge over vast areas of brush, the earth seemed to shake. Deering thought an earthquake had struck but Nash-Boulden reassured him that it was "just the fire." They, along with many others, witnessed what was probably the most stupendous firestorm on record in California. As firefighters watched in awe and amazement, the fire swept 12 miles airline, on a front nearly five miles wide, from the Sespe Gorge to Mutai Flats, up the canyons and ridges of the Sespe River to the Pine Mountain Ridge and Devils Heart Peak in one hour. Perhaps 35,000 acres of brush and timber burned in just sixty minutes! By that evening, the fire had blackened 80,000 acres with no end in sight.

No part of the fire was accessible by road. Seventeen fire camps were originally set up, 12 of which were supplied by pack stock. It took 58 packers with 130 mules and horses to supply the fire camps. Bulldozers dragged supplies to some camps, opened a few roads and built a few miles of fireline but were not very effective. Experimental models of two-way radios were used but as Nash-Boulden said, "Once in awhile, you would get a message through but twice in awhile you wouldn't." Over 3,000 men and women fought the fire, although there were never more than 1,200 on the lines at once.

Between September 10th and 18th, crews were repeatedly driven away from the fire by shifting winds and fast moving fire runs. Three fire camps were destroyed and a fourth saved only by desperate backfiring. On the 13th, Reyes Peak Lookout was destroyed by the fire and on the 18th, Santa Paula Lookout was saved only by backfiring. On the 17th, a woman packer, Mrs. M. M. Hunt, saved a fire crew when the fire overran them, by keeping them under wet blankets in a creek.

Despite wildly erratic fire behavior and the huge size of the fire, there were only two serious injuries, a broken leg and a back injury. This is partly due to outstanding leadership from Nash-Boulden on down to crew leaders. He told his camp bosses, "The first thing to do is to make yourselves safe. Nobody's going to get burned on this fire if I have anything to say about it." Also, this was the first large fire in California on which most of the firefighters were members of trained crews.

The fire ran wild for ten days and grew from 80,000 to 120,000 to 200,000 acres. Strong winds and low humidity continued until the 18th when a high fog rolled in from the Pacific. Almost immediately the winds shifted to the south and humidity rose to 30 percent. Within the next 36 hours, the great fire was cold trailed and controlled. Over 500 miles of line was out. The final toll was 219,254 acres, all but 10,700 acres of which was national forest. From Fillmore to the outskirts of Santa Barbara, an area 32 miles long, averaging 11 miles deep, had been blackened. The Matilija Fire burned more acres than did all of Region 5 in the next five years (1933-1937).

Despite the best efforts to reduce costs, the fire cost $123,000, a huge sum in those early days of the great depression. The mobilization for the Matilija Fire was the first time a multiagency control center was used for a major fire. The Angeles National Forest dispatcher coordinated the crews, supplies and equipment for the fire and acted as zone dispatcher for the first time in Region 5. The fire showed the need for standard training and terminology, the need for better record keeping and for better forest maps.

MY FIRST LOOKOUT JOB

Jack Smith

I was assistant warehouseman at Tiller Ranger Station in 1933. There were two new CCC camps on the district and during the start-up program the district furnished substantial work tools and equipment to the camps. We operated three pack strings and there was a lot of push to get new lookout houses built to increase direct visibility coverage from lookout stations. I hauled supplies by truck to stations and fire camps. It was a pretty good job. The ranger arranged for my personal car to be used by the law enforcement man—the car to be maintained and served in the CCC camp shop. This gave me additional income of some $25 or $30 per month without out-of-pocket expense to me.

In mid-summer the Ranger, Gene Rogers, came to me and said he wanted me to be the lookout at Callahan emergency lookout station, which was okay with me. I bought an order of groceries, and with my packsack and bedroll was on the way to Callahan the next morning, where the packer would leave my gear and go on to Green Butte. Upon arriving at Callahan, which consisted of an 83 foot ladder up a tree and a 12xl2 foot ground cabin, we started to unpack, but first had to kill a three foot rattlesnake which had possession of a spot in front of the cabin door. We hooked up the telephone to make sure it worked and the packer left. Then, a hike more than a mile down a steep trail to the spring. I dug out the spring, waited for it to clear so I could fill a five gallon man-pack and a couple of water bags to take back to the top, and then I was set up and ready to do my job.

But first I had to climb that 83 foot ladder up an old growth Douglas fir, which on the first trip seemed endless. When you got to the top, you pushed open the trap door and crawled through onto a 6x6 foot crow's nest. The base for the firefinder, of course, was the top of the tree which had been topped at about 12 inches in diameter. But the firefinder was down in the cabin and it had to be hauled up with a rope, put in place, and oriented. I needed a telephone in the crow's nest, but the district was short of equipment, so I would have to fill out a fire report and climb down the ladder to report a fire. I had been there only a few days when I spotted a fire in green timber on Three Horn Mountain. I reported it and climbed back to the crow's nest, only to do this six more times. A fire bug had gone around the side of Three Horn and set a string of seven fires some distance apart. However, with the abundant manpower in CCC crews these fires were promptly controlled.

A few days later the smoke from the big Tillamook Fire came in with the northwest wind and covered everything like a blanket. Visibility dropped to less than one fourth mile. I walked many miles every day on fire patrol, as did just about all the other lookouts on the district.

The smoke stayed with us and visibility was very poor. I left the lookout station for a few days to fight a 50 acre fire that a patrolman had found after it had burned for several days. It was a smoldering fire in old growth timber. I had a crew of CCC enrollees and the fire was dangerous because of falling trees and snags. Now and then an old tree would burn through and fall with very little warning. When this fire was out, I went back to the lookout where I stayed until the fall rains started. It was considered normal to get a rain around September 15–20th, but they didn't arrive. In our evening telephone visits among the lookouts, we witnessed many signs of rain, such as, "Hoot owls hooting on the north side of the mountain," and "Coyotes barking in the daytime." "Rain crows calling at night," etc., but none of the signs we knew about seemed to work.

I ate a few gray squirrels to extend my food supply, but was down to a can of tomatoes and a package of macaroni. I cooked them together and ended up completely out of grub when the rains came on October 13. I put the firefinder away, snapped the Forest Service padlock on the cabin door and headed down the mountain.

During my stay I had six or eight visitors, including Oscar Howser, the forest central dispatcher, but I never could get a visitor more than 20 feet up the ladder.

A HUGE HOH HUNT

Sanford Floe

In February 1927, I was transferred to the Port Angeles District of the Olympic National Forest. The Port Angeles Ranger District covered the area from Deer Park west and south to the divide between the Hoh and Queets River, about 660,500 acres.

In October 1933, the Hoh Valley was opened to elk hunting after some twenty years of closure. There was an estimated 2,000 people in the area. This was four or five times the number expected and created a "Coxey's Army" situation. Biologist Jack Schwartz, two experienced guards and myself were the local forest officers in the area. The Washington State Game Department had five or six protectors there also. In addition to these, the following observers from other agencies were on hand: Leo K. Couch, representing the U. S. Biological Survey; Mr. M. P. Skinner, a former National Park naturalist, now representing the Boon and Crockett Clubs; Mr. David Madsen, Supervisor of Wild Life Resources for the NPS; Foster Steele representing the regional forester.

The first night a former Forest Service employee visiting among the camps accidentally shot a hole in the gasoline lantern pressure tank in a tent. He and two others had to be hospitalized for burns. In rapid succession a hunter was shot and killed at the head of Owl Creek, miles from a road and it took the combined efforts of all forest and game department people to get the body out. A party of hunters tried to ford the Hoh River in a truck, drowning one of them. A hunter shot himself in the leg about six miles out in the timber and his partners were so exhausted they could not guide us back to him. We did find him, however. Someone shot a white horse being used to pack out elk meat. Then, we had four or five inches of rain in twenty-four hours, marooning hundreds of people on the wrong side of the river from their camps. Flood stage continued several days. All kinds of craft were pressed into service for ferrying people but the best were Indians with dugout canoes. There were hundreds of other small incidents. The usual percentage of lost people, car wrecks on the congested road, disputes over who killed the game, camps placed too close to the river bank washed away when the river rose while the owners were out hunting, etc. One night we were awakened by a pounding on the station door. A somewhat drunk hunter said his partner had not returned to camp. Questioning him got little helpful information. Finally, "grasping at straws," I asked how his partner was dressed. In reply the man said he had on a red hat. Since everybody in the area was wearing a red hat I took considerable ribbing from my partners for not being able to identify the man immediately from his useful information.

A DOUBLE WINNER

Clark Gleason

It was April 3, 1933, when Dave Ilch and I drove from Berkeley to the Feather River Branch, California Forest Experiment Station. There Earl Morrow provided us with bundles of bare root pine seedlings for planting in the Kennett smelter fume area on the Shasta National Forest. The planting went well but it was hot and dry that week. Came one day's end and we drove into the town of Kennett (now submerged). Glory be! It was the first day of the end of prohibition! Well!!! We sat down for a beer. "Gentlemen," said the barkeeper-owner, "I want you to know that this is the first liquor served over this bar since prohibition." It surely tasted good after that long hard day and we took the owner's word for his "first time" statement. Oh! Our trees took hold OK; they were followed by plantings at higher elevation that didn't get drowned when Shasta Dam was built.

THROUGH ROSE COLORED GLASSES

David O. Walker

In 1933, I helped build the new lookout tower and garage on Peavine Mountain on the old Lakes District of the Mt. Hood National Forest. Then I worked as a faller on the road between Beaver Butte and Mount Wilson. In 1934, O. J. Johnson asked me to man the lookout on Peavine Mountain. I found the lookout very interesting. The day started at 5:00 a.m. The first two hours I worked around the grounds and buildings. Then back on the tower for roll call and breakfast. The rest of the day was spent in the lookout unless it was fogged in or the fire danger was very low. I had a lot of fun and good exercise landscaping the area surrounding the buildings and turnaround.

In 1935, Mrs. Walker and I went to Peavine Mountain on our honeymoon. It was a wonderful place. Marcelle spent most of her spare time with maps, field glasses and the firefinder until she really knew the topography of the land around us. When we wanted company, I would walk around the catwalk playing my harmonica through an amplifier I had made from a large fruit juice can. Before long, we could see John Conroy's camp tender and herder on their way to spend the evening. It was quite evident by their speech that Ireland had been their birthplace. The camp tender's concertina went fine with my harmonica and we gave the herder some fast tunes for his clog dancing. He was very good. Marcelle usually served cake or cookies which we all enjoyed. Their lilting manner of speech reminded me of my father and our Scotch neighbors in Idaho.

I brought our season on Peavine to an abrupt end by tangling with a ladder and bucket of paint. The ladder won! My left shoulder had been dislocated and broken. The bucket of paint had crowned me with its contents. It was 5:30 a.m. and I had a little trouble getting to my feet and finding my way up the stairs to tell Marcelle that our honeymoon had been sidetracked by a ladder, ably assisted by one half gallon of gray paint. I was a mess!

In 1937, we were stationed at Bear Springs. I believe my rating was fire guard. I was in a position at this station to take a more active part in the work of the Forest Service—fighting fire, counting livestock, some cruising, scaling and meeting the public. The guards who came in contact with the forest visitors had to wear dress boots, breaches and a half Norfolk coat when around their station. After trying to explain to them the difference between a ranger and a guard I found it less confusing to answer their question, "Are you the ranger?" by stating, "The district ranger is at Clackamas Lake, but I am in charge here. May I help you?"

A few of our forest visitors seemed to be under the impression that our station was a combination restaurant and grocery store where they could buy what they needed. In most cases if it was some small item we were able to help them. One lady wanted to buy the food we had just cooked for our dinner and became quite upset when we refused to sell. Another demanded the use of our toilet because she refused to use the one by the registration booth even if it did look clean. She stated flatly, "I will have nothing to do with a rest room which is used by the general public!" But ninety-eight percent of the so-called general public were just fine and certainly made up for the others.

Ranger Lynch was a quiet prankster and some of his ideas certainly livened things up and broke the monotony. He usually had some ideas on the proper way to welcome newlyweds. The year before we moved to Bear Springs, it was Fire Guard Ray McGinley who moved into the guard station with his new bride. Our greeting was enthusiastic. We descended on their home like a flight of locusts and filled it to the point of overflowing. We started crying, "I'm hungry! I'm hungry!" Ray was embarrassed. His bride was very upset. We could have eaten them out of their supplies in a few minutes. We kept it up until we were all chanting, "We're hun-ger-y! We're hun-ger-y!" Ben Illige, who was boss of the ERA crew had them busy in the campground setting out a feast that would open the eyes of our newlyweds. When the lanterns were lit, we invited the honeymooners to cross the bridge over Indian Creek and be the honored guests at a wonderful party.

Clarence Gilbert and I slipped away. We wanted to make sure their home was comfortable. We met some pranksters just leaving. Ray came over and met us on our way out. We undoubtedly took the blame for everything. All we had done was to plug the cookstove pipe at the elbow.

The next was Bill Phelps, the lookout on Mount Wilson. When Bill brought his bride out from town, they moved into a lookout which had been papered with pinup girls. This had taken several hours of my time while I was filling in at his station. In searching through the magazines, I had found a four page folder of a beautiful girl wearing a bikini and a big smile. I pasted this on the ceiling over Bill's bed. We had a lot of fun kidding Marge Phelps about Bill's pinup girls.

Some of the events at Bear Springs had tragic endings. One of these was the water fight in and around the stable at fire school. What might have started as a boyish prank turned into a full scale battle with no holds barred. Part of the force were sleeping on steel cots in the barn stalls. Some were billeted in the barn loft, some in tents and the balance in buildings near the office. Someone slipped into the barn, tied two buckets full of water over the stalls so they could be tripped from outside. The tripping took place about 11:30 p.m. and all hell broke loose. Thinking the men in the loft had been the instigators, the men downstairs dragged in the fire hose, hauled it upstairs and proceeded to flush the others out of their beds and sleeping bags. They in turn retaliated by emptying two full barrels of water by the bucketful into the stalls below. The driver of one of the fire trucks pulled out for his station in the middle of the night, fighting mad and soaked to the skin. I still remember our

Assistant Supervisor Foster Steele's words the next morning, crisp, clear and to the point. No, my ears didn't get red. I wasn't in on the perpetration or the receiving.

Fire is always a tough and relentless enemy. Some of the men I worked with on fires will always be very close to me, as real friends who were there with what it takes when needed. Roger Sherman was with me on the North Wilson Fire. It was a small tough fire on the edge of an old blowdown. We would have plenty to work on if we failed to control it. We worked all night but there was still a tall hemlock with a snag top inside our line. Although we could see no sign of a lightning strike on the tree and no smoke was visible at its top, we decided to make a bed and fall the hemlock into it well away from the blowdown. We finished our first cut and had started on the back cut when wisps of smoke started rising from its snag top. We didn't have long to wait. Thirty feet of the tree's top broke apart. Limbs and all came hurtling down at us. As a rule, the safest place is close to the tree but this time it didn't seem to be the answer. A sky full of debris came plunging down. We ran from the tree to escape. Roger tripped and landed face down. I yelled, "Roll, Roger! Quick!" He did. A second later a huge limb stood like a dagger in the spot he had vacated. Burning bark, rotten wood and limbs were piled around the base of the tree and we had to cool them with dirt before we could finish falling it. It took us most of the day to mopup. As we started back, the rain poured down and soaked us good on the way to our cars on East Gate Road.

My separation from the Service came in October, 1942. Mrs. Walker and I had gone to Larch Mountain for twenty-four hour duty during the war years. We were to live in the lookout house on top of its 110 foot tower. We were to divide the time between us, watching for and reporting planes to the filter center.

The tower's welcome to us was far from heart warming. The wind and rain came in around the lookout windows. Water lay on the floor up even with the threshold. The sway of the tower would almost empty a pan of water left on the stove or table. All our wood had to be packed up 176 steps because it was too windy to pull up and over the guard rail. I caulked around the windows with dish towels, using a hammer and screw driver.

This kept the wind and rain out but we were fogged in most of the time. We could hear planes but we couldn't even guess their distance or direction of flight.

A flock of geese, headed south, almost crashed into the lookout windows but were able to stop in mid-air, three feet from the glass, tails pointed straight down, wings braking them to a stop where they hung in balance, honking about the obstruction. Then back in formation they continued on their way.

On the third morning of this adventure, I stepped out on the catwalk. It was a sheet of ice! My feet went out from under me and I sailed across the catwalk on the seat of my pants. When I stopped most of my body was hanging out from the catwalk but I was still clinging desperately to the bottom guard rail with both hands and the point of my chin. Marcelle was sound asleep inside the lookout. With my chin holding part of my weight, I couldn't have yelled anyway. When its 110 feet to the ground, you can put in a lot of effort to stay on top. When I could feel the catwalk under the seat of my pants again, I pulled myself up to a sitting position and sat there looking out between the guard rails. This time I lacked my usual enthusiasm as I looked down at the Columbia Gorge. I thought, "There is surely something better in life for Marcelle and me!" My promises to her about life in the great outdoors had turned to ashes. Ten years on the Mount Hood, plus 176 icy steps up the ladder, a little more money and then this!

I worked my way back inside the house and sat there thinking it over. I woke Marcelle to tell her, "I am quitting the Service!" I said, "The catwalk and steps are covered with ice. Stay inside until it thaws!" She answered, "I know you like the Forest Service, please don't quit because of me." I said, "I can get another job, but not another wife like you! I would never forgive myself if you were hurt!"

I called Ranger Baker to tell him we were leaving but we would stay until our replacement arrived. I knew I had to act quickly while the steps were icy or I might change my mind. They put a woven wire safety fence around the catwalk on the Larch Mountain tower, but in spite of this, the wife of the man who replaced me fell and broke her arm.

The Forest Service has every right to be proud of its personnel, past and present. Mrs. Walker and I were very happy we were a part of it for a time. I only wish that every young married couple could be fortunate enough to have lived through at least part of our experiences. I am sure they would leave with a better understanding of themselves and the fine people around them. My opinion of the Forest Service personnel was very good. They were a few hard working men doing a very fine job for a salary so low it wasn't even an inducement for a cowpuncher to change jobs. Twenty years later I found out what held them and why the Forest Service is so hard to leave.

FOUR TALES FROM THE ARROYO SECO

Buel B. Hunt

THE GHOSTLY ARSONIST

In the very early days of the great depression there were many different programs put on to employ the needy. One of these was a road being constructed (about 1933) out of Millard Canyon, which is a branch of the Arroyo Seco that flows past the Pasadena Rose Bowl.

The foreman was Jack Cowan and the district ranger was Bruce Coulter, a rough and tough old Scotsman, who ruled with a positive iron hand over the Arroyo Seco Ranger District before it was divided in later years.

Those were the days when the Forest Service manual was a loose-leaf binder that would fit into our saddle bags. Those were the days when the district ranger knew every inch of his district and the people and use patterns on the territory.

The powerful high tension line from Big Creek on the Sierra ran across the front country of the Angeles above Altadena and the south face of Brown Mountain, where Jack had his men building the new truck trail by hand labor.

On a very hot summer day, Jack reported to Bruce Coulter that there was a fire high up on the face of Brown Mountain where the high tension line was. Bruce (who knew every inch of his district) said, "No, that's impossible. No one ever goes up there. There hasn't been any lightning and there are no trails in that heavy brush covered mountainside." So, no one went up that hillside to confirm Jack's report.

Finally, after about an hour, the fire had grown so big that the Los Angeles County lookout and forestry department reported it. So Bruce decided to take the old Moreland fire truck and drive up Jack's road toward the fire which was spreading rapidly.

About three hours later, Ranger Coulter came back to our office, and was he a mess! He was black from head to toe and the front of his blue jeans was burned out. He was utterly exhausted. I said, "Bruce, what happened?" He said, "That damn fire's spreading all over the mountain, get busy and call out the CCC crews." I said, "But Bruce, what happened to your pants?" He said, "Oh, the damn hose caught on fire, so when I was dragging it out, it caught my pants on fire—and look at my shirt and hands!"

Why did the fire occur high on that brushy mountain? Bruce was right! There was no one there but the heat of the day had so expanded the spans of the high tension line that they sagged far down to the ground where there was a very fast growing "tobacco" tree of a very succulent nature. The electrical corona from the powerline shot down into the brush and had started that fire which burned for several days on the western face of Brown Mountain.

RED BOXES

When Bruce Coulter was ranger on the Arroyo Seco Ranger District, stretching from the Ridge Route to the Mount Wilson area, we didn't have all the roads and stations which were later developed. We covered most of our district on foot by means of trails, some of which were too steep for use by horses. This brought about the "Red Box" presuppression system, so a crew needn't be burdened with tools, canteens of water and lanterns while hiking to a going fire. Red boxes were placed in strategic locations. Those tool boxes were of various sizes and had tearable boxcar seals through which the standard Forest Service padlocks were inserted. On patrol we would change the water in the canteens and make sure the kerosene lanterns remained filled. Back in those days, people were trusted; mountain cabins were unlocked and supplied with food for hikers who might get stranded in bad weather. And, it was this trust and realization that made the red fire tool boxes safe for storage.

A SONG TO REMEMBER

The Arroyo Seco Ranger Station, when it was located at the junction of the Arroyo Seco and Dark Canyon north of the Rose Bowl, was involved in the great flood of 1938.

Before the ground got saturated and the huge downpours began (12 inches in 24 hours at Ophid's Camp), I was going back to the headquarters. My pregnant wife and I had been buying groceries and a new portable typewriter, which we had placed in the rumble seat. (What fun we old-timers had in rumble seats.) The big concrete dip in the road which crossed the rising Arroyo Seco stream had about 14 inches of water in it. We had to get home for supper so I eased the old Model A in. However, rocks and gravel struck us and the engine killed. Wow—and me with a pregnant wife in the middle of the rising stream!

Well, I used the old Model A to patrol in, so I had a lot of stuff in the back end. One of the articles was a "Yellow Ribbon" tow line. With it I tied it onto the front axle, waded through the rising waters and tied securely to an old oak tree. The next priority was to carry my wife across the roaring waters toward the ranger station. Then I went back for groceries. By the time I was ready for the new typewriter, the water force was too risky.

The next day I went back to look at the Ford. It was still there, tied with a "yellow ribbon to the old oak tree," but another car had been washed downstream. The raging waters stretched across the whole canyon! Arroyo Seco and the Dark Canyon overflowed their banks and the whole country seemed to shake with the roar of boulders and trees washing towards Devils Gate Dam. It was so scary that we felt our houses were going to wash away. For three days, Ranger Bill Durham and my wife and I occupied the office which was further from the flood waters. We survived and so did the Model A.

EARLY DAY RADIOS—SX SETS

In the late thirties, when Bill Durham was district ranger on the formerly extensive Arroyo Seco Ranger District above the Rose Bowl and spreading to the Ridge Route, the Forest Service came out with SX and SPF radios, both of which required the stringing up of wire antennae. Not only was this a disadvantage on a fast moving fire in steep mountains, but the locations were critical (SXs "line of sight") and the techniques of getting a message across, "dialing in," etc., weren't understood by many of our key people.

Consequently, many of the old-timers didn't want anything to do with "those new gadgets" and they continued to send messages up and down the firelines by hastily written and often inexplicit notes.

But, can you believe this, the real misunderstanding and frequently heard comment by crew leaders on the fireline: "Well, I'm not going to use one of those damn SX sets because they work on short wave and I'm too young to get sterile so that I can't have a family!"

ADVERSE POSSESSION

Fritz Moisio

Federal acquisition of forest lands under the Weeks Law did not come about in Missouri until the state passed an enabling act. This was accomplished in the forepart of the '30s. Soon after, purchase units were set up in the Ozarks and by 1934, the Forest Service had the acquisition program well on its way.

By 1938, sufficient forest land had been acquired to formulate the Mark Twain and Clark National Forests.

To me, the most interesting aspect of the acquisition job was the search for adverse possession. This involved a field re-examination of each tract, just prior to finalizing the purchase, to determine presence of any use that might obscure the title. In most instances adverse possessions were cleared directly by obtaining disclaimers from occupants or users other than the landowner. However, there were cases that required exceptions to be worked out between the title attorney and the owner to provide an acceptable title. Here are some of my most interesting cases.

On the Springdale tract near Bull Creek, two miles southwest from Chadwick, I discovered an unoccupied three room dwelling. It was in keeping with the local hill folks' architecture except the entire basic structure was made from black walnut lumber. This included the siding, studs, joists, rafters, foundation posts, floors, doors, roof covering and porch. I felt I was walking away from a small fortune, but what the hell, it was more exciting to be a happy ranger.

Continuing the search I went south along the banks of Bull Creek and found an old house place of long ago on a gentle knoll. The only remaining evidence was the fireplace and chimney. The top two thirds of the chimney had toppled down, however, the lower part appeared solid. I paced several chains north along the knoll and there I discovered a cemetery. Some of the old-timers had talked about a Civil War skirmish having taken place in the area, but to my knowledge, no mention had been made of the cemetery. Several graves were marked with native stone and the others with handscribed white oak boards or crosses. Many of the markers were still readable. Both the Union and Confederacy were represented and in rank included troopers, corporals, sergeants, lieutenants and at least one captain. As I looked to the west the sunset was filtering between the leafy branches of the oak, walnut and sycamore trees, making the old stone chimney and fireplace stand out as a glorious monument. There was no doubt in my mind that these guys had valid title to their resting place. It was signed, sealed and delivered at Gettysburg when Abe Lincoln said, ".......we are met to dedicate a portion of it as the final resting place of those who here gave their lives that that nation might live........"

My next search took me southeast of Swan Lookout Tower to examine a 120 acre tract. The seller was an absentee resident and not familiar with the land. His deed contained an exception for an existing grave but the description was so vague that it would require more luck than skill to find. However, I found it by running a compass and pace grid course over the most likely area. Instead of one, there were two grave mounds, one adult size and the other about the size of a small child. The child's grave was unmarked. The adult's had a flat upright sandstone with a dove chiseled into the upper part and underneath the words, "Hanged by the Bald Knobers." I wondered if the hanging had taken place in the immediate area or elsewhere. Also, how come the child's grave was unmarked or do you suppose they buried the master's hound there? I checked with the Christian County Courthouse for information and made inquiries at Harris Brothers' General Store and elsewhere in Chadwick without success.

Somehow, I got the feeling that I was digging into the past of too many kinfolk and that everybody wanted to leave it all buried. But, I did learn the "Bald Knobers" was a committee of unauthorized vigilantes who rode the hill country during the Civil War and post war period.

The limestone soil formations of Taney and Christian counties contained open grass covered balds with sparse scattering of post oak trees. These balds were a favorite place for Ozarkians to meet during moonlight nights to have fox hunts. The participants rode to the top of a bald knob where they tied their horses and built a warming fire while one member took off on his horse, dragging a fox pelt over the surrounding countryside. When the rider returned, the hounds were turned loose to seek the trail of the sly fox and the hunters settled about the fire telling tall tales, solving issues, smoking, chewing and sampling from an encircling jug. All this time the hounds were baying and yelping in chase of the ersatz fox and finally the winner was foretold not by sight but by sound. These guys really were not hunters but acoustical experts on a social outing. Evidently, during the Civil War era, these bald knobs were favorite gathering places for the vigilantes who called themselves the "Bald Knobers."

My last adverse possession search was made near Garrison. I had just crossed a branch of Garrison Creek when I discovered a moonshine still within a small opening. My idea of a still had always included a copper setup with gleaming boiler, tubing and containers. This vision was shattered upon finding 14 rusty oil drums filled with corn mash, connected together with galvanized pipe. There were lye cans scattered about indicating the corn had been cracked with lye. Two barrels had dead mice floating on top and a third one had a dead squirrel. Now I understood why most of the hill folk preferred "store boughten" whiskey. My attention was diverted by the sound of approaching voices. It became imperative to make tracks, so I hid behind a mound just beyond the mash containers. Although covered with leaves, the ground felt lumpy. I gingerly explored down with my hands, hoping to avoid any copperheads. I discovered the lumps beneath were sugar sacks—the moonshiners' sugar cache! Gees, what a helluva place to be, and just then all these guys burst into the opening with a lot of commotion and hollering. To my relief, I recognized Luther Schumaker, Treasury Division Alcohol Tax Unit agent and "Snake," his one eyed deputy who supervised the moonshiners in destroying the still before taking them to Springfield. Needless to say this adverse possession did not require a disclaimer.

THE REMOUNT STATION

M. M. "Red" Nelson

In the mid-1930s, the Forest Service in much of the West, found that supplying fire camps on large back country fires was becoming a problem because the number of pack strings that might be rented for use on fires was becoming fewer and fewer. The answer was for the Forest Service to maintain some remount stations so as to be sure to have pack animals to go on large fires. Region 1 out of Missoula, Montana, established a remount station with brood mares and jacks and raised mules for their supply of remount pack animals. Later in the 1940s, their mule production was great enough so that they were even supplying mules for some ranger stations in Oregon. We used mules from such a shipment on the Umpqua NF. Region 6 decided to establish a remount station on the Siskiyou NF for use in the Northwest. In 1935, or spring of 1936, Ken Blair, the Redwood ranger, and Frank Folsom of the regional office made a horse buying trip to eastern Oregon for animals to establish the remount. I know it was in operation in 1936 because, as an assistant ranger, I was assigned as service chief on the large Sandy Fire (one of many being handled by the FS during the late fall conflagration situation). We determined that we needed to establish a fire camp a good distance back from any road. Otto Lindh was division boss in charge of that camp. My responsibility was to keep the camp supplied. I heard about the new remount station and ordered packers and pack animals to pack to Otto's camp. A couple of months later, I was made ranger with responsibility for the remount station.

It was then that I began to really learn about that remount. The animals bought for it were, in general, range horses and had not been broken to carry a pack. The remount packers had a big job just breaking horses. Some were impossible to break, so the packers and I were allowed to "horse trade," always trying to upgrade the pack strings. That was in the days before the Siskiyou rated an administrative officer on the staff. Later, Ernie Shank was assigned in such a job and discovered in the manual that "horse trading" was not allowed without getting out bids and other paper work that made it not worthwhile. By that time, however, we had a pretty good string of working horses.

When I took over the ranger district, I also assumed considerable responsibility for horses and mules! I had my district string of seven or eight mules, plus saddle horses and also five or six burros used with trail maintenance crews. I also had full responsibility for the regional remount station. I had about 65 head of horses maintained at my ranger station for use in packing on fires any place in the Northwest (also used some in No. California). In addition, each fall all of the string of mules and the saddle horses from the other five ranger districts were assembled at my ranger station. I was responsible for getting winter pasture for the hole herd, about 110 head. We usually got the pasture in the Medford area and drove the whole herd to the pasture and returned in the spring. It was a two day drive each way.

I recall one year when spring was late in coming and the rented winter pasture played out. We rented another pasture for a month but it was located nearly 40 miles away. One of my packers and a summer fire guard and I went over early one morning, caught us some riding animals and started the drive to the new pasture. The first 10 miles were at a full gallop. The route required crossing a ridge of mountains. As we neared the top of a logging road, the road divided into many branches and we had different horses taking different branches. It was a big roundup job before getting them headed down the other side on a narrow trail. It was well after dark and raining hard before we got them into the new pasture. Needless to say, part of me was too sore to comfortably sit down the next day. After returning the whole herd to my station in the spring, the other districts would come for their pack strings.

During my first year at the station, other improvements were made. The old standard ranger's barn was greatly extended on both ends, making a very large storage area for hay, and mangers divided by stall partitions were put in so that all the animals could be fed at the same time. We also built four pole corrals, one for each quarter of the barn. Also three large red pack animal hauling trucks were provided for use in sending pack strings to fire assignments. Special loading ramps were built to facilitate fast getaway.

There were many interesting experiences in operation of a remount. Pasturing the stock in winter was one already referred to. When at the station, the stock had to be fed twice a day. They ate a lot of hay in a season. I had to buy hay and some of that was done in the packer's "off season," meaning that I was the only one available for the unloading and stacking of the hay bales in the barn, a back breaking job! After each horse was broke to pack, they had to be shod before being used on the rocky trails. We hired a special horse shoer to help the packers with that job in the spring. Then it was up to the packers to replace shoes as needed. Horse shoeing went on for several weeks in the spring. Some horses did not like it and a special sling rig was used to hold up the horse and a special method used in tying up the flailing hind legs until the metal shoe could be heated in the forge, pounded out to fit the shape of the hoof and then nailed in place. Spring always had something interesting at the horse shoeing shed. If the animal had not yet been branded with the "US" that was also done at shoeing time while we had a fire going in the forge to heat the branding iron white hot. The bad odor of burning hair was common.

One year, we had a young jack burro and decided to try to raise a couple of mules using mares in our pack string. We built a special corral for the jack to keep him away from the jennie burros. We put mares in with him but he would have nothing to do with them. My packers claimed it was because he had previously been with jennies and liked them better. They could be right, for the next spring we did have a jennie come in foal with the cutest little burro colt you ever saw. In years after I had left the remount, it was reported that unexpectedly some of the mares in the remount did come in foal. That was a mystery until it was learned that one of the employees had been secretly experimenting with the use of artificial insemination—pretty new stuff in those days!

While I ran the remount station, the pack horses were used on many fires in all parts of the region. It was in days when dropping supplies from aircraft was just beginning and usually used only for those camps that had no trails near the camp. In 1938, we had one packer supply camp with 125 head of pack and saddle stock on the Illinois River. Use of pack animals was not without accident. I recall when one of our remount horses was killed when it fell over a cliff while packing to a fire on the Galice District. Later that year, when we had missing items in our annual inventory we usually said, "Lost when pack horse went over cliff." That was OK until someone in the office figured one horse couldn't possibly carry such a load.

THE ORIGINAL ERA

Ted Niehaus

The ladies may be surprised that we had an ERA back in the thirties. However, it was for men only. The initials stood for "Emergency Relief Administration." Camps were established for down and out men. The crews included everything from retarded people and derelicts to pretty good men.

In early June of 1935, we had a lightning storm on the Plumas. I was dispatched with ten ERA men to hunt down a fire somewhere in the wild headwaters of Nelson Creek. Only Mt. Hough Lookout was manned, so we had one bearing on the fire and a rough estimate of distance. We started up the Nelson Creek Trail from the bridge on the LaPorte Road, carrying tools, an SPF radio and one lunch per man. In those days it took two good men to pack a radio and batteries. A packer with three mules was already at the bridge waiting, with supplies, to follow us.

Things were routine for a couple of miles, then a new lightning storm blew in. We had reached a small creek where a miner was panning gold. The clouds were getting noisy and the miner's wife was standing in their cabin door begging her husband to come in. Suddenly a bolt of lightning struck a huge ponderosa pine about 50 feet behind the miner's back and about 100 feet from my crew. Large limbs popped out of the tree and flew ten or more feet before falling. The tree swayed and whipped like a limber fishing pole. The concussion knocked us half way to our knees but no one was hurt. The miner came out of the creek with the longest strides I ever saw. Someone asked him, "Did you see that?" He replied, "Nope, nope, but I felt it," and he continued on a beeline to the cabin.

About a mile further we reached a high point from which we could see the fire in the distance. We were not to see it again until we were there. The point was atop a near vertical cliff into Nelson Creek about 100 feet below. I stopped my crew for a rest. I was a little concerned about two or three older men who were not too steady on their feet. Suddenly, there was a splash in the creek and my heart sank. Excitedly I said, "What was that?" Now there was a young retarded fellow whom the other men called "Nuts and Bolts." Well, Nuts and Bolts said, "It was me. I threw my lunch in the creek. I ain't carrying no lunch up this mountain for nobody." "Good thinking," said I, "Let's go."

In about another mile or so, we crossed Nelson Creek at a point where two Polish miners had a cabin. Stopping to fill canteens, the crew made quite a lot of noise. It was now 11:00 p.m. One of the miners came out in his long johns to investigate. His comment was, "What amata you fellow. Time to zlip now." We explained that we were going to a fire but it did not change his opinion of our sanity.

In another mile or so I could smell a little smoke carried down by normal nightly down-drafts in the canyon. I stopped my crew and walked a little further to locate the strongest smell of smoke and decided the fire must be directly above. I nailed a note on a tree for the packer and called up my crew. We went about a half mile uphill to the fire. It was about a half acre in pine needles. The time was 1:00 a.m.

The first thing my crew did was to make a bed of pine needles for Nuts and Bolts. Then they gave him part of their lunches and put him to bed like a baby to get him out of the way. We surrounded the fire in a couple of hours. We set up our radio with the long antenna strung in trees. We could receive no signal but learned later that our transmissions were copied in Quincy.

During the early hours I went down to the trail two or three times to meet the packer but he did not show. At daylight I went down again and met a miner with a week's supplies on his pack. He said he had met the packer headed back out. There had been a rain squall from a lightning cloud and the packer had decided the rain must have put the fire out and we had gone home by another route. I told the miner I had ten hungry men depending on the packer for breakfast. He said he could not feed that many but he would like to have me join him for breakfast. He made delicious hot cakes as long as I could eat them.

Returning to the fire, I told my men that no breakfast was in prospect and we had better mopup the fire and start home. I did not think it wise to mention that I had had a sumptuous breakfast.

We cleaned up the fire by 11:00 a.m. and started back. The Quincy dispatcher, having received our radio messages, had started the packer back and we met him about half way out. The crew unpacked the mules in record time and we had a feast.

A truck was waiting at Nelson Creek Bridge and took us to the SO in Quincy. The driver stopped to let me out and was ready to take the crew to camp at Feather River Experiment Station. We noticed that Nuts and Bolts and one other man had gotten out and they said they wanted to talk to the dispatcher., There being no hurry we let them go. I thought they would complain.

The dispatcher told me I had made quite an impression on the crew. In the first place, we had not seen the fire for miles and "that little ranger just marked a tree and took us to the fire and did not miss it a damn inch." The other thing that impressed the men was that I had gone up and down that hill several times during the night on a supposedly empty stomach and was still going strong. It all proves that sometimes one can make points by keeping his fat mouth shut.

WHAT FIRE?

Harvey Mack

On the evening of July 10, 1935, forest Fireman Ed Curtis was watching the traffic on Waterman Canyon Gear Road to Lake Arrowhead, when he saw a car go into the forest, throwing sparks and smoke out of the hood vents. Several things flashed through his mind, the most important being to stop that car.

Knowing he would never catch up with the car if he took the large Moreland pumper, and also knowing he could not make a call to the dispatcher to report where he was going and still catch the car, he "piled" into his own "trusty" chariot and made chase.

He rapidly overtook the car and flagged down one Glen Welsh of San Bernardino; raised the hood of the car and with a few well directed handfuls of dirt, placed the fire under control and out in one motion, much to the surprise of the owner who, until that very moment, did not know that his car was on fire. Thereby hangs the tale of one more fire put our before it actually started.

(Two of these Moreland pumpers were built for the Forest Service in 1931. They were large, open cab, powerful, but slow vehicles. One was assigned to the San Bernardino Forest and the other to the Sierra Forest. They did not prove to be a practical design for forest fire use.)

A PISTOL PACKING RANGER

Phil Lord

Editor's Note: If we are to believe Phil Lord, the Lassen NF officers of his day were a special breed. As a range and wildlife staffman, Phil may have felt it to be another example of survival (on the Lassen) of the fittest. Or maybe it was just the water. –E.J.

District Ranger Rueben P. Box was a man of many parts. Not the least of his many skills pertaining to the outdoors was his proficiency with either rifle or pistol.

Now, not everyone knew it but when Rueben Box was in the field, which was most of the time because he detested office work, he packed a pistol. It was a genuine Luger and he carried it in a shoulder holster under his shirt. Shooting, as anyone who has checked on Forest Service history knows, was at one time a part of the ranger examination.

Rueben was on the Hat Creek District when the following episodes took place. Rueben and I, and a young California State Fish and Game employee were out in the Hat Creek lavas looking at winter deer range. We jumped a coyote. The fish and game man grabbed his rifle, swung off his horse and started shooting. He ran the gun dry and never cut a feather. By the time the coyote was quite a way off and really in high gear, Rueben turned his horse sideways, reached in his shirt, out came the Luger and pow! The coyote went end over end and lay still. The bullet had hit him in the right ham and evidently he was just starting to turn a little to the right because it had gone clear through. Rueben looked at the dead coyote and said, "I must have jerked the trigger. Usually I hit them right at the toot of the tail." He said it serious like and that fish and game kid said, "Gee, Mr. Box, I'd sure hate to have you shooting at me." Later Rueben told me, "At that distance I was lucky to hit him at all."

One spring Rueben was riding with several grazing permittees. A cottontail jumped up and ran and someone said, "Let's see you hit him, Rueben." So Rueben did. He was fast on the draw. It wasn't till the permittees started laughing that Rueben realized it wasn't cottontail season. He told me, "All I could do was go over to Fall River Mills and turn myself in to the judge. He fined me twenty-five dollars and he didn't suspend it either."

TWO HAP ARNOLD STORIES

Don Bauer

The advent of the Forest Service and the military using aircraft for fire detection is documented very beautifully in a book written by District Forester Coert DuBois. World War I had ended and the military was down. There was no budget to keep their things going, and the Forest Service was in about the same boat. It happened by chance that Coert DuBois wandered into the Waldorf Bar, which was a gentleman's bar in San Francisco in those days. He chronicled this story.

These two rather lonely officers, one from the Forest Service and one from the Army Air Corps were having a drink and ran into each other; they were reciting their relative problems. DuBois was trying to get funding and get a system for organized fire protection in California which was keyed to first of all detection. This meant lookouts and they were expensive; then tying those in with a network of telephone lines. These were tremendous challenges. I stand in awe at how they were able to do as much as they did in those days with as little as they had. Anyway, prompt detection of forest fires was a key point in the whole picture.

The Air Corps officer was Hap Arnold. They hit upon the idea, after about the third drink. DuBois said, "Hell, you guys got airplanes that you don't have gas or money to fly, what would be wrong with them flying around looking for fires if we gave you the gas. We could work together." This was great. It would give flying time to the Air Corps people who were all grounded because there was no money to fly the airplanes. So the great experiment in aerial detection was born. March Field was the airbase and rangers would be detailed from different districts to come up and fly the observation, usually for a week at a time. Hap was one of the pilots. He took the word back to the Air Corps command and they put together a cooperation agreement where the Forest Service would provide observers and share in the cost of fuel and aircraft operations instead of building lookouts.

The CCCs were organized with the Army as the service and support agency, and the Forest Service was essentially a technical agency that worked the enrollees during the day. They would be taken out of the camp in the morning and brought back in the evening, doing forestry work during the day. The Army commander for this area was a major in the Army Air Corps named Hap Arnold. The military had budget problems too, during peace time. One of Major Arnold's duties was to inspect the camps. He wanted to keep his flight time up. He had to fly four hours a month to keep his proficiency as a pilot up. He ingeniously solved this. He would fly his little Army airplane and deliver the mail to different camps by hanging out the side and dropping the sack of mail with a rock in it. Vista Grande was notable for the fog that will often sock it in. He was up looking for the camp, buzzing around. The guys on the ground could hear him. They knew he was right overhead. He hit the top of a tree and knocked a wheel off the airplane while he was dropping the mail. The wheel, for many years, was on the mantle of the fireplace in the CCC camp.

HIAWATHA TALES—THE MANISTIQUE RANGER STATION

Ed DeGraaf

The development of a new ranger district and ranger station may not sound very romantic but those of us who had the opportunity in Region 9 during 1934-1935, enjoyed an unusual experience. The new districts took place throughout the region. My particular experience took place in the Upper Michigan National Forest.

It was my good fortune to start the new Manistique District effective January 1, 1935, and in turn develop the new ranger station. The district opened up with an office in a small second floor housekeeping apartment on the main street in Manistique. There were two rooms of about 12 x 16 feet in size, with a toilet room down the hall (shared by others). One room was for the clerk and included a small quantity of forms and office supplies. The other room was the ranger's office and included such maps as were available with a place for manuals and handbooks which were nonexistent. What was on hand for all business needs was a purchase order book that the ranger shared with the clerk. With this start, the new district "flew by the seat of the pants," according to the best instincts of the ranger and his clerk. For the first couple of years, there were not enough manuals and handbooks to serve all the new districts in the region, so the RO would send out an occasional instructional letter on policy and control with high hopes that the young rangers would stay out of jail.

The ranger's office room had one feature that was of unusual interest as long as we occupied the rooms. The door from the ranger's office room into the hall was matched by another identical door that opened into a small closet. When someone would have business with the ranger and turn to leave the room, there was a 50 percent chance that they would try the wrong door and step into the closet. This shook them up and they would take leave by the right door in consternation.

Although timber sales had not yet started, the district did not lack for work. The region's organizational policy was for the ranger to supervise the CCC camps and work programs, as well as a booming acquisition program. One of the early office problems was to order parts for the many vehicles and heavy equipment we had on the district. We had to obtain parts catalogues, as well as substitution lists for each kind of vehicle, tractor, grader, etc. The camps had to have the catalogues, as well as the ranger's office. The supervisor's office could not provide them, so each district had to scrounge their own. This was necessary so the field could order the parts correctly and the ranger's clerk could check the identity, numbers and consolidate the orders into requisitions and hope for the best. For sometime, each district was equipped with a mixed fleet of vehicles—Dodges, Fords, Chevrolets, Internationals, etc. Finally, someone decided to put vehicles of one make on one district to simplify ordering parts and to facilitate maintenance and repair.

When the station development was completed, a detailed audit was made by the regional fiscal office to see if building limitations had been exceeded. After a detailed review of all invoices and charges, it was determined that the house costs had exceeded the limit by some $200. This was "verboten," so we had to reach a way out of this predicament. Finally, the auditor hit on a solution. It was to place all venetian blinds and FS locks on the property records. This totaled over $200, so a young ranger's neck was saved.

One of the features at the station was a tall flagpole that was a neighborhood attraction. It was a tall, selected jackpine that had been brought in from the woods and dressed on the site, where it seasoned for six months. The final raising was a real ceremony and from then on the flag was flown each day. I do recall one special occasion when some of us had been out socializing with some of the neighbors the night before and kept rather late hours. Due to the exuberant behavior the night before, the flag was accidentally raised in an upside-down position the next morning. Since this is the "International Distress Signal," there was an immediate response. The telephone started to ring and many of the neighbors, some of whom had been in on the party the night before, had to know what kind of emergency had arisen. Needless to say, the flag was quickly lowered and raised in the correct position.

The station site was basically beach sand on top of a limestone formation with a sparse growth of juniper and local brush. The landscape plans called for development and plantings to embellish the attractive location. One of the outstanding needs was for good topsoil. It so happened that the county needed to develop a new gravel pit on the east side of town, not far from the station site. This called for stripping an overburden of good loam topsoil. The county had no use for this topsoil, so it made an offer to the citizens of Manistique that anyone could have free topsoil if they brought their truck to the pit being developed where a power shovel would load the truck. I checked with the county and the city to see if the Forest Service was eligible. They readily agreed and said the topsoil was available on a first come, first serve basis. With the unofficial concurrence of the SO, the road graveling fleet of dump trucks was brought in from the Cooks CCC Camp and put on the topsoil haul. Since it was late winter and an inconvenient time for individuals to get topsoil, no local citizenry were interested. It resulted in the ranger station getting all the topsoil (several hundred yards). The county was glad to get rid of it and the city fathers were pleased to see the station development.

Walt Sauer was especially interested in fish and wildlife and had a talent for public relations. He was married to a good looking school teacher in Manistique and soon was called upon by various educational groups to give a talk on birds and wildlife. He made such a good impression on his first appearance, as he described the hardship the birds and animals had to survive the severe winter, that the audience took up a collection to buy feed for the creatures. This was an unexpected surprise for Walt, who checked in the next day with a favorable report on his program and hat full of money for bird feed. He wanted to know what to do with it.

For lack of better advice, I told him to buy some feed for the birds and put the rest in the bank for similar needs in the future. What I did not realize was that this became part of Walt's program and he kept putting the money contributed by the audience in a FS bank account. He later transferred to the

Federal Fish and Wildlife Service as manager of the new Seney Refuge. The money stayed in the savings account, and by the time I left Manistique, must have amounted to several dollars.

In addition to the regular staff, there also were a couple of special employees that cannot be overlooked, Axel Mortenson and Herb Poupor. They came to us from Cooks' community via the ERA (relief) program. They were two of the best woodsmen I've ever known and had a local reputation as rascals for successful poaching and game violation. However, they became two of our best employees for firefighting, construction and field work. They became expert on running lines and recovering corners as field surveyors. Their work time was limited to 16 days per month and when the fire hazard was high, they patrolled the south half of the district where a lot of bad fuel conditions occurred. They made an excellent pair of fire guards and could handle almost any emergency.

We equipped them with a pickup loaded with tools and pump cans. This pickup was painted a bright fire engine red that showed up some distance on the berry plains and in the slashings. A regional inspection picked up the red enamel paint job and I learned that all FS vehicles were supposed to be green. My red paint job on the fire patrol pickup that cost $35 got me into trouble. However, I retained the red job as it was excellent for fire control work.

As the ranger station was completed we were happy to move from the meager office of two rooms to our new office building. By this time, the station also included two large warehouses, an oil house and the new residence. Technically, the house was not complete, as a length of sewer pipe in the basement was not installed. Since I was single and had excellent living arrangements for $10 a month in town, I was in no hurry to have the house declared completed and start a payroll deduction for rent. The Supervisor, Paul Wholen, was very tolerant of the sewer pipe stall and frequently reminded me that I needed to complete the house and move in.

However, after a regional office inspection that caught up with the situation, Supervisor Wholen took me aside and gave me an ultimatum, either take over the house and move in, or get transferred so they could put a married man on the district who would use the house. So, I took a transfer to another district, where new experiences awaited.

TRAINING A FOREST RANGER

Everett Jensen

I reported for duty as an assistant ranger at the McCloud office of the Shasta Forest on a cold, blustery day in January 1935.

There was no one in but I heard an irregular thumping and banging in the nearby warehouse that sounded interesting. I waded through several feet of snow to the door and looked in. There was Ranger Dutch Sullaway, red of face, flailing away at an old automobile tire with an axe. "We ran out of firewood," he explained, "But these old tires burn OK." He handed me the axe and I found myself banging away at another tire until Dutch decided we had enough fuel for the rest of the day. For some days thereafter, my mornings started with producing enough tire sections to keep the stove going.

INSPECTIONS

Fritz Moisio

Six months after my assignment to the Willow Springs Ranger District on the Mark Twain NF, I experienced my first general integrating inspection. The inspector was H. Basil Wales, Assistant Regional Forester, R-9. Assistant Forest Supervisor, George Ferrari, represented Galen Pike, Forest Supervisor, who was away on other business.

The theme of Basil's inspection was, "How well does the ranger know his district and people?" We left the ranger station in the morning in an apple green Graham Page. I was the chauffeur while Basil and George occupied the rear seat, the

First FS woman lookout Hallie Daggett at Eddy Gulch Lookout, Salmon River RD, Klamath NF, California – 1913.

Mail and file room, supervisor's office, Cleveland Forest, California – 1911.

post of inquiry and observation. Many people resided within the ranger district boundary, as the acquisition program was still underway. There was also a large number of squatters waiting to relocate upon Forest Service acquisition of these lands from nonresident owners. Through the CCC program we had already developed a very good truck road system. I knew most of the people, but towards evening was having trouble placing every occupant and became shy of names. On approaching the district boundary, I tried to speed by the last house without success. H. B. asked, "Who lives here?" I replied, "Pete Schedoni," and could hear George having trouble with his sinuses. After proceeding about a quarter mile, H. B. asked, "Say, what the hell was Sam Chaney's name doing on Pete Schedoni's mailbox?" There was nothing I could do but let my pants hang down.

My next inspection episode occurred on the Ava Ranger District, Mark Twain National Forest. This time the inspector was from the chief's office. By the time he reached my district his six gun hung low in a tied down holster. You see, up north on the Superior Forest they had let him walk a greased seaplane slip and thus he slid like greased lightning on his pantseat into cold Lake Superior. Then, on the Nicolet Forest the personnel had prepared their fire cache in A-1 shape, mounted on a very recently painted red truck. The grapevine had worked so well that it was almost impossible to uncover a weakness in this outfit. Like all inspectors, he was suspicious and when he started to leave everyone gave a sign of relief, but he stopped, sniffed, turned and walked back to the truck. Upon reaching the track he released and lifted the hood and behold, the engine was gone! So now, it was my turn. He seemed like a real nice guy, but still I was gun-shy. I was somewhat relieved on seeing Al Hall, Director of Central States Experiment Station and Marvin Smith, Assistant Forest Supervisor of the Mark Twain and some other friendly people in the inspection party.

The inspector observed our fire danger meter on the office wall. He asked, "Mr. Ranger, do you follow this meter explicitly?" My answer, "No, I use my judgment and make adjustments accordingly." Before I could explain, he turned to Al and said, "We spend thousands of dollars developing a scientific instrument and some punk ranger has to louse it up with his judgment." Al took exception to this. All I had to do was hold their coats while they went at it. I regretted when the inspector left without the knowledge that this highly scientifically developed meter would at times indicate an extremely high fire danger with a foot or more of snow on the ground.

PATROLMAN'S BETTER HALF TAKES ACTION

Harvey Mack

Mrs. Emett Hart, wife of the patrolman stationed at the Hook Creek Campground on the Arrowhead District, discovered what appeared to be a fire about a quarter mile down the creek. Her husband Bert, was away on a fire. Upon her investigation, it turned out to be a campfire in a non-designated area. In addition, the man was busily engaged in fishing for trout by the light of a Coleman lantern which was hanging from the branches of a tree. It being nine o'clock p.m., Mrs. Hart laid down the law to him; put out his campfire; caused him to move back to the campground and furnished the district ranger with all of the information necessary, resulting in a fine of $25. Just what was expected of a 1935 Forest Service wife!

BILLY'S PEAK

Carl Wilson

On applying to the Shasta Forest in early 1936, I was informed that I could be a fire patrolman if I could provide my own vehicle. Without a car, I was relegated to a lookout position. It was the end of my second year at the University of Idaho, and I was eager to get any kind of a summer job with the USFS, particularly in California. Little did I realize what was in store for me.

I arrived at Mt. Shasta in mid-June and reported for duty to Assistant Supervisor George Sergeant. Earl Backman was the supervisor. I was told I had been assigned to Billy's Peak which sounded sort of exciting! I rode to Trinity Center with one of the forest engineers and was told to stop en route at a civilian conservation camp and get a physical exam. An Army doctor gave me the exam and said I was "fit as a fiddle" and qualified to work for the USFS. It wasn't until later that I realized he hadn't checked my eyesight.

I reported at the Trinity Center Ranger Station to Ranger Bert Cary and Dispatcher (later FCO) Harold Peterson. Several days later, after arranging for groceries at the Trinity Center store, we left for Billy's Peak, driving to the mouth of Coffee Creek where we loaded the mules with groceries, LPG bottles and my suitcase. Before the packer had a chance to throw the hitch on one mule, the animal decided he didn't like the load and bucked my suitcase off into a rock pile. One side of the case was crushed and my new alarm clock smashed. That should have been a warning to me! Finally, the mule was repacked and we headed up the six mile trail to the peak. Several sweaty hours later we arrived at the end of the horse trail and the beginning of the people trail up the steep rock ravine to the summit. Although I don't have acrophobia, my blood chilled upon arriving at the top of that rocky pinnacle known as "Billy's Peak." I wondered who Billy was? After we pulled the two mule loads up the hill on the winch, "Pete" Peterson and the packer left me in my new penthouse in the sky.

Life went on, as it usually does, and I survived the first month, including the twice weekly two mile round trips to the spring for water. I survived loneliness by talking with the patrolman at Ramshorn and "rubbering" on the Callahan Ranger Station line when George Hendricks called his stations. After a month my initial food supply ran out and the packer was tied up on fire business in the Sacramento River Canyon. So, I began using leftovers from previous years. My main staple was cornmeal mush, liberally laced with tiny green weevils. After the first week I began to have fantasies about juicy hamburgers, hot dogs, apples and anything FRESH! Finally, after two weeks of bare subsistence, supplies arrived like manna from heaven!

Meanwhile, I discovered several conflagration type fires, which turned out to be exhaust smoke from a dredge and dust from bands of sheep. I missed a fire under my nose on Coffee Creek. In addition, a small fire burned in punky logs near Granite Peak for several weeks before I could see the smoke. Harry Churchman, observer on Bonanza King, had all the "firsts" thanks to my myopia. On the bright side, I was probably one of the pioneers in the "let burn" program.

I did stay on the lookout until the end of October, which is sort of late for the Shasta. Despite the "hardships" suffered on Billy's Peak, I decided that I wanted my Forest Service career to be in the California Region. So, in 1937, I returned.

A CITY GIRL GOES TO THE WOODS OR I MARRIED A FOREST RANGER

Mildred G. Nelson

My life with the Forest Service started in 1936, when I married M. M. "Red" Nelson who only months before had received his "permanent appointment" as a junior forester. He was assigned as assistant ranger on the Naches District of the Snoqualamie National Forest. Some say in those days, when you married a Forest Service man you also married the Forest Service. There is some truth in that, but I have no complaints. Both have been good to me. I was no young kid when we were married, being a registered nurse and working at my profession. I wasn't, however, wise to the mountains or backwoods because I was raised in the Los Angeles area and was a "city girl." So coming north and locating out on a ranger district meant a lot of new experiences for me. As I think back there were experiences that would be impossible to have in this day and age. One example: All of my belongings were packed up for moving and came north with me on the train. As the train crossed some of the desert lands south of Yakima it suddenly stopped out where there was nothing in sight but sagebrush. It stopped to pick up some Indians that had flagged it. It made me feel like I was riding a stage coach waylaid by Indians. I was glad my mother was with me.

We were married in Seattle in mid-August on a weekend. In those days it was unheard of for a Forest Service employee to have any leave during the fire season. Red felt lucky at even getting the Saturday morning off. That means a pretty short honeymoon before being back to work at the ranger station on Monday morning. In fact, one might say our honeymoon was at the Naches Ranger Station.

The ranger station was located on the eastern side of the Cascade Mountains in pine country about 40 miles upriver from the apple growing community of Naches and about 60 miles from Yakima. It was on a large flat with the American River on one side and mountains on the other side with the Naches Highway running through it. The ranger's office and three "year long" residential homes were on the mountainside. On the riverside were the warehouses, workshops, corrals, barn and some small "summer" dwellings and a building for work crews labeled "The Bunk House!" At one end of the flat was a CCC camp with 200 boys and their foremen and Army camp personnel.

It is natural that a bachelor assistant ranger did not occupy one of the dwellings even though it was labeled "Assistant Ranger." He lived in a room above the office and ate at the CCC camp. It was also natural that if he married in mid-fire season he still didn't rate one of the yearlong dwellings until fall. What he did rate was the only unoccupied building which was the bunkhouse.

So the "Bunk House" became our honeymoon cottage. It was designed for a crew of eight. It had a good size kitchen with a very large wood range, a table to seat eight, running water (in summer), sink and cupboards. The main room was surrounded with built in lockers so each man could have one. When we moved in, that room was filled with cots and mattresses. The third room was the bath. It had an open shower, wash basins, a urinal and a toilet like I had never seen. It was designed for cold weather winter use when water would freeze. So, it had no usual water trap in the bowl and no normal water holder. There was a metal tank up above that filled with water only as you sat. The seat stayed up at an angle until sat upon, which then opened a spigot to fill the tank above. When you got up from the seat the strong spring caused the seat to fly up. That shut off the water to the tank and also opened a valve to flush the toilet. It worked fine, but first you had to learn to hold down the seat with one hand while you slid off. Red kept telling me how much better it was than just having a "biffy" out back. A year later another newlywed couple lived in this "Bunk House" and his mother-in-law came to visit. She swore the young forester had fixed the contraption just to irritate her.

The wood range was another thing that took some learning for a city girl. There was no electricity at the ranger station. I had to learn to light a Coleman lantern, clean chimneys, make best use of candles, and learn NOT to USE an electric toaster, oven, refrigerator, washer, vacuum, iron, stove, etc. The wood range took a lot of practice. I let the fire go out three times while I was baking my first cake. I also burned many a slice of bread making toast on top of the stove. Red did not complain, because he said, he was in high school before he learned there was a way to make toast without first burning it on top of the stove then scraping it over the sink. Later he did learn from the fire camp cook, that if you sprinkled salt on top of the range then toasted the bread, you could do a fine job. It worked too. Ranger Ray Hampton and wife Ramona gave us a gasoline iron as a wedding gift. It scared me, but Ramona had me over to her house to train me how to use it. She got it all pumped up after putting some gas in the container used to heat the combustion unit. Then she lit it. There were flames three feet high. Nonchalantly she opened the back door and threw it out in the backyard. My NEW IRON! But she saved burning down the house!

Country shopping was a new way for me. There was no refrigeration, not even an old ice box, so I had to learn to make out with canned goods or things that did not easily spoil. There was some help by community use of a root cellar built like a cave back into the hill. It kept things a bit cooler in summer and from hard freezing in winter. The nearest store was at Naches, 40 miles away, but most real shopping was done in Yakima, even further. Usually once a week one of the four or five families at the RS would furnish a car and drive to Yakima with some of us wives going or sending our order list with those that did go. It took some time for me to get over going to such stores as Penneys, and seeing old Indian women sitting on the floor in the isles with their back next to the warm glass showcases. I did not get into Yakima Indian village to see just how they lived but apparently even the kids know about modern conveniences. The story is told (true) of one small Indian boy needing to go to the bathroom while his family was in the Sears store. He nonchalantly lowered his pants and used the toilet on display.

As Thanksgiving approached, I sent my grocery order to town with some of the other wives and asked them to bring me a chicken. Imagine the surprise of my "city girl" face when I took it out of the bag. It was a whole chicken (as ordered). When they said "whole" they meant "whole!" All I can say is that it was dead and the only thing missing was some blood from having its throat cut. We worked for hours getting it ready to cook. It took a reference book, time and patience before it was defeathered, dressed and stuffed. I learned chicken anatomy first hand.

Social events at an outlaying station were not many—perhaps a dance at some distant grange hall every month or so. The district personnel had to make their own entertainment. Having a newly married couple on the flat made one such opportunity. It was called a "shivaree." That is where the people all wait until you are well settled in bed then all come calling on you en masse with much noise, pan beating, bell ringing and horn blowing. I had never heard of such a thing but Red was wise to it; so guessing when it might occur, we turned the gas light off early (but stayed up in the dark! Sure enough they came—a jolly lot! We were prepared with beer and refreshments. Our first party was a success!

We had not been established at the RS too long when one day the "crank-your-own-number" phone rang with my number—two longs and one short. It was Red calling to say, "The Indians are coming! Come over to the office quick!" I did not know what to expect but followed the orders. It was a large group of Yakima Indians, mostly on horseback but with a few horses and buggies. They had been up on the ranger district in the high country picking their annual supply of blueberries. A colorful parade going back to their reservation. I shall never forget one old old Indian woman on a swayback horse who looked as if she had spent her life on that horse and they grew old together.

We had been married six weeks when one Friday evening the ranger sent us into Seattle to deliver something to the supervisor's office on Saturday morning (a work day until noon at that time). We did the job then seriously considered staying to attend a university football game. But, Red said it was still fire season so we better get back. We returned and had not been back more than an hour when the ranger got a call telling of a most serious fire situation on the Siskiyou Forest and all of SW Oregon. They were sending CCC crews from most all camps in the region. Red was to go also, but not with the CCCs. He was to take our car, drive back to Seattle, pick up Ranger Paul Piper and head for Oregon. He soon left his bride to go firefighting. The next morning, I think it was Sept. 26, the battery powered radio carried reports that the Oregon coastal town of Bandon had burned to the ground the night before. It was a bad period of dry east winds that came after many loggers and ranchers had started their usual fall burning. There were many large fires; it looked like a long siege and it was.

I was stuck in a strange backwoods area without even a car (no matter about that because at that time I did not know how to drive). The other people at the ranger station were especially good to me and made me feel at home. Fire season ended on the Snoqualmie and one of the dwellings next to the office became vacant so I got lots of help moving from the "Bunk House" into a real house—even had some furniture. Red kept me informed as he went from one fire to another getting a world of experience in running crews, scouting fires, running fire camps, etc. I knew he would not ask to be relieved as long as he could be getting more experience that he thought valuable. So, when I had some friends visit me from California, I joined with them and they drove me down the Oregon coast to the town of Coquille. It was then the GHQ for all of the fire area and Red was there, by that time running the service and supply end of the firefighting. I remember the devastated look of what had been the town of

Bandon. I guess every home had a fireplace. All that was left was a sea of standing chimneys—all else was burned to the ground all the way to the edge of the sea. Lives had been saved by taking boats out into the ocean.

I also learned about gorse! It looks something like Scotch broom with its yellow blossoms. It grows bigger and in very low humidity becomes very volatile. It was growing rampant in and near Bandon and was the culprit that made it impossible to save the town once it caught fire in the high winds. I stayed at Coquille with Red the last week or more as he finished up shipping out men and supplies and writing purchase orders for the last of supplies, phone service, warehouse and office rent, etc. He was the very last of the imported Forest Service overhead to leave the area.

We had a leisurely drive back to Naches and arrived six weeks after he had left. We had been married six weeks, then he had gone firefighting for six weeks. In another six weeks we would be leaving to return to the Siskiyou. Red was to be the district ranger on the Page Creek District at Cave Junction (now called the Illinois Valley District). Before Red left Coquille, the Supervisor, Glen Mitchell, had come to him and offered him the ranger job.

DOUBLE DUTY ON THE FIRE TOWER

Edward C. DeGraaf

When the Manistique District took over the fire protection on the South half of the Hiawatha Forest it included a few widely separated fire towers that had been manned by the State of Michigan. One of these was the Steuben Tower located on a high hill in some rough terrain between the village of Steuben and the highway to Munising. Due to its strategic location the observer could detect any spot fires started by the old M.&L.S. or Haywire Railroad line, as well as any fire occurring in the forest land that stretched for miles in all directions. The lookout at

Steuben Tower was manned by Gene Scott who had handled the job for years for the state before the U.S. Forest Service took over.

Scotty was known far and wide for his sharp detection and ability to pick up small smokes. Over the years he had driven over every road and had learned the country thoroughly. He had been crippled by polio in his youth but it did not handicap his lookout ability or detract from his knowledge of the woods. I made it a point to visit Scotty frequently in order to get well acquainted and I found him a very likable individual.

It was not unusual in visiting Scotty to encounter his ex-wife who apparently made frequent visits to see Scotty, especially right after payday. They were not on good speaking terms because invariably she would be at the foot of the tower trying to talk to Scotty by shouting at the top of her voice while Scotty stayed up in the tower and answered back as he saw fit. She was afraid to climb the tower and he knew better than to come down. He once told me that during his marriage he heard more than enough from her.

While Steuben was Scotty's favorite he did duty on occasion on other towers such as Cooks and Tie Hill. In the early summer of 1936, Scotty was on Tie Hill when the Bay DeNoc slashings in the Sturgeon River swamps caught fire. The fire weather was the most severe I have ever observed in the lake States. In order to plan for control, it was necessary to determine where the head of the fire was running. I went to the Tie Hill Tower at midnight to check with Scotty as he could see the entire area. The weather conditions at that time of the night were severe, humidity was real low, it was 90 degree temperature with a 35 mile an hour wind from the south. From the tower he could see the fire crown all night through mixed hardwood stands that bordered the swamps. The next day the winds died down and the fire no longer crowned in the green hardwoods, and control was ultimately accomplished.

At the time we had taken over protection from the State there were a number of prominent Manistique citizens who made it a point to recommend that we be sure to include Scotty as lookout for Steuben Tower. It so happened we were glad to get someone of Scotty's ability and knowledge of the country. I later learned from these same well-wishers that in the past Scotty had operated a still each season concealed in a thicket north of the tower and was able to do this very successfully in combination with lookout duties. He had a number of preferred customers in Manistique who were concerned with his welfare now that prohibition had been repealed. As far as I know, his moonshining was discontinued when the protection of the area passed to the USFS. This coincided with the repeal of prohibition and the return of legal liquor.

NOBLE FIASCO AT CUYAMA

Walter Puhn

In April 1937, I was transferred from the Eldorado National Forest as assistant ranger to the Los Padres National Forest as district ranger of the Cuyama District. In looking over my new assignment I found stacks of empty, new, five gallon "honey cans" piled to the ceiling of the district warehouse. Also piled were sacks of fire retardant chemicals. These items, I learned were to be used in an experiment to determine the feasibility of controlling forest fires by bombing from aircraft.

In August 1937, a crew under the direction of David Godwin, Assistant Chief of Fire Control in the Washington office, came to Cuyama to conduct the experiment. The airplane, as I remember, was a Nordyn Norseman owned by the Forest Service. Operation was to be conducted from a dirt landing strip which bad been graded on a flat across the Maricopa-Ventura highway from the Cuyama Ranger Station.

I learned that my part in the operation was to start and manage test fires on a sage and piñon pine grass flat near the mouth of Quatal Canyon about three miles south of the landing strip. I was to do this with my spike camp suppression crew of civilian conservation corps blacks (before integration) stationed at Cuyama. Little did they know, nor did I, that years later I would have the authority to call out powerful fleets of aircraft to bomb hell out of fires in various parts of California. Listening "between the lines," I gathered the thought that the experiment was as much to prove, or at least indicate, that aerial bombing was not a feasible means of controlling fires as it was to determine whether there was any practicality in such bombing. However, the principal purpose was to have some answers for the many critics who were demanding the Forest Service attack fires from airplanes and the Army Air Force cooperate in doing the job.

At the appointed time I started the test fire, which was no problem in Cuyama Valley in August, and then tried to keep it somewhat contained with my good crew of firefighters until the Nordyn Norseman arrived with its load of filled honey cans. The plane would come in low (it seemed to be only 110 to 200 feet) and loose a few cans that would burst open on hitting the ground or the stubby piñons and splash water in or around the fire. The water would quickly dissipate from Cuyama's heat and low humidity and was without any noticeable effect on our going fire.

As soon as the plane passed to make a turn for the next drop the crew would rush in to slow up the briskly burning fire. We would just about succeed in getting it contained when we had to run for safety and not be part of the target on the following drop. It was apparent from the first drop that aerial and ground firefighting are not compatible at the same time on the same piece of ground. After a few repeated performances of trying to contain the spreading fire, escaping from the drops and rushing back for more containing, we ground forces were nearing exhaustion and had formed our own conclusions on the feasibility of fighting fire from aircraft.

During a refueling stop, we managed to get the fire well contained as well as get some needed rest. When we got the word the plane was taking off for another series of drops we looked for its arrival but only saw a cloud of dust arising between us and the landing strip. We soon learned the Nordyn Norseman had crashed and the experiment had ended. Neither the plane nor the occupants were totaled. Pilot Harold King had suffered some cuts and bruises. Again, "listening between the lines," I got the picture that after takeoff one of the crew had accidentally kicked a valve controlling the fuel tanks and in doing so gave the motor only enough fuel to get the plane in the air.

I cannot remember how much time was spent in actually experimenting with water drops. My guess is that from start to disaster all was done in the course of a day. I do know that stacks of honey cans and sacks of fire retardant remained in the warehouse. In my belief the experiment lasted long enough to prove the bombing of fires from airplanes was not yet practicable. The water drops at Cuyama had little effect on the test fire. The strikes inside the fire gave only a momentary cooling of the flames, while the strikes outside the fire were well evaporated by the time the flames reached them. I learned it is quite useless to bomb a forest fire with water unless you can hit directly on its burning edge. The Cuyama test was an easy setup where the plane could come in low and slow without the usual problems of rugged terrain and uncertain air currents. I don't know whether the "inner circle" came to conclusions other than mine.

Perhaps the Cuyama experiment had something to do with the increased use of aircraft on fires. Within a year or two it had become standard practice on the Los Padres, and I presume elsewhere, to service fire camps with cargo drops in hard to reach places. Los Padres, with its difficult problems of access, especially relied on water drops to service camps and crews on the line.

LEARNING THE HARD WAY

Everett Jensen

It was late May or early June after a winter of heavy snowfall. I was following the trail, homeward bound, and had reached the ridge top above the head of South Russian Creek. Here a long-ago glacier had created a cirque and I looked down the steep, snow covered slope to the basin below. Far down and across the basin to the north I could see the trail free of snow where it dipped into the timber. To reach that point by trail I would have to continue onto my left and follow in a loop as the trail spiraled down. It would be possible to head down hill across the loop and cutting across country on foot was one of my favorite sports. I walked down slope on the snow. The sun had a fair shot at the snow here and I sank above my ankles at each step. This was going to be easy.

I had reckoned without my host. The slope gradually steepened, the sun's angle was less favorable in terms of melt, and there was more and more crust. I remembered reading somewhere that Eskimos, who live close to these things, had several dozen words, more or less, for different kinds of snow and ice. My boots weren't making much impression now and I stopped to take inventory. Downhill progress had taken me toward the lip of a cliff which seemed much too close and by now I knew that I could expect more ice between me and the break. I had visions of myself hurtling out into space biding by the laws governing free falling bodies.

How had I managed to trap myself this way? I wasn't equipped or trained for this type of mountaineering and I didn't much care for it anyway. Off to my right, perhaps 100 yards distant, the cliff seemed to terminate in a snow slope that reached unbroken to the bottom. I had better try for that. I kicked as large a hole with my heels as possible and with both feet using the hole I turned carefully, face down. I reached out six inches with my left leg and kicked a toe hold then another twelve inches for a second hole. I shifted my feet, first left then right and proceeded crabwise across the slope. There was no interest in speed. This was the way it was going to be even if it took all afternoon.

Eventually I reached the goal and found the going easier. This section of the cirque received more afternoon sun and as I progressed down I was once more sinking over my ankles. At the bottom I felt relief, of course, but also a feeling of guilt at my carelessness. I looked back at the snow free face of the cliff with the mat of boulders at the bottom and realized that it had no feeling whatever.

THE RED CAP FIRE

Pete Hanson

It was the Red Cap Fire in 1938 on the Klamath National Forest. When I landed at Orleans, the ranger station walls were bulging: Bob Deering, Roy Headley, George Gowen, Jay Price and many locals. I was sent into the lower elevations of the fire to scout the situation. I've never had quite that experience, either before or since. Fire was everywhere but I could walk anywhere I wanted to go. The fire was not moving, just smolding; I could not keep a match lit nor could I get a breath of good air. I had to sit down frequently, there just wasn't any oxygen left in that atmosphere and no breeze. I have since wished that I could have had that so-called air analyzed.

Back at Orleans my next assignment was coordinator with my headquarters at Black Lake Camp, high up on Red Cap Prairie. I never had contact with any other camp with which to coordinate. It was quite a steep climb from where Supervisor Tom Jones left me at the trail head. I was accompanied by a military officer but left him behind in the climb. I never saw him again.

As I remember, Black Lake had about forty men assigned. When I arrived in camp I learned they had all gone down into Red Cap Creek at the head of the fire. Now, I'd estimate it would take a good four hours to climb up from the creek to camp. Also, the crew had not fired out the line as they descended. That afternoon a mass of fresh air forced its way into the canyon and all those smoldering fires at lower elevations literally exploded. As I remember, the blowup occurred early in the afternoon and for quite awhile I sat at the head of the fire trail. I guess I was praying. The fire roared and threw a smoke ball in the air that resembled an atomic blast as we visualize it today. There would be no stopping it until it consumed the entire basin and reached the snow barrens of the Salmon-Trinity summit.

There was no logical reason to expect the forty-man crew to make it back to camp and safety. There I sat and apparently my prayers were answered. I saw the bobbing heads of forty men with my old friend Charlie Crews prodding the laggards. His uncanny fire sense had told him to quit work in late forenoon and start his hike up the hill. If ever there was a man deserving recognition for lifesaving it was old Charlie, fire foreman from Hayfork. I reported this event but I'm afraid it landed in some waste basket.

We had a little T or S radio and I got an order to put the crew up to the head of the fire. By now they were sleeping and I wasn't about to send them.

BROWNIE: A HORSE OF ANOTHER COLOR

Russ Rogers

Editor's Note: Russ Rogers has submitted a nostalgic yarn that defies easy classification and that is troublesome for an editor who prefers the safety of well traveled ruts. Should it be "pass" or "file zero?" But wait. Alexander the Great had his Bucephalus, Robert E. Lee had his Traveler and the decision is that Russ Rogers should have his Brownie. If that be treason, make the most of it. –E.J.

I would like to say a few words about Brownie. Brownie resided for almost 30 years at Brush Creek Ranger Station on the Plumas NF. Toward the end he mostly just stood around and did nothing. Brownie was a horse, the district ranger's horse.

I first became acquainted with Brownie in 1938, when I came to Brush Creek in a contingent of 50 newly enlisted CCC boys from Los Angeles. After two weeks of working on the road crew under Frank Church, Walt Graham and Jay Peterson, I was picked to work for the ranger in his fire warehouse. I hung axes and sharpened tools. Deene Stowell was the district ranger.

Deene would often ask me to go down to the corral and give Brownie some oats. I would do this and Brownie seemed to appreciate it. The time came when Deene needed to go to a ranger meeting in Quincy for three days. He asked me to feed

Brownie while he was gone. Brownie liked oats so I gave him oats—lots of oats. (What did a kid from a beach town in Southern California know about feeding a horse?) In the afternoon of the third day, there were sounds like the firing of a shotgun coming from the vicinity of the corral. And then "whap," Brownie raced to the other side of the corral. This was repeated several times. It was then that I realized what had happened. Brownie had gas and I was responsible. Brownie did recover and I got smarter regarding the feeding of a horse.

I have known three people who, at different times, were district rangers at Brush Creek: Deene Stowell, Spike Slattery and Bill Turpin. None ever said that Brownie was a great horse.

I left in 1939, after a year at Brush Creek. On the last day, I gave Brownie some oats—but not too much. When I left, Brownie was standing in the corral. Years passed, nineteen years in fact. In 1958, I returned to Brush Creek as the TMA. On the first day, I wandered about the station trying to reconstruct the past. The CCC camp was gone except for the rec hall which was being used as a school. Pine and fir saplings were growing where the barracks had been. I walked down to the corral. There, in the middle of the corral, stood Brownie.

By then Brownie must have been at least 25 years old. He was seldom ridden. I would like to say that Brownie died peacefully of old age while eating oats, and that there was a solemn burial with a drum roll and a 21 gun salute. It didn't happen that way. About 1962, Brownie was sold to a guy in Chico who was in the business of buying old horses for reasons that I prefer only to surmise.

I know that Brownie was not a great horse. Secretariat he was not. There will never be a statue erected in honor of Brownie. But he was there every day for nearly 30 years. That is worth something. Happy grazing in horse heaven, Brownie, and eat all the oats that you like. Gas is no problem in horse heaven, or so I have been told.

WHEN THE FFF WAS UNLEASED ON THE KLAMATH

Al Crebbin

I remember back in 1938, when the Forest Service's fire suppression policy changed almost overnight from one of "financial deficiency" to one of "financial affluency." As I passed through Orleans on the way to the Slide Creek Fire, I could feel the tension of headquarter's personnel, including Supervisor Tom Jones. A fire team composed of Washington and regional top firemen was on its way to the Klamath to put a new suppression policy into action.

As I climbed out of Bluff Creek on my ten mile drive and 15 mile hike to the Slide Greek Fire in the Blue Greek drainage, I could see the gigantic mushroom of smoke from the Red Cap Fire 30 miles to the east. I was very happy to be out of the organized confusion sure to prevail in Orleans. The Slide Creek Fire was 2,500 acres in size, located between the Slide Creek and Nickowitz Creek forks of Blue Creek. It was manned by twenty-five "down river" Indians. Jimmy Peters was in charge of the men, all old friends of mine from my topography mapping days on the lower Klamath. It was a typical Blue Creek fire job. Slide, Nickowitz and Blue creeks were still flowing enough to stop the fire, given a little assistance and a lot of luck. The main control effort was to line and backfire the head of the fire from Slide to Nickowitz creeks; no small job for 25 men and absolutely no hope of reinforcements judging from the amount of smoke continuously billowing into the sky over the Red Cap Fire. We had radio communication with Fred Wilder on Orleans Mountain with an "S" set that only transmitted Morse code. Unfortunately, there was no one within 30 miles of the fire camp that could send or receive in Morse code. The first and only pack train to the fire had left a fair supply of grub and we hoped another pack train would arrive from somewhere before starvation set in.

Nothing ever stopped Tom Jones. When he could not get a radio message to me about the new policy, he sent the word out by messenger. I am sure the new policy was a great shock to Tom; straightening out old bent nails, using poles for construction and pinching government pennies until they squeaked, for over thirty years. His policy message read, "Al, do not spare a cent to control the fires. The Treasury doors are wide open." I understood the policy perfectly but mused there was nothing I could do about it way out in Blue Creek without communication. How wrong I was. In less than two hours, tools, food, blankets and a voice radio were dropped by air on our camp. One old Indian well expressed the awe we all felt when he looked up into the sky and said to himself, "It is just like food being dropped from heaven by the 'Holy Critter'."

The next day Ranger Milt Morris of the Eldorado showed up, leading pushing and dragging 50 pooped, ill shod and almost worthless Sacramento winos; lots of quantity but absolutely no quality.

I soon learned that the affluent have many more problems than poor folks.

THE MISSING BID

Phil Lord

Retirees who worked in the Forest Service during the late 1930s will remember how scarce funds were on ranger districts. They will know how rangers had to scratch for money to give their trained short-term fire people enough out-of-season work to keep them coming back. Ranger Wes Brokenshire of the Mineral District of the Lassen National Forest was in this same leaky boat.

One year Wes was allotted money to build two miles of fence. The drawback was the project could not be force account but had to be put out for bid. Wes knew that when he advertised in Red Bluff there were several fence building outfits that would undoubtedly bid. One of Wes's lookouts, Hubert Grissom, by name, was a good fence builder. Wes thought that if Hubert could get the bid he could make up his crew by hiring other short-term Mineral District people. But could he get the bid? There were two drawbacks: One was that Hubert would be bidding against professional fence builders and the other was that he was pretty well out of touch. At that time, Hubert was running a trap line in Mill Creek and didn't plan to come out until trapping season was over. Most any part of Mill Creek was a long way from anywhere. The end of trapping season would be just about right to start the fencing job which was to be in the front country, but Wes had to have the bid in the meantime.

At that time of year, it was a long, hard day's ride in from the Sacramento Valley. Wes knew what to do and he did it. He drove to Bill Facht's place, which was as far into the foothills as you could get by car, and stayed all night. Before daylight next morning he was headed for Hubert's camp, riding one of Facht's best horses. It was a long, wet, cold, muddy ride but Wes made it just before dark. Hubert was glad to see him, of course, and was very much interested in the fencing job. After supper Hubert did some figuring and then made out the bid form which Wes folded and laid on the table. Next morning when Wes was ready to leave, the bid form was missing. They could not find it. So, Wes finally said, "Well, I know what your bid was and I can make out another form at the station and sign your name to it." That's what Wes did and Hubert won the contract. But Wes was bugged. What became of the original bid? He told the story many times.

About ten years later some Indian people who lived in Mill Creek came in and asked Wes about the old cabin where Hubert had stayed that winter. The roof had fallen in but they said there still was some pretty good siding on it (it was double) and they would like to get it. They had asked Wes because they knew the cabin was on national forest land. He told them to go ahead but to pile up the rest of the cabin for burning. Ten days later, one of the Indians came into the station and said, "Wes, when we tore down that shack we found this government paper in a rat's nest between the walls." He handed over the missing bid form not too much the worse for wear.

THE ILLINOIS VALLEY AIRPORT

M. M. "Red" Nelson

In 1938, the Siskiyou Forest had a really bad "fire bust" which was not unusual in the 1930s. At the time I was district ranger on the Illinois Valley District (then called the Page Creek District).

Some of us began to think that proper use of aircraft might lessen our problems of firefighting in that isolated country. Yet we did not have an airstrip in all of the Illinois Valley. We thought we should have one. Even the thought of the Forest Service having, or getting funds, to build an airport was unthinkable. It was even hard to get funds to maintain the few roads and many trails. But, where there is a will there is a way. We went to work finding a way. I picked out an area in the Illinois Valley that seemed like it might make a suitable airstrip. It did not have too much in the way of large trees but it did have lots of rocks (mostly loose). Some of it was land owned by BLM and some of it by private owners (not too valuable because of the rock and no valuable timber). I went to a fellow named Elwood Hussey, who was owner of much of the needed land. He was also a kind of one man chamber of commerce and promoter for the new town of Cave Junction. He agreed that an airport would be very desirable so I got him to agree to donate to the Forest Service a part of his land needed. But, there were other private landowners involved who would not donate. Naturally, the FS had no funds to buy such land. I again consulted with Hussey and we worked out a land exchange between the other owners and Hussey for the needed land. Afterwards Hussey made further donations to the FS. I also applied for and got a permit for use of the BLM land.

Now it seemed we had the land available for an airstrip. How do we get it started? I had an unique situation on the district. The Oregon Caves National Monument (about 500 acres) was within my district. It was operated by the National Park Service. That organization had a CCC camp assigned for work on Crater Lake National Park. In the winter months they would move their 200 man camp to the Grayback Guard Station on my district. They had some small projects at the Oregon Caves and did some projects, such as sign making, for use at Crater Lake Park, but they were in need of many projects on Forest Service land to keep their crews employed. That is where I came in. I furnished the projects. Since the CCC camp was that of the Park Service, the work program did not have to be approved by the RO and not even the forest supervisor. It was purely my district projects, although I kept the SO informed. A young engineer named John Ulrich was head of the NPS CCC camp. He and I would get together and develop the work plan for projects. Since the projects were for the Forest Service, the Crater Lake Park superintendent, for whom he worked, had no particular interest in the work done for the FS, except the usual accomplishment reports. John Ulrich and I had excellent rapport. When I included building the airstrip, John agreed it would be a fine project. Together we surveyed the location, designed the airstrip and he started his CCC crew to clear the brush and few trees and the mammoth job of removing rock. As I recall, the start was in the winter of 1938-39.

I left the district at the end of 1939, so did not see it completed. I think the CCCs got it to a state that planes might use it, but perhaps an enterprising ranger, Harold Bowerman, who followed me, may have gotten some other type of help. I do know that in later years when the jumper base was established, R-5 furnished part of the funds and they were on call for fires in Northern California forests. After I became fire chief in R-5, it was decided to pave the airstrip (that after 1952), and I budgeted R-5 fire funds to help pay for it. It was then that I learned I had not done too careful a job when I originally got deeds. Title check prior to the investment showed cloudy title. It delayed the paving job.

BUGS HELP SMOKECHASERS

Donald Peters

It was late June or early July of 1939. The daytime temperatures had been very hot for several consecutive days and the buildup of thunderheads indicated an oncoming lightning storm.

Finally, all the contributing elements combined and all "hell" broke loose. Within a two or three day period, the Heppner District experienced approximately forty lightning caused fires. A small amount of rain had fallen over the district but it was just enough to moisten the flash fuels and to raise the humidity sufficiently to keep the fires from spreading rapidly. Most of the strikes hit ponderosa pine snags of which there were many in those days as this was during an era before timber sale contracts required the felling of snags in a sale area. It was also before the practice of salvage logging.

I was stationed at the time at Ellis Guard Station and was sent to several of the fires whose smoke could only be seen from Arbuckle Mtn. LO. After getting those that could readily be found, we had a problem with others which were only emitting intermittent puffs of smoke. In those days radio communication was not available to a smokechaser and one had mainly to rely on his compass and the lookout's azimuth, verical angle and landmarks.

Occasionally an intermittent puff of smoke would give forth and thus enable the smokechaser to find his fire. Upon felling the tree, its rotten trunk would burst into flame as it hit the ground, and unless flames or smoke grew larger according to the lookout's judgment, no further aid would be sent the initial attack fireman. He stayed with his fire until it was out! And woe to those who didn't!

Other tree fires, which were not receiving sufficient oxygen or where the rotted wood was too damp to permit the fire to burst into flame, presented problems. However, upon close scrutiny—especially shortly after daybreak or late evening at times when the temperature had cooled off—the smokechaser would observe swarms of gnats hovering around the area where the fire existed. Having this clue to go on, further observance of the tree would oftentimes yield a puff of smoke and the source of the fire could be determined.

ALASKAN REMINISCENCES

Art Glover

One afternoon late in 1939, I was working on the all purpose transportation plan in my Portland office in the Mercer Building when Frank Heintzelman, Regional Forester of R-10, came in and without any preamble stated that he needed an engineer to take charge of all engineering and Civilian Conservation Corps activities in the region and wondered if I would consider a transfer. I took the job.

Alaska has two forests, the Tongass in the Southeast and the Chugach in the Northwest areas. Both of them consist mostly of islands, all heavily timbered with hemlock and spruce, some red cedar in the South of the Tongass and some white Alaskan cedar in the North. The Chugach has practically all hemlock and spruce with large areas of scattered small birch, none of it merchantable. The total coast line of the forests, including all the islands is estimated at 2,500 miles. The timber was estimated at 80 billion B.M.—total for both forests.

The ground cover of practically all timbered lands consists of thick moss with a thin clay subsoil. Under this there is usually solid rock for an indeterminate depth. This formation proves the common theory that Alaska is a very young land, geologically speaking, and only a few centuries since the ice age covered the entire country.

The CCC work had been under way for some time with an enrollment of about 1,100 Eskimos and Aleuts in the Chugach area and men of the native Indian tribes in the Tongass area. Small camps were established from Point Barrow in the North to the South tip of Prince of Wales Island in the south. Beyond this to the south is British Columbia.

Projects consisted of fire control in the flats of the many rivers in the Northern part of the territory, including the immense flats of the Yukon River; construction of roads and trails; boat and pontoon plane mooring floats; restoration of old totem poles and constructing new ones, all done by the Thlingit Indians in the Southern areas. Then there were water and electric power projects for many small Indian villages in the

Southeast area; improvement of stream beds used by spawning salmon; creating recreation areas around principal cities; building small boats on lakes difficult of access except by plane. These were to be used by trail crews and fishermen. One excellent log cabin was built on Lake Hasselborg's shore, 30 minutes flight from Juneau, to be used by fishermen, recreationists and trail maintenance crews.

I was involved in the supervision of all these projects which required a lot of flying to get to them. The one motor planes were equipped with pontoons for landing on lakes and rivers.

On lengthy projects such as road location and construction, dam construction and pipe lines, travel was on one of the launches owned and operated by the FS. They were equipped with living quarters for from six to ten men, including the skipper, cook, deck hand and permanent employees. These boats could be used for indefinite periods provided they were well provisioned.

The regional forester was anxious for me to develop a transportation plan for the region as soon as possible. This entailed much travel in order to link up the trail system on the many islands by planning and building more mooring floats at the terminis of the road and trail system. The Washington office approved using FRD funds on these projects with CCC labor whenever possible. Mooring floats were built to allow for tide flows from zero feet to twenty-seven feet, so it was necessary to install ramps from the floats to the dock runways above. These 60 foot ramps were equipped with rollers on the float end and were permanently fastened at the upper level. At low tide the ramps would often be on angles of 45 degrees. With the ramps cleated it was unnecessary to install ladders on the piling supporting the upper deck.

The second World War started shortly after my arrival. When the Japanese were active in the Aleutians it appeared they might also invade Juneau since we had no coastal defenses. The regional forester gave instructions to build an escape trail from Juneau to the Canadian border, about 40 miles east. This was to allow women and children to escape in event of an attack. Supply caches were located at strategic points along the trail which traversed some very rough country.

Fortunately, the trail was never used for the purpose planned but provided access to the back country for fire and maintenance crews.

As the war progressed in other parts of the world a heavy demand arose for clear, strong, straight grained and resilient timber for certain parts of aircraft. Since R-10 had an abundance of old growth spruce at different locations, the Department of Agriculture instructed us to prepare a logging plan as soon as possible. Logs were to be shipped to the Puget Sound area, sold to the highest bidder and then manufactured into the sizes required by the aircraft industry. The accepted plan was to establish operations at 10 locations known to contain heavy stands of spruce timber and arrange with experienced Oregon loggers to man these camps on a contract basis. The FS was to supply all necessary equipment and pay the contract price for all logs delivered at a central base.

Since the largest tract of timber was located on Kosciusko Island, a few miles west of Ketchikan, it was decided to have the main base up there on Edna Bay. All other camps were to be built on large rafts and tied up at strategic locations of available timber. Floating camps could be moved on short notice from one timber area to another. Edna Bay camp was also planned for truck logging which involved constructing about 20 miles of two lane well graveled road, mostly through solid rock. Housing for all camps was prefabricated to FS specifications in Seattle on contract, then shipped by barge to individual camps where they were set up in a few hours.

The Edna Bay logger used the skyline system for yarding but in the small floating camps regular ground skidding was used to the spar trees. A 100 foot high A frame firmly fixed on the raft, which also carried the swing donkey, permitted dragging the logs into the water where they were made into log rafts of about 300,000 feet B.M. These rafts were to be towed to Edna Bay where logs from all camps were to be made into ocean going log rafts of about two million feet each and towed to Puget Sound for sale as planned. These rafts were cigar shaped, about 30 feet in diameter and 200 feet in length with half under water and half above. About 10,000 feet of five-eighth inch wire rope was required to bind log rafts securely enough to stand a month of towing through good and bad weather. Due to the unwieldiness of the rafts, top towing speed seldom exceeded two and one half miles per hour by 1,500 hp tugs. None of the wire rope could be salvaged for further use because of the corrosive action of the long exposure to salt water.

By the time 47 barge loads had been shipped north and 42 rafts of logs had been shipped south, the aircraft manufacturing companies started using metals instead of wood. Suddenly the Washington office instructed us to clean up all timber that was on the ground, to cancel all contracts and sell all equipment locally or in the Puget Sound area. We had no difficulty selling everything, as the war was still on and equipment could not be purchased on the open market without a high rating. We realized 80 percent of the original cost which was considered very satisfactory.

THE COMING OF THE SMOKEJUMPERS

Roy Mitchell

During the fall of 1939, an experimental parachute jumping project was conducted on the Chelan (now Okanogan) National Forest. The object of the experiment was to determine under what conditions man could be safely transported to or near a fire in inaccessible mountainous areas and also to determine what protective clothing and equipment was necessary to land safely in timber and on other hazardous terrain.

Parachute jumping was not new. Neither was the proposal to use parachutes to transport men to fires, as various personnel in Region 6 and other regions had discussed it for several years; however, the proposals for this mode of transportation "jelled" in the following manner:

During the summer of 1939, the Washington office assigned to Region 6, for fire experimental work, a conventional commercial five place, high wing, cabin type airplane piloted by Captain Harold King. Prior to being assigned to Region 6 the plane was used for a short time in Region 5.

When first used in Region 6, attempts were made from the Portland airport to drop water and chemicals on dummy fires from various containers, but the small volume that could be carried by this type of plane and the sighting methods were so unsatisfactory that there was real difficulty in hitting the target effectively and favorable results were not obtained in these tests.

Melvin L. Merritt, Assistant Regional Forester in charge of operations, discussed the problems with Captain King and Jack Campbell, who had charge of Region 6 fire control at that time. Suggestions were made that if the plane could drop a man, or men, to a fire quickly after it started and while it was small, perhaps the plane could thus serve its most useful purpose. At any rate, it was felt that dropping water or chemicals in the volume that planes available at that time could carry was ineffective and not worth the cost.

One day soon after, Jack Campbell and Captain King went into M. L. Merritt's office and said they would like to have permission to drop men on fires from planes, pointing out that there were obvious and serious risks to such a procedure. M. L. Merritt said he was willing to have them try it if the regional forester and Washington office would approve.

It so happened that Roy Headley, Fire Chief from the Washington office, was in Portland, so Merritt, Campbell and King went at once and put the proposition up to C. J. Buck, Regional Forester. He approved if Mr. Headley would agree. Mr. Headley gave full consideration to the matter and after lengthy discussion approved the experimental project and said the Washington office would finance the project to be undertaken if Region 6 would arrange to have Mr. David Godwin, his assistant chief of fire control, present and assist during the trials.

As a first step, Mr. Merritt asked Captain King to secure the necessary parachutes. Since no parachutes were available from the military, he accordingly, in cooperation with Shirley Buck, Region 6 purchasing officer, prepared specifications and bids for supplying an initial order of parachutes. These were sent to several manufacturers. A few days later Beach Gill from the relatively small Eagle Parachute Company, Lancaster, Pennsylvania, personally came into the office of fire control and said he not only wanted to bid on the chutes but also would like to assist in getting the operation started by furnishing a few experienced jumpers as well as chutes.

Mr. Gill's bid was accepted. The contract providing for parachutes, two jumpers, protective clothing and a rigger (parachute packer) was drawn up. One of the first things done was to prepare protective suits and equipment for the jumpers.

Exhibition jumps were quite common at carnivals, fairs, etc., but protective clothing for a jumper to land in hazardous cover and terrain, and equipment making it possible for a jumper to release himself from the parachute harness and descend to the ground if the chute should catch in the top of a tall tree, had not been developed. This initial phase of the project required more time and funds than was first anticipated, but all the things that were done to make this hazardous experiment as safe as possible were time and money well spent.

It was decided in the regional office to conduct this experimental project on the Okanogan National Forest.

The project got under way October 5, 1939, after much had been done in the way of preparing equipment. It was completed about the middle of November.

Four different types of parachutes were used on the project. One was the seven foot by seven foot burlap cargo chute used to determine wind drift prior to the jumper leaving the plane. Another was a condemned military chute weighted with about 150 pounds and dropped from the plane to determine problems which would face a live jumper if he landed in tall trees, a snag or rugged terrain, and to determine damage to the silk canopies when landing in trees, snags, etc. Many dummy drops were made with these chutes prior to any live jumps in timber types. The two types of live jump parachutes, furnished by the Eagle Parachute Company, were the 30 foot canopy backpack and an emergency attachable chest pack which, I believe, had a 27 foot canopy. The chest pack chute was to be used only in case the 30 foot backpack chute failed to open. Both chutes were so constructed and rigged that the jumper could turn right or left by pulling on the right or left guide lines and would always face in the direction of drift.

These chutes were very rugged and were not easily torn when a jumper landed in timber or got hooked on a snag. Although a chute would occasionally suffer a small tear in the silk canopy when caught on a snag, or sharp limb or a tree, the crew was surprised at the small amount of damage caused in this way. The chutes received more damage when they were being retrieved than when the jumper landed in timber.

The original protective suit for jumper use was made up in Portland and was a one piece suit made of heavy canvas and padded. It had a stiff leather collar sewed to the canvas and this collar stood up about 10 inches high. It was designed with the idea of giving the face protection when landing in timber or in case of being dragged on the ground by wind if landing was made in open country. One of the first few jumps made with this original suit resulted in the jumper's chin and one side of his face being skinned very badly. The ground crew at the jump spot thought the jumper was seriously hurt from the looks of all the blood on his face and suit. Fortunately, only numerous small blood veins were cut; however, this was the worst injury received by any of the jumpers during the project. The cause of this injury was analyzed and it was determined that the jumper's body was in such a position when the chute opened that some of the shroud lines caught under the stiff leather collar, pulling its edges across his chin and face.

The jumper's suit was redesigned at Winthrop and was made a two piece suit of lighter material, as the heavier canvas material in the original suit was not flexible enough. The redesigned suit was padded at the knees, shins, hips, abdomen, shoulders, elbows, etc., to protect the jumper. Webbing was sewed in the crotch of the trousers and also down to the cuff. The trousers of this two piece suit were fitted with suspenders and were worn over the jacket. A football helmet protected the skull. To this helmet a wire mask was hinged and fastened by a leather strap and buckle. This protected the skull and face and yet did not interfere with visibility.

The jumper's outfit also included a back and abdominal brace, a wide leather and elastic belt, to guard against possible back and abdominal injuries during parachute opening, boots with eight inch or twelve inch top, ankle braces, athletic supporter and gloves.

On one leg of the trousers a pocket extending from about the knee to ankle was provided in which a rope could be carried that would enable a jumper to let himself down to the ground if he should land in a tree.

In the first tests it was found that when a jumper landed in a tall tree it was difficult for the jumper to extract himself from the parachute harness due to the bulky suit. A detachable riser with snaps was developed which permitted the jumper to release himself from the parachute and descend via rope.

There were about 60 live jumps made during the 1939 experimental project.

Tests were also made in dropping a parachute jumper's firefighting outfit to him. This outfit was wrapped in burlap or canvas and included items such as pulaski tool, shovel, mess kit, first aid kit, canteen, rations, etc. This outfit was dropped by the seven foot by seven foot burlap chute.

After these tests were finished in the fall of 1939, all the crew felt elated that the project with so many hazards was completed without serious injury and that the tests proved that smokechasers could safely land in inaccessible mountainous areas. The success of this project prompted the Service to undertake "smokejumping" on an extensive scale.

THE FORTIES

THE TRINITY NF 1940 – 1946

A. G. Brenneis

The Trinity, up to 1940, was basically a custodial and protection forest. It had only a few very small timber sales. It was a long way from railhead at Redding and Arcata. The main highways were not geared for heavy log and lumber hauling, and there was a good supply of timber closer to the market place than that of the Trinity Forest. The Trinity Forest had the major network of roads and trails and also the telephone lines in the area.

In 1940, gold was a major industrial activity in the county. The two bucket line dredges, Mary Smith and Junction City, were hard at it and several smaller outfits and doodlebugs were still going strong. On the forest, there were three Civilian Conservation Corps camps in operation—Minersville, Harrison Gulch and Hawkins Bar.

The forest began its cruising of timber and development of a timber management plan for the Indian Valley working circle in the Hayfork District. The cruising was done by a party from the regional office under Oscar Evans and Bill Bacon. Andy Schmidt and Lee Thomas did the bulk of the work on the management plan which was approved by the regional office. This was the first approved management plan for the Trinity Forest.

Late in 1940 or very early 1941, things began to happen. Gil Snyder and the Western Telephone Company bought out the Weaverville Exchange and began to modernize the system. In time, the company expanded service area and took over several Forest Service lines to improve commercial service.

During the war and particularly after V-E Day, the forest received excellent cooperation from the military. They furnished troops for major fires, when available, cargo dropping aircraft and aircraft for observation. War was beginning to make its ripples in the USA and in California. In several ways, the Forest Service, in cooperation with the military, was getting set "just in case." The CCC camps were being cut down as some were sent to help build new military camps such as Camp Roberts.

A system of aircraft warning stations were put on paper all along the West Coast. These would be manned in event of war as an early detection system for enemy aircraft. The Trinity had 15 of these stations ranging from Kettenpom, Horse Ridge and Virgin Buttes on the west side to Weaver Bally on the east. Planning included people (two to a station), servicing, necessary buildings and communications. A practice alert was set for December 15, when all stations in the entire West Coast network were to give the system a "dry run" to see if it was prepared to function. December 7 beat the plan "to the punch." By shortly after noon on that day, word was received that the system was "Go." By that night, the teams were lined up, some were on their way and a few lookouts were manned. The system worked!

Every station needed large fuel supply tanks and each station needed an additional building to use for cooking and sleeping quarters for the "off shift" lookout. These were cut out and mass produced in the Weaverville carpenter shop under supervision of Mary Goodyear, packaged and taken to location by truck, or in many cases, by mules. Part of the packing was done by private contract, where available, as the Forest Service string couldn't quite handle the massive job.

The system, manned in part by man and wife teams and in others by two men, worked effectively for the duration of the war with very few complications.

Another surprising and almost unbelievable part of the war experience was the Japanese fire balloon program. This was designed by the Japanese to send huge laminated paper balloons with a fire bomb to the US on the jet stream. These balloons were to set fires on touch down on the North American continent.

Eagle Rock emergency lookout tower, Yreka RD. Klamath NF, California, built during teen years.

Forest protection agencies had been briefed, in great secrecy, about this program and we had all been alerted to the possibilities. The idea was to set many fires, create smoke blankets, draw workers to the firelines and generally create apprehension about what else might be in store.

One of these landed intact on the Trinity between Hayfork and Hyampom. It was reported by the South Fork Lookout and picked up by Ranger Roy Deals and his crew. Fortunately, the explosive device did not function or there could have been problems. We think it was one of the early "invaders" to be captured intact and turned over to the military.

NATURAL CAUSES

Jack I. Groom

Hunting season was never my favorite time of year and an experience I had as a ranger on the Touchet Ranger District left me puzzled by the behavior of some of the hunters—one I have never forgotten. During the 1940 season, the weather was very stormy as often happens in the high country of southeastern Washington. Late one night some hunters knocked on my door at Godman Ranger Station and told me one of their party had become lost and they were unable to find him. They explained that their party had left the road's end west of Oregon Butte, and that one hunter was unable to keep up with the rest of them so he asked them to go ahead and he would catch up with them at camp. All of them were walking with a pack horse carrying their supplies. After camp was set up and the one hunter had not arrived, they retraced their steps but when he was not found and it became dark, they came to me for help.

The next morning I brought my horse by trailer to the road's end where the hunters had started walking the day before. Near Oregon Butte, the trail forks and the hunters were camped several miles south of that point. So I took the north branch of the trail thinking that the lost man had perhaps missed the turnoff. I had gone about a mile when I found a man lying in the middle of the trail, obviously dead. Some of the members of the hunting party had agreed to meet me at Oregon Butte, but they never showed up. I went back to Oregon Butte, and telephoned the sheriff in Dayton to report the hunter's death and asked for advice on what to do next. The sheriff asked me if I would bring the body back to the road's end where he would meet me. I agreed to try but was definitely not sure how I would accomplish this task.

Back at the location where I had found the man, I tied my horse to a tree nearby. While I was trying to make up my mind how I was going to handle this situation, I took numerous pictures of the scene as I had found it. These were to be used in a report the sheriff had asked me to make. The dead man was a rather large man, certainly larger than I was. Still I thought, if I could manage to get him across the saddle, he could be secured there with my rope. I realized when I looked over at my horse Tommy, that I might have one heck of a time convincing him to do this particular pack horse job. Tommy was a very good saddle horse but he was set in his ways when it came to unusual situations. Obviously being anywhere near this dead man was not to his liking. Looking at me Tommy seemed to be sending the message, "Let's get out of here!" My exact thoughts at that moment.

Having reached the conclusion that I would have to go back to the road's end and try to get some help, to my total relief I heard someone coming up the trail from the north. It turned out to be the answer to my prayer. Here was the camp tender from a sheep camp with a saddle horse and a pack horse. This pack horse made no fuss about the load we put on his back and we were soon at the road's end where the sheriff and the coroner were waiting for us. Some of the members of the hunting party were there to answer numerous questions about what had happened. (The coroner later ruled the man's death to be from natural causes—whatever that means.)

After asking the sheriff to notify the dead man's family in Dayton of his death, the members of the hunting party went right back to their hunting, never allowing an incident such as the death of a hunting partner to interfere with a long planned hunting expedition!

THE FIRST ENVIRONMENTALIST (OR ONE OF THE FIRST)

Phil Lord

Editor's Note: Cowboys have sometimes been called "cowpokes," and I assumed this was because they sometimes poked cows. Phil Lord is teaching me that they also take humorous pokes at people, including themselves. In the following story, you may want to count the number of pokes Phil takes. –E.J.

The term, "environmentalist," has not existed in its present usage for too long a time. At the time of this incident, I doubt that it was in use, at least I hadn't heard it.

Andrew G. Brenneis was supervisor of the Lassen, Wesley J. Brokenshire was ranger on the Mineral District and I was the grazing and wildlife staffman. The three of us were to make a grazing and wildlife inspection on part of the extensive area of brushy front country on the west side of the Lassen. The RO was worrying about deer and livestock competition for winter forage in the area (yes, they worried about this and that, then, as they do now), and the forest was supposed to take some type of positive action. At that time, the Lassen had some yearlong cattle permits in the area, hence the need for an inspection. Grazing inspections then were a little different and I think a little more conducive to obtaining basic information than they are now. There was no skipping from one C&T plot to another in a staff car; there were not many roads and anyway, C&T plots had not yet been invented. We planted our posteriors on the hard side of a saddle (the soft side being next to the horse) and tried to cover as much of the country as we could. Granted, it took a little longer, but I think we came up with more and better information and, as the fellow said, "What's time to a horse." Someone may question what we did with this information, but I am ready for that question too.

It reminds me of a time on a Saturday night at North Fork on the Sierra. A dance was in full swing and, as usual, there were some drunken Indians there. There were also, as usual, some drunken palefaces. One of the Indians staggered up to Roy Blood and said, “Roy, we got some horses loose and some tied up.” Roy looked at him and said, “So what?” The Indian, somewhat set back said, “Well, it’s damn good information.” Anyway, on these inspections, we felt we were getting some “damn good information.” Sometimes, we could use it, sometimes not.

Coming back to my story, Andy, Wes and I had started early and had covered a lot of country by the time we headed westward down a long, brushy ridge. The going became fairly difficult and Andy said, “Wes, we must be getting close to the forest boundary.” (I don’t think that part of the boundary had ever been surveyed and posted; maybe it hasn’t yet.) So we stopped and Wes got out a map and looked and finally allowed that we might be just a little outside the forest. It was about noon and we had lunches of a sort, a bacon sandwich and a small can of fruit apiece. We decided to eat so we tied up and got out our lunches. Now, this can of fruit bit needs a little explanation. Roy Blood, who was at that time handling the Lassen funds, had allowed us a per diem for this trip, something like $1.50, with a percentage off for the use of a government sleeping bag, of course. Wes and I, when buying the grub for the trip, had decided to go all out so Roy wouldn’t suspect us of knocking down on this generous per diem. We would be extravagant, hence the cans of fruit.

We sat down on a kind of rim rock with a steep canyon just below it (what we called a “hot chute” when I worked on the Angeles) and ate our lunches. Wes and I finished with our little cans of fruit at about the same time and one or the other of us said, “Two bits mine goes further than yours,” and we dropped them off the rim. As the cans clattered down, Andy lit on us with all four feet and we right quick learned about how wrong it was to throw away cans or any other refuse in a pristine area. Now, anyone who has never been chewed out by Andy Brenneis just has no idea what a chewing out is. (Don’t get me wrong, I liked working for Andy; you always knew where you stood. He could be just as generous with praise if you deserved it as he was rough with the chewing outs.) He really worked Wes and me over about throwing away those little cans. I remember the last thing he said was, "You two don't need to think you can go hog wild the minute you get outside the forest boundary." Andy was an environmentalist before it was popular by that name anyway. As we were tightening up our cinches, getting ready to leave, Wes muttered to me, "Well, at least he didn't make us climb down after them."

SABOTAGE?

Ira W. Roberts

While I was fire dispatcher for the Cleveland National Forest, I had also been filling in as PA on the Descanso District during 1940 - 1941, after Bill Clark had decided to go back in the Navy to complete a 20 year hitch.

We had been preparing for a one week test of the Army's proposed aircraft detection system when we got the news that the Japanese were bombing Pearl Harbor. Very shortly afterwards, we were ordered to activate all aircraft warning stations immediately.

That was at approximately 3:00 p.m. on the afternoon of December 7th and we notified the personnel who were to man the stations to report there immediately and we would meet them there with all the necessary equipment to make the stations operational and habitable. But, we didn't reckon with the activity which had gone on behind our backs. All of the observers made it to their stations on time but none of their equipment made it.

After loading their supplies onto a truck and two pickups, we started out. But not one vehicle got more than a few hundred feet from the Descanso Ranger Station yard. We soon found the reason why. Every carburetor was full of water. After cleaning the carburetors, we tried again, but the same thing happened. We then checked the gas tanks and found that they all had water in them. After draining them and refilling them from the underground storage tanks, we figured we had it made. But not so. The main storage tank was also diluted with water. We then resorted to filling wash tubs with the diluted gasoline and skimming the top to get undiluted gasoline for the vehicles. That worked and we finally got the supplies off to the lookout stations but not until well after midnight.

The next day we had Standard Oil Company come out, dig up and test the storage tank. A large quantity of water was found in it but no leaks were found where water might have entered accidentally from rain or other underground sources. There is no doubt that all of the water found in the motor vehicle tanks originally came from the underground storage tank.

RIDING THE CREST OF THE FLOOD

Story from Dick Chatham

Editor's Note: Harvey Mack excerpted the following story from the March 21, 1941, California Ranger, a copy of which had been sent to him by Dick Chatham. The story is written in the third person, presumably by the editor of the California Ranger, Region Five's newsletter at the time. –E.J.

During the recent storms, R. A. Chatham, equipment inspector from the Redding shops, had a hair-raising experience while traveling on the Red Bluff-Eureka Highway a few weeks ago.

While rounding a turn in the highway at the entrance to Button Canyon he was much surprised to see a four foot wall of water descending upon him. He set the brakes on the vehicle and proceeded with alacrity to make an exit. Discovering that in his confusion he couldn't get the door open next to the bank, he had to make his exit from the opposite side and climbed up onto the hood. Here too, he found it hard sticking on, since the water was flowing over the top of the car. Then the car began to wash downstream. Finding it impossible to cling to the top, he decided he would have to swim. He plunged into the water and started downstream, ducking logs, trees and driftwood. Then he found himself being carried swiftly toward a submerged barbed wire fence, and he was quite busy for the next few minutes endeavoring to make a landing upon a post instead of being caught on the wire, where he may have met disaster. He was fortunate enough, however, to reach a post and it held long enough for him to collect himself and swim ashore. The panel sedan was washed approximately one quarter mile downstream and, strangely enough, it never turned over, remaining upright at all times.

In an accompanying letter, Dick told how a one-legged man took him in and dried, fed and lodged him until he could get out. They recovered his panel truck, tore it down and cleaned out the debris and silt and returned it to service.

RESEARCH ON THE MODOC

Joe Elliott, Jr.

Editor's Note: Joe tells of a time of troubles in Marlboro Country. There was agitation to have the Modoc NF transferred from Agriculture to Interior. Joe Elliott, Sr., RF Show's troubleshooter, was delegated to head off disaster. What was the major problem—public relations? range management? We get some insight into how the situation was analyzed: Joe Sr. asked for and got two public relations experts, one range expert and an engineer. Instructions were to overlook nothing that would ward off criticism. Joe Jr. writes with his usual attention to necessary detail. However, taking the situation as a whole, one troubling question remains: Were George and Paul wearing their cowboy hats? –E.J.

Just prior to Pearl Harbor, two unrelated events occurred that have a bearing on this report. While not too many people are aware of it, there was a sagebrush rebellion of sorts starting in 1941. This was a limited rebellion, as I recall, that involved the Modoc, the Fremont in R-6 and I believe a forest somewhere in the Rocky Mountains. The purpose of this rebellion was to have the forests involved transferred to BLM where the stockmen felt they could better control grazing policies through their advisory boards. With the advent of World War II, the rebellion gradually came to a halt.

The first event: In June of 1941, my father Joe Elliott, Sr., was transferred from the Sequoia to the Modoc as supervisor. His job was to try to pacify the stockmen and to improve relations with the local press, some of which had become very anti Forest Service. As an example, for a time any Forest Service vehicle on the streets of Alturas was referred to as a part of the pine tree parade. For this job, Regional Forester Bevier Show said my father could pick the crew he wanted to take with him. It turned out to be a modest one consisting of George James, who I believe was the fire officer, Hamilton Pyles, who was the forest engineer and a couple of fair to middling district rangers by the names of Floyd Iverson and Paul Stathem. While emphasis was given to range management, every effort was to be made to do the best possible job on overall forest administration. Nothing should be done that could embarrass the Forest Service in any way, and all actions should be models of efficiency.

The second event: This one had to do with the delivery of a Bosworth trencher to the forest for test and evaluation. Bosworth trenchers had been around for awhile, but this was the latest model. I think it turned out also to be the last model. It consisted basically of an internal combustion engine mounted over an axle and two wheels with a flywheel about ten inches in diameter, facing forward and at right angles to the route of travel. Equally spaced around the perimeter of the flywheel were eight to ten chains about eight inches long. The machine was guided by two plow like handles. To the handles were attached controls to put the chains in motion and a throttle to control their speed. The theory was that this machine would efficiently beat out a fire trail in connection with grass fires such as cheat grass. One day, probably in 1943, George and Paul had been over to Klamath Falls with the boss, where the latter developed a very bad headache. Doctors associated it with a minor stroke. Accordingly, they decided the best place to leave him was in the Japanese Relocation Center at Tule Lake, where they had excellent medical facilities. Having left him, they had every expectation that it would be some days before they would see him again.

About this same time the Modoc experienced one of its famous cheat grass fires, some of which have been known to cross parts of the forest in a day's time. George and Paul felt this would be an ideal time to carry out some research with the trencher. It is not known how much line they built, but they did manage to kick up quite a dust storm. Paul was operating the controls pushing as best he could on the handles. George had a harness-like device over his shoulders to which a rope was attached from the trencher. Thus, he could help pull the contraption as he also acted as line locator and fire boss (which I suspect he probably was, in fact). All of a sudden Paul realized that George was no longer pulling so he shut things down to see what was going on. As the dust settled, Paul saw a ghost like figure that he thought looked somewhat familiar and gradually it took on more and more the profile of the boss—in fact, it was the boss.

Apparently, while our heroes were in their manmade dust storm, the fire had now proceeded about two townships ahead of them and was gaining. The boss looked at his two star pupils, then at the trencher and once again back at his pupils. He shook his head a couple of times in disbelief and figured coming out of the hospital and all, he must be suffering hallucinations. Additional checks convinced him there were hallucinations involved but not his.

Now, since I was not there, I do not know exactly how the conversation went. Knowing my father's normal mode of operation, he likely told them, considering the problem the Modoc was in at the time, that they might like to consider the possibility of putting the plaything away and organizing to suppress the fire, using more conventional methods of the period. Of course, it is possible there could have been more zip in this meeting. Based on the ensuing fire action, it appeared that my father's suggestion had prompt and complete acceptance. Thus, the research talents of George and Paul were terminated, at least while they were on the Modoc. After the fire action had been fully completed, very personal invitations were extended to come in and close the door.

P.S. I never learned of the events that allegedly led up to those so-called "come in and close the door" sessions from my parents; likewise they never knew I had heard about them from others. One side had participating knowledge; the other hearsay only.

AIRCRAFT WARNING SYSTEM MEMORIES

Harvey Mack

At the beginning of WWII the Cleveland, like other Southern California forests, was heavily involved in the operation of the aircraft warning system for early detection of possible Japanese aerial attack. On the Trabuco District we had observers at Santiago, Bolero, Margarita and El Cariso lookouts and stations. Maintaining communications was a major problem as we normally did little maintenance during the winter months as all of the facilities were closed down until the next fire season.

After a major storm, all of our lines were out and it was imperative that communication be restored as soon as possible. Since our winter staffing was minimal, Ranger Rex Albright had me meet him at the ranger station immediately. We loaded our tools and supplies and headed up the Indian Truck Trail clearing debris, checking the lines and making repairs. As we arrived at the Main Divide, we could see we had real problems as the ice storm had raised havoc with the line and the poles were coated with ice. Rex was not much of a pole climber and neither was I. We came to a corner pole with cross arms broken and six twisted lines. Rex decided that this was my pole to climb. I had never scaled an ice laden pole before but, since he was the boss, I decided to give it a try.

I put on the spurs and started climbing and soon had "burned" the pole three or four times. I took off my spurs, threw them in the back of the pickup and sat down, telling him that if he wanted the pole climbed he could climb it himself. Rex's reply was for me to climb the pole or maybe I should look for a new job the next day! I climbed the pole. When I got to the top I violated a few safety regulations and fastened my safety belt over the damaged cross arm so I couldn't burn the pole again. I have no idea what Rex would have done if my spurs had kicked loose and I was left dangling in midair. After I had finished the repair and gotten back into the pickup, Rex said, "See, that wasn't so bad."

Somewhere along the Main Divide there must still be boxes of insulators, pins and tie wires that I hid along the line for such repairs. Now the entire line is gone and they may never be found.

That same winter following another severe storm, the AWS observer at Margarita Peak called in to report that she was experiencing severe abdominal pain and needed medical help as soon as possible. I picked up another employee, loaded my gutless 1937 Plymouth pickup with tools and timbers and took off on our rescue mission. We worked most of the night and into the morning shoveling off slides, building temporary bridges over washed out roadbeds and picking up the timbers after we had crossed one for use on the next washed out area. It was cold, wet and tiresome but we finally made it. When we climbed into the tower, not knowing what to expect, we were greeted by a smiling woman who said she was feeling fine, "It just must have been gas!" She also wondered if we had filled any part of her grocery list! Oh well, just a part of an average day's work.

Oh yes, when I got back to town, I checked my mailbox and—there were my orders to report for military service in less than a week. XX@@##%%¢¢&&.

AN EARLY DAY CLEARCUT

Buel B. Hunt

We old-timers can remember those early motion picture days when Brian Donlevy was the cowboy villain and Roy Rogers was the most popular cowboy singer. I met both of these stars when movies were being made near the old town of

Kernville. This town was the center of activities for us at the Isabella Ranger Station; shops, bars, Saturday night dances, street fights, etc.

Most of the timber on Ray Stevenson's "super-sized" ranger district lay south of what is now Lake Isabella. My job was focused mainly on that area. Among other discoveries, I found a whole ridge top of huge old growth ponderosa pine, clearcut and abandoned on the ground.

Brian Donlevy had a mine in the vicinity of the clearcut and he was showing me gold deposits with a fluorescent lamp when I asked him about the clearcut and abandoned timber.

"Well, Buel," he asked, "Did you see anything peculiar about that?"

I said, "Yes, the trees were felled many years ago but there is no sign of slash of limbs."

"That's exactly it," he said, "No limbs were left because a congressman had a nearby mine. He needed resinous limbs for his boiler. He didn't steal the timber. He just cleared up the fire hazard."

LET ME OUT!

Jack Groom

A simple request for assistance in getting a car back on the road turned out to be a little more than just routine one day when I was ranger at Paisley, Oregon, in 1942.

There was a young man living on a small ranch between Paisley and Summer Lake and adjacent to the Fremont National Forest boundary. Some said that he had been injured while serving in the Army, but most everyone thought of him as harmless, although very difficult to understand sometimes. Someone who knew him well said the man would tolerate no personal questions and in fact might become violent if he thought he was being pressured.

There was no way he could make a living on that small ranch, so he bought trees from the Forest Service and from these, cut wood and posts which he sold and delivered to people in Paisley and Summer Lake. I had never had any trouble with him and he always left his cutting area neat with the brush piled. He never objected to the trees I marked even though they were full of knots and often had much rot. Trees that would be valuable for lumber sometime in the future could not be sold to him.

One day he sent me a note which read something like this, "Ranger Groom, please come and help me. I have run my car off the road down below my cabin and I also need a couple more trees marked." So, Sheldon Moss, my district assistant and I gathered up all the cable and blocks we could find and drove up there. His Model A Ford was sure enough off the road in a real bad spot but we managed to get it back on the road. We couldn't take it to him because he had removed the key. So we drove to his ranch. His house was very small with three windows set fairly high, each covered with a screen, and only one door. After knocking on his door and getting no response, we walked out to the barn but he wasn't there either. We gave up and went back to our pickup and then heard someone shouting, "Help me, let me out." Back at the cabin, there he was talking to us from one of his windows. He said he couldn't get his door open, so we went to the door and after careful examination found that the door was nailed shut from the outside!

I got a hammer out of the pickup and pulled out all of the nails. The door swung open easily, indicating that there was nothing basically wrong with it. He offered no explanation, just wanted to go right out and get the trees marked for cutting. After marking the trees we told him his car was back on the road and we would take him down there. He didn't thank us, just said he would go down and get it later.

On our way back to Paisley, Sheldon told me that when he saw how that door was nailed shut, the hair on the back of his neck raised right up and he broke out in a sweat. My feelings were somewhat the same, not knowing what to expect. Later, in talking this incident over, we felt sure he could not have gone in through one of his windows to get into the cabin because they were too high off the ground. Sheldon thought perhaps he had a tunnel from the house over to the woodshed nearby. I wondered if he might have a secret panel in the cabin wall someplace. Neither of these theories really seemed logical because he could have used a tunnel or secret panel to get out while we were at the barn. To us it seemed improbable that anyone could have or would have nailed his cabin door shut while he remained inside.

Considering his peculiar temperament and the intimidating glare in his eyes, there was absolutely no way either Sheldon or I would have asked the rancher why his door was nailed shut. For that reason I still don't have the answer.

CLAIMS, COURTS AND COSTS

Don Bauer

The Clark's Ranch Fire was in about 1942-43. It was during World War II, and Swede Nelson had just gone from the supervisor's job to state forester in Sacramento. Bill Peterson, who was assistant supervisor on the Angeles, was brought over to be acting on the San Bernardino. He was fire boss on the Clark's Ranch Fire, which had a very interesting critique and report on it.

It was caused by a Navy plane that was flying out of San Diego. They would fly out to the desert to do bombing practice, then return to San Diego. This particular Navy bomber had the bombs affixed to the wings. One bomb only partly released and he tried to shake it off over the desert. It wouldn't dislodge, but he had to get back to San Diego before he ran out of fuel. As he was flying over these mountains, the bomb dropped off and started the fire. It was quite a fast moving and damaging fire that burned up and slopped over the ridge.

It had two very interesting things come out of it. The fire was started on land that was owned by Bear Valley Mutual Water Company, but what they alleged was federal negligence; flying airplanes with hazardous objects on them and starting forest fires. So they filed a claim against the United States for the damage they had sustained as a result of their watershed being destroyed by federal negligence. We had just come out with a watershed fire damage appraisal study that assigned dollar values to acres burned, and they used that same technical study as the basis of their claim against the United States. We thought this was sort of unfair because nobody had ever sued for watershed damage. Watershed damage had historically always been assessed at one dollar per acre. But the study showed that many watersheds, especially here in Southern California, were extremely valuable.

So the claim was awarded and it established a principle in law that assigned a basis for watershed values that had been sustained in the courts. It laid the groundwork for the cost benefit ratio calculations we later had to do to justify such things as fire control budgets, airplanes, fire labs and all of that. So the Clark's Ranch Fire, although it was devastating to that particular area, did lay the groundwork for great advances in the intensification of fire control efforts.

THE WORLD WAR II HOME FRONT

Edwin F. Smith

Editor's Note: We owe Lyle Smith, Ed Smith's son, for resurrecting this appealing Christmas newsletter from wherever it had been buried. Forest Supervisor Ed writes to his "boys" out there on how well the girls back home have taken over in former male territory. –E.J.

We thought you boys that are scattered to the four corners of the globe might be interested in knowing how us folks at home try and carry out our job with the tools that have been left us, as the best of human tools were taken from us by the necessity of war. We have had a lot of fun and a lot of grief.

Starting out, from a fire standpoint, between the use of the males that were too old or were rejected by the armed forces, and taken on as cooks, lookouts and what not, women and boys 16 and 17 years old, we came out at the end of the season in pretty good shape, having had about 76 fires which burned a total of 84 acres, and we feel proud that we could carry on and not destroy your home plant even though the mechanics of doing this had a lot of tears and humor connected with it. Believe me fellows, it seemed most strange to go out on a lookout point where one had always met men, either single or with families, and find nothing but women, but these gals did a fine job in more ways than one. I think you would appreciate the fact that even Milt Morris kept himself shaved and his hair combed and did more personal inspection this year than heretofore. You all know about how many short-term people we use every year and it might be of interest to you that there were 27 women used in protection jobs this summer outside of our regular office force —and—by the way—this office force, too, consisted most of gals, as our administrative assistants also went in the armed forces (even though one night wonder how they could use them) and so Margaret Ward took over that job and we had a nice harem consisting of our old line people as well as war widows. In this connection, both from the protection and office standpoint, I assure you that it helps out us old boys with limited number of males in our part of the country so that quite frequently a nice looking girl will convince herself that some of us old boys' hair is getting dark and this makes it nice, too, at home, for with this competition our wives treat us much better when they believe someone else might have designs upon us—do you follow me?

One thing further along this line is that some of our bold guys like Joe and Bob Flynn who always gave us the impression that they were so almighty important, I might add this little item—that both of your jobs were filled with high school girls so I am sure you will never be again conceited enough to believe that you are indispensable!

Believe me, our old timber went into lumber for war work this year. Every mill ran to its full capacity insofar as man-power could be assured, and the cut from the Forest that you so ably protected over the years was quite enormous, and is still continuing. It makes us all glad that we had this product so necessary during these vital times.

As to the cows and calves and sheep, we are holding up the total numbers; in fact, are a little above what we normally have, which is a fine thing. People tell us you do get the "red points" one way or another, but it is pretty tough to see a nice fat steer, knowing in our minds that we won't get any of it! That is part of our selfish nature and we do hope you enjoy it. This is where you have it on us.

The recreation business was, I assure you, not flourishing, and those of you that used to sneak away from your stations and spend pleasant moonlight hours with some forest nymph (who had migrated from the city) would sure be out of luck, as most of them are now known as "Rosie the Riveter," or you may find them as WAVES, WACS, SPARS, etc. Very few summer homes were used and camping was limited. You know we Californians are really getting into this game and putting out at last. The whole thought of the 16 and 17 year old boys used in fire crews was "well, next year I won't be here because I will be in the armed forces," so their interest was not too great but we find quite a difference in their thoughts when they used to hang around Mac's and drink ice cream sodas and think about the gals that were there or elsewhere. Young boys grow up quick these days.

Do you remember when we used to celebrate Labor Day, 4th of July, Christmas, Washington's Birthday, Lincoln's Birthday and all those holidays? Well, this is changed, too, as the calendar now covers Christmas only and you will find us going on just the same all day Saturdays, Thanksgiving, etc.

You remember Johnnie Davidson. Well, the other day he stepped out and was married to Rosie at Pacific so cupid is still with us.

Joe Baumann and Marie finished up their stone mansion this year after five years of effort so they are very happy.

Bob Dasmann left for the Navy and so we have a new ranger at Lake Valley.

At the present time all the boys (and Jessie) are taking a day or two annual leave and trying to kill themselves a pheasant and I assure you nowadays that even blackbirds and kildees are looked upon for helping out with red points. I am sure we all wish them success.

Probably I had better ring off this little historic memorandum, but we do want you all to know we think of you always. We have our Honor Roll hung in our office and we wish you every possible success and our greetings for the holiday no matter how meager it might be for some of you. GOOD LUCK.

RECRUITING FIREFIGHTERS

Bob Gray

Pickup firefighters were much more common during World War II than in later years of organized crews. We recruited men from anywhere. Sawmills and logging operations were the best source, but hunters and the man on the street would do. An agreement with sawmills worked well. If twenty men were needed, a call to the company would usually produce results. The request would be made to the mill superintendent who would first select the men whose absence would least affect the mill production. Or he might shut down one shift and send the whole crew. The Forest Service tried to cooperate with industry by taking only what was needed and by releasing men and crews as soon as possible. Many mill workers and wood workers enjoyed the diversity of being on the fireline, though not all shared this enjoyment.

One weekend the local mill wasn't running, so we were unable to get a crew from them. District Ranger Dutch Sullaway decided we'd better recruit from the street and from the McCloud Hotel where most of the bachelor workers lived. We had signed up a few men but the word must have gone faster than we had. When we knocked on one door in the hotel a voice answered, "I'm sick in bed." Dutch opened the door, and sure enough, the man was in bed. I was totally shocked when Dutch took hold of the covers and yanked them off. The sick man was fully clothed, including boots. It seems Dutch knew this fellow and told him, "You fooled me last time but you can't fool me twice." The "sick" logger grinned and reported to the Forest Service as requested.

During the hunting season, Elmer Hall the woods boss, would gladly provide men for firefighting but only if we guaranteed an equal number of hunters to match his men. Our system was to stop hunter vehicles along the road and tell the hunters to report to the fire at a given point at a certain time. Their identity was assured by checking licenses and I can't ever remember any hunter not showing up. We usually kept them only one shift, and if they had legitimate reasons, they were excused from duty. Medical excuses were checked through their doctors by telephone. Most of the hunters took their "induction" in good spirit. No, they weren't the best firefighters in the world, but that's the way it was.

Some Forest Service officers were empowered to recruit firefighters and enforce fire laws by being appointed as deputy fire wardens for the State of California. Most rangers, fire management officers and prevention technicians had these appointments without which recruiting and enforcement were illegal.

Seldom did anyone question this authority, but one time I was glad I had it. There was a fire reported near Covington Mill on the east fork of Stewart Fork of the Trinity River. Fire Control Officer Larry Thunborg at Trinity Center made initial attack and nearly had it hooked when winds caused it to get away. I was coming from Weaverville, and about one quarter mile before reaching the fire there were two TD-24 dozers working a strip mining gold operation. When I asked them to bring their dozers to the fire the men declined, saying, "We're making $60 per minute on this job." I was taken aback, but not for long. Knowing I had no authority to draft their equipment and that I did have authority to draft the men, that's what I did. Well, most men would rather fight fire with dozers than with shovels, so they brought their dozers and did an excellent job. I served as tractor boss, line locater and swamper for the two tractors, guiding them as they built line through tall pole sized timber, sometimes thinking the reason they worked so fast was to drop one of those trees on me. They built a mile of line in record time and were released back to their "$60 per minute" gold mining operation in about three hours.

Three or four days later when I went by to pay them for their time and services, the men declined to accept payment for themselves or their tractors. Maybe they *had* been making $60 per minute.

TENAJA STATION STORIES

Gilbert E. Belcher

In the spring of 1943, Richard Bell, who I knew in the Future Farmers class at Corona High School in Corona, California, asked me if I would like to work the summer in a remote forest station as a fire suppression crewman.

I filled out an application and asked for the Tenaja Station, as that was where Richard would be. Later I was notified that I had a job. I was 16 years old.

At Tenaja Station, on the Cleveland Forest, I met Blackie Williams, the forest guard. Blackie was recently married and had arrived earlier to open camp. Blackie's father-in-law had given him an old Dodge Brothers car. It had been a touring car, the back seat section had been removed and a wooden bed made it a pickup. Once a week or as needed we went to Murietta for food, mail and a paper. It was a dusty ride, many gates, a thirsty radiator and a lot of wilderness. I had to pinch myself to be sure I was part of this big adventure.

It must have been hard to find good workers for summer fire season jobs during the war. The firefighting agencies across the nation had enjoyed the labor of the CCCs, but lost them in 1942 during the war. Many of the regular work force joined or were drafted into the military. Many good paying jobs opened up. Few men were available to fill the firefighting jobs at low pay, no days off and in some remote location.

Blackie had two documents he seemed proud of and showed them to Richard and me. One was his dishonorable discharge from the U. S. Marine Corps, the other was his discharge from San Quentin State Prison. Not anyone's choice of experience needed to be a forest guard. He did do a good job, as I remember.

The Vale Cattle Company owned a large area east of the station. They employed Mexican cowboys who dressed the part of "The Wild West." They carried guns and rode the fences to look for rustlers. Meat was rationed, sold on the black market and in big demand. So those who raised animals had to keep a close watch as a lot of stock was hauled off or butchered on the spot.

We bunked in an eight man barracks. It was small and held together by very large hooks and eyes. It was a portable building, made by the CCCs. It was on the south side of the property, near the creek and among many granite boulders. The boulders had many mortar holes made by Indians grinding acorns less than a hundred years before. This site had also been a homestead but all that remained were a few square nails we picked up.

A wood rail type fence surrounded the station which we painted tobacco brown and hoed the firebreak around the fence; no small job. The routine was one day you worked on station projects and the next day you cooked, cleaned barracks, disposed of trash, heated water for evening showers and cleaned lanterns. There was no electricity. A camp stove had a pipe coil inside which went to an upright tank. Thermo siphon heated water in the tank for our daily shower. A screened in cabinet with a door was outside. A large pan was on top with burlap in the pan hanging down the outside of the cabinet. The pan was kept full of water by the cook. He siphoned water out of the pan and down the sides to evaporate and cool the inside of the cabinet. It was good for fruit and vegetables but not meat. We used canned milk and a lot of canned everything.

A small campground was between the barracks and the creek. Because of our remoteness and the war, no campers came in 1943. The toilet was a first class USFS building. A small can of lime was sprinkled in the hole after each use. Between the garage and the corral was a slight drainage and a pump house. Inside was a Fairbanks Morse one cylinder gas engine and a positive pump. Water was pumped to a redwood tank on a wooden tower. There was a small garage, storeroom and workbench. The fire truck was a 1935, one and one half ton Chevrolet with 200 gallons of water and a 50 gallon a minute Viking positive midship pump, truck number 542. It had about 2,500 miles on it—still a new truck. How proud we were of it.

A guard school was held for a week before the camps opened and one learned a lot about his role as a forest guard. When riding on the back of the fire truck we wore hard hats and eye goggles. Safety was stressed in our training. We learned how to rehandle wood chopping tools and to sharpen them, skills I was able to teach others and used for many years. A great deal of time was spent training with the truck, its hose and tools, and of course, polishing the critter.

We learned to climb telephone poles and repair the line. We cut some brush under the line but didn't make much headway with a three man crew. At night you could hear, but very faintly, KFI radio on the telephone. Everybody listened in.

A weather station was in front of the guard station and a flagpole. Twice a day the temperature, humidity and fuel moistures were taken, recorded and called into headquarters.

Sunday was a goof-off day and much time was spent laying in the big hammock near the garage. The hammock was made of heavy rubber conveyer belt, three feet wide and long enough for two of us to lay facing each other with our feet dangling down. It was supported by the many oaks on the station property.

I don't recall any fires. We started to one and got turned back at Murietta. There were no radios in the fire trucks, so at every phone you called in to see if the orders still held. I had saved $150 and paid my way through the next school year. It was my first government job with many years to follow.

Somehow I muddled through nine months of the eleventh grade. I was now 17 years old and had my first car, a 1932 Ford V-8 roadster. It was a hot rod. I recruited Donald Bradley to work at Tenaja with me as Richard Bell was off to the Marines.

My mother was working at a small grocery store and Don and I were able to buy three months worth of canned goods and flour products. These were bought at wholesale prices and saved us a lot of money.

The forest guard was A. J. Jones. He had been in the peace time army, getting out just before the war. Also, there was a Dee Parvin who was a patrolman. The camp had a government horse with USA branded on its rump and on each hoof. It was a very wise horse, an army mind of its own. When Don or I mounted it, it would head for the nearest tree and leave us on a low branch or the ground. Parvin was an old buckaroo and he became the master.

There were some unbranded wild cattle in the thick brush which couldn't be herded out, so Dee would rope one and drag it to camp. He had a stakeside truck and would haul the cow to Los Angeles and sell it.

Dee didn't stay the whole summer. He left for town one day with some cows he had drug out of the brush. Having passed through two or three gates, he had a mishap and several cows fell out of the truck. He couldn't catch and reload them. A rancher came upon him and recognized his cattle and started to ask questions. Dee Parvin, wife and truck just faded away. The Army horse got a transfer.

Fred Varney and Clair Miller came one day and showed us how the CCCs cut brush with brush hooks. They were very good. They could cut brush faster than A. J. Jones, Donald and I could throw it off the line.

A contest was held for the best camp. All camps worked hard to be #1. The winner got a green pennant to fly under the stars and stripes. Tenaja won it!

Two weeks before the Tenaja Station had opened, a Lockheed P-38 crashed just west of camp beside the road to El Coreso. Clair Miller, the FCA, was the first to arrive and the pilot was sitting on the wing trying to bandage his bloody head. One propeller had broken off and skipped off to the north. It hit the ground and cut a deep slice about every 50 yards. Don and I spent many hours after work looking for the runaway prop. It may still be there.

The camp routine was much the same as the year before. The road grading crew arrived and stayed a week. They graded roads all over the Cleveland Forest and had lots of stories to tell.

The camp was more liberal in 1944. I was allowed to walk to the Santa Margaretta Lookout, arriving there about 8 p.m., stay the night and be back at 7 a.m. I went two nights a week. The lookouts name was Yokum. We called him "Little Abner." Yokum drank, and it could be detected on the telephone. One morning he put the flag up upside down and the Marines at Camp Pendleton came charging up to see what the National distress was.

Clair Miller made lots of trips to Tenaja and the lookout. The lookout could see the dust coming his way and would clean up the cab and hide the booze and do all the cover ups drinkers do, so as to appear on the ball.

Yokum finally took a walk and we went up to clean and paint and haul off the trash. The lookout had an underground water cistern. It filled up during the winter. A hand wobble pump put the water up to the cab. The new lookouts were very religious and asked A. J. Jones if any alcohol had ever been in any of the containers, as they could not ever use them. A. J. said no! Good thing we had walked out the brush as far as one could throw and picked up the bottles.

Clair Miller came up one day to train us on the new SX radio. It was a small wooden box with two folding lids which hinged up like a picnic box. It bad a small head set and a button to push down to transmit. A wire antenna was held aloft and could transmit and receive several miles in line of sight. We couldn't have been more impressed if it was the Hope Diamond.

A Marine artillery shell landed outside of Camp Pendleton in the San Mateo Canyon. I don't remember how I got there, but my first job was to fill 400 canteens from the creek. A fire camp was set up but there were not any hand tools. We waited two days until tools arrived from a fire an the Palomar District. All tools were then sharpened. It was my first field fire camp experience. I was assigned 125 Marines and told to hike to the fire and find Fire Boss Johnny West. A lot of the Marines took off for LA and many of them threw away their tools. West gave me a drawing of the fire and where to start building the line. Here I was, with zero fire experience and doing the job of a crew boss.

We had our canteens, K rations, but no flashlights. Our fireline ended up in the creek, but it was dark. I told the Marines to walk downstream on their own, as without lights we had no organization. The bottom of the canyon was a jumble of boulders and very slow going. I found a Marine who had jumped down and broken both ankles. He was in much pain. I did a lot of hollering but could not attract any help. I got him on my back with my arms under his legs and him hanging onto my neck. We stumbled around for hours and soaked his feet in the creek. It was getting light when we got to camp. The Marine left in a Jeep. Someone at camp determined I was used up and sent me back to Tenaja. While overseas in 1945, I talked with Marines who had been on that fire.

1944 was a busy season for everyone. I met many other employees, made a lot of firelines and saw a lot of other forests. I had spread my wings!

A fire had started on the San Bernardino Forest at the base of a mountain and ran up to Crestline. I was assigned a crew of Army soldiers and two Pacific Marine pumps to work down a creek and wet down both sides. We came out at the Arrowhead Springs Hotel, which was then a Naval hospital. The next morning I was told to drive a pickup loaded with barrels of gas and two riders to the Kenworthy Guard Station on the Idyllwild District. I told him I was a crew man and he said again to drive the pickup. I finished the season as a driver. This was the fire that started just east of Hemet and went to the desert—70,000 acres I think. I was gone for three weeks and wore out a pair of boots. I knew then that my life's work would be firefighting.

I turned 18 on September 29th. I knew I'd be getting my greetings from the President and I was getting messages from high school to return to school. When I got back to Tenaja, A. J. Jones was gone. David Roberts was in the high school class ahead of me. At this time he was in the Marines at Camp Pendleton. He was denied liberty (a pass) and went over the hill. I was alone at Tenaja and he stayed with me for 10 days. David had a large motorcycle. We were like Tom Sawyer and Huck Finn. We rode every truck trail, slipped off to Fallbrook and Murietta for ice cream and newspapers, hiked the nearby mountains, swam in the creek, played cards and had a swell time. Dave knew he was in deep trouble and went back to Camp Pendleton. He was, of course, slammed in the brig. Marines were very hard on prisoners. He went through some unpleasant routine for days, which made you hate the service and vow to never again get in the brig.

David Roberts and Richard Bell were on the same ship and fought the Okinawa battle in the Fifth Marine Regiment, First Division.

I soon returned to school. I wanted to enlist, but my folks like most folks, followed the war news and knew Hitler's days were over, as our armies sped across Europe. Japan was losing island after island and most of its ships. One could walk around town and see the gold stars in the windows and parents would press to keep their sons out of the war.

I was drafted in November and told to report in January, 1945. I elected to be a Marine like Richard and David. I was in PLT.2 San Diego boot camp, January and February. In March I looked at Santa Margaretta from the Pendleton side. In June I was wounded while serving with "B" Co., 1st Bat., lst REG., lst Marine Division at Okinawa. I was discharged April 1947, in time to go to forest guard school at Julian, California.

A RESIDENT LOOKOUT

Don Bauer

A question came up on what some call Windy Knob or Arrowhead Springs Lookout. It was built right at the beginning of the CCC period, maybe a little before, to provide detection. It was on land owned by the Arrowhead Springs Hotel people and the reason for its architecture was they wanted a type of architecture that would not detract from the residential character of the area. It was built as a lookout. Actually, there was a position called lookout-fireman, where it would be a lookout and if conditions were right the dispatcher could dispatch the guy to a fire. I'm told that the first forest fire pumper, a giant Indiana fire truck, with a 500 gallon tank on it, operated by Chuck Fuller, was stationed there. It was used as a lookout-fireman station for a number of years.

Then, during World War II, Swede Nelson was selected to be the new state forester. Swede did not want to resign from the Forest Service but the regional forester twisted his arm and said there are things that foresters have to do for the good of forestry and sometimes it's necessary they resign. So Swede went on to become the state forester.

The assistant forest supervisor on the Angeles at that time was a young man named Bill Peterson. Pete was detailed over here to be the acting supervisor while Swede's state forestry appointment was going through the state legislature. He had no place to stay, so the lookout was made available for him to stay at Arrowhead Springs. He thought it was a pretty good place to live and he did live there with his wife and two children until his appointment was confirmed. Then Pete was transferred to the Plumas in 1948.

Norm Farrell came down as supervisor and, looking at the price of housing and all, allowed that the old lookout was still a good place to live and he moved in and named it Windy Knob. When some fire studies were made, he had made more initial attacks personally as supervisor from that location than two thirds of the suppression crews on the forest.

After Farrell moved out, Sim Jarvi moved into it for a short time. It wasn't big enough for his family and he moved out to Rialto in 1953. At that time I, as forest fire control officer, took a strong position that the structure had been built for fire purposes and it should be made available for a Cajon District fire person. That's how we recaptured the site for fire control, and I believe Gene Kemble was the first one to move in there after that. He lived there for several years. After Waterman Canyon Station was built, our need or justification for a fire facility at that location no longer existed so we took action to terminate the lease and return the land to Arrowhead Springs, which by then, was Campus Crusade for Christ.

A HOLE IN THE GROUND

Ralph Bangsberg

Editor's Note: Ralph has titled his story, "A Hole in the Ground," a commonplace heading. Considering the nature of his assignment he could easily have come up with a more eye catching title. We can only conclude that it was really tough to get into and out of the American River Canyon. This conclusion also explains why the sheriff and two rangers involved were willing to delegate the job. Or did they have other reasons? –E.J.

I was assistant ranger, employed on the Eldorado National Forest in 1943. I couldn't understand why I was putting in my day painting signs at the Georgetown Ranger Station when I'd requested the transfer over here in order to have my own ranger district; after all, I had ten years experience, seven in Wisconsin as acting ranger and also superintendent of a 200 man CCC camp. I was bored. Another hot day in July.

Then, the ranger stopped by the shop and asked if I would like a change to do something else the next day? Without asking what, I said, "Yes!" I soon learned it was a strange mission. A local man had escaped from a sanitarium. He was holed up in a shack on the American River below the community of Foresthill on the Tahoe National Forest. He was thought to be dangerous and probably armed. I was to meet another forestry man and the two of us would go in to bring him out.

I left very early to avoid the heat, arriving at Foresthill just as the sun was topping the pines. Jim and I left immediately. If ever there was a piece of country that could be called a "hole in the ground," this was it! The trail was just short of straight up and down. We slid and stumbled 2,000 feet into the canyon. Suddenly, we came across a shotgun shell that had been reloaded with Babbitt and jagged chunks of metal. Now we proceeded with more care and less noise so as not to cause alarm to our quarry. The last of the trail was dangerously steep and we almost crawled. Then, we came upon an opening in the brush. Almost straight down below us was a tarpaper shack. On the roof was a man attempting to repair some shingles and beside him was his shotgun.

Jim and I retreated a little way to discuss what to do. After much thought, we decided in order not to make too much noise, we would go in two different directions. I would work my way down river and the roar of water would muffle any noises I might make. Jim gave the signal later and we closed in quietly toward the shack. Wonder of wonders! He climbed down off the roof without his gun. He was a strange man, tall, at least six feet with a white beard and the largest hands I had ever seen. Not a word was spoken as he entered the little cabin with us behind him. He seemed quite docile and to sense why we had come, while proceeding to pack a bag with everything from a can of beans to some vicious looking knives. We quietly persuaded him to leave these things, assuring him we would return and get them later.

Well now, how to get out of this hole in the ground? We couldn't return the way we came, it was too steep. We must go down river three or four miles to a road where, hopefully, the sheriff was supposed to meet us.

We left the cabin just as we'd found it. I believe the man expected to be back before dark and he would have had had we not been constantly alert.

He had a way of slowly slipping away, so gently you hardly knew he was moving—like a shadow. He spoke good English and continually stopped to point out places on the canyon walls where there was gold. (Of course, there wasn't.)

Now we found ourselves in very rough canyon and no trails. We used the old water flumes of gold rush days but they kept crumbling underfoot and were more dangerous than the rocks to travel over. What a challenge for the 49ers building this flume!

We heard a noise and came upon a mining claim operated by two men near a small cabin. I realized they had guns in hand, waiting for us. They seemed to know this man and didn't trust him. After some small talk and a little rest, we moved down river on a fairly good trail, finally reaching the road.

The sheriff put our captive friend in his car and said, as he left in a whirl of dust, "I'll radio for someone to come get you." Tho' terribly tired I was no longer bored! I felt a certain sadness for this giant of a man who longed for the freedom to relive his memories of the days of gold.

A WORLD WAR II TIMBER SALE

C. Glen Jorgensen

Timber sales were given high priority during the war. Under the War Powers Act, the Forest Service was permitted to sell timber without competition, although intent to sell a non-competitive sale had to be published in advance. This story concerns the trials, tribulations and confessions of a Fremont NF sale to Weyerhaeuser in the Horseglades area (section 21, T. 34 S., R. 14 E, WM). It is not remembered under what authority the Horseglades Sale was made, but when Weyerhaeuser thought they might run short of logs during the winter, the forest took emergency action to mark the timber for cutting (tree selection).

Ed Cliff was forest supervisor and Jim Thompson was timber staff officer. All timber sale work was projected out of the SO in Lakeview, Oregon. On a Tuesday morning, early in January, 1943, Verus Dahlin, Jay Hughes and I were sent to the Bly District to mark the section of timber which lay adjacent to the company railroad into Camp 6. We did not go on Monday because we had to wait for Jay to get back from Diamond Lake siding where he had moved Forrest & Annette Jones in minus 30 degree weather!

We encountered a heavy snowstorm going over Quartz Mountain. As was customary for project crews, we stopped at the Bly Ranger Station to tell Ranger Ross Sheapard where we were going. Ross told us that because of the heavy snowfall, we could not drive to Weyerhaeuser's camp. In that event, Verus (our leader) said we would go to Beatty and catch the company speeder into camp. Ross said the railroad was also snowbound. With this news, we planned on going back to Lakeview and wait for better weather. Before we got around to departing, the State Patrol stopped in and said the highway over Quartz Mountain was now blocked. We were stuck in Bly. Nothing to do but relax and wait, so the three of us moved our gear into the warehouse (nowhere else in Bly to stay, then or now). "Swede" (forgot his real name) had a double barrel stove hot and he was doing tool maintenance work. In those days on the Fremont you couldn't leave foresters idle very long before they were playing poker. So it was with us. We found the green front store, got a few snacks, drafted Swede, and we were in business. Other than stoking the stove and occasionally replenishing supplies, we played poker all day and most of the night. By the next day, we were fit to be tied, but still could not travel in any direction. We agreed it was better to snowshoe into Camp 6 than sit around, so we drove as far beyond the Obenchain Ranch as we could buck the snow, put on our packs and took off.

We had some 11 miles to go, all uphill. Nothing for timber beasts! It was not long until we put on the snowshoes. In an even shorter time, we were into the routine grind of going single file and alternating the lead job of breaking trail. The snow was fluffy and we would sink in over a foot with the bear-paw snowshoes. The day wore on. We stopped often to rest. Our rest stops became longer. The night life was taking its toll. With still a few miles to go and darkness approaching, we were showing signs of being dog tired, but kept encouraging each other onward. We were getting somewhat irrational. (None of us had heard about hypothermia or ever experienced shear exhaustion.) However, camping in a snowstorm without food or proper gear was not appealing. We struggled under our packs as the grade flattened. Our endurance barely lasted as we dragged into Camp 6 after dark. I cannot remember anything about that first night in camp.

Section 21 was mostly surrounded by Weyerhaeuser land so the section lines were well blazed. The railroad paralleled the west section line. For some reason, we started marking strips parallel to the railroad. It was bench type land and the topography was flat. The timber was a beautiful park like mature ponderosa pine type with occasional patches of lodgepole. It was enjoyable getting around on snowshoes. We marked with marking axes, blazing the tree and stamping the blaze with "US." No chance for stump marks in the deep snow. It was customary for pine markers to torch a dead snag for lunch fires.

One day, Verus said it was about time to eat and whoever finds a snag to yell and we would build a lunch fire. Verus found a good snag and I went over to help him fire it. In a few minutes we had a nice fire going and were getting ready to sit down to eat when we realized that Jay had not arrived. We were puzzled. There was no logical answer to him not being at the snag. Then we heard a faint h-e-l-p. We investigated and found Jay down in a deep hole alongside of a broken off tree that stuck above the snow about four feet. With the intentions of building a lunch fire, Jay had tramped solidly around the stub with his snowshoes, then he took them off and tried to tramp some more alongside the stub when he slipped down and was unable to get out without help. We then realized the snow was really deep. While in the hole, we handed Jay his marking axe, which he shoved down handle first and determined that the snow was eight feet deep at that spot. This figure was later confirmed by the Highway Department, as they reported eight feet of snow at Sun Pass on old Highway 97 north of Chiloquin.

It was snowing hard on another day. I was on the outside marking strip some one half mile or more east of the railroad as the end of the day approached. We always marked up to the section line before stopping. Before getting to the line, I ran into a patch of lodgepole with scattered overmature pine that had to be marked. In the process of going from tree to tree, I must have gotten a little disoriented. Darkness overtook me without finding the line, which I would have followed back to the railroad. I quit for the day and headed west. Camp was near the SW corner of the section. Weyerhaeuser had a diesel light plant that ran continuously. I soon observed that the sound of the diesel engine was getting faint. I took a sounding and headed towards the sound. Again the engine sound became faint. I took a new sounding and followed my ear. After an hour or so of trying sound orientation, I came to the conclusion that my ear was not going to lead me out of the woods. It was light enough to see the snowshoe tracks of the previous day's marking. The tracks seemed to go every which way. I knew the tracks should have a north-south orientation, except the weaving from tree to tree. Finally, by establishing the general directional pattern of the tracks, I made it out to the railroad some distance from camp and avoided the utter embarrassment of lying out in the woods all night when I could hear the light plant.

We finished marking the sale by midafternoon on Saturday, packed our gear and walked back to our rig. The snow had been plowed with a bulldozer leaving hard chunks of snow and ice that made walking very difficult. Upon arrival at the car we found it buried. We had to use our snowshoes to shovel away the snow so we could get the back door open. Then with shovels, it took us a couple of hours to dig out and get the car turned around. We arrived back in Lakeview around 2:00 a.m. Sunday morning. Probably the most exhausting week in our entire career.

As most emergencies go, Weyerhaeuser did not cut the timber that winter. In the meantime, I was transferred to the Bly District as assistant ranger with timber work 99 percent of my duties. The following summer Oliver Erickson, ARF for timber management, Mr. Campbell, Weyerhaeuser boss in Klamath Falls, Ed Cliff and Jim Thompson made a tour through the district looking at timber activities. For some reason, they checked the winter marking job on Horseglades Sale and they did not like it. In due course, I received instructions to blue out (unmark) trees of specified Keene tree classes. On a hot August day, Bob Appleby and I went to the sale to change the marking. I had mistakenly told Bob that it would be an easy job in the open park like stand on flat ground. When we arrived at the sale area, it was unrecognizable. The understory was dog hair reproduction. We had to almost crawl through the reprod to find marked trees, and then you had to look up, way up, some 14 feet to see the mark. When we finally found a tree that met the instructions for bluing out, we realized that we were not properly equipped. One of us said we would have to get a ladder, the other said either that or climbing spurs. We had little enthusiasm for the task to do, considering the unpleasant conditions. On the way back to Bly, we stopped at Camp 6 and learned that Weyerhaeuser planned to start logging in the next few days. Bob and I discussed what we should do about the problem. We rationalized that we had plenty of other work to do, the war effort called for production, not frills, and once a tree is cut, there is no second guessing the decision. Accordingly, we gave the job low priority.

Evidence of my insubordination was soon missing, as Weyerhaeuser cut the sale out during the month of August. Forrest Jones administered the sale and did all the scaling with woods calipers. During that month, 20 million board feet were cut and scaled on the district. Sales activities involved two Weyerhaeuser sales, a sale to Ivery Pine Mills and a Ewauna Box Exchange cutting. All the sale administration and caliper scaling was done by Jones and myself.

WWI surplus Pierce Arrow touring car converted to FS fire truck for Sierra NF, California – early 1920s.

Louis Flock at Brock Cabin Ranger Station, Oakridge RD, Willamette NF, Oregon – 1920s. Left to right: Bess, Ruby and Bonnie.

The Horseglades Sale was closed with no further incidents. Bob Appleby went into private consulting work, Forrest Jones went to other timber work on the Fremont and I was transferred in November to Heppner on the Umatilla as assistant ranger. Everything was done except one thing. I had not told Ed Cliff or Jim Thompson of my insubordination in not carrying out their (illogical) orders. Not being one to mar my integrity, and wanting to clear my conscience, I finally confessed to Ed Cliff. It was done under prudent conditions. Ed was about to retire, I had 30 year's service and the discussion was held during happy hour in a Portland hotel room. The story closed in good humor.

CARGO PLANES: FIRE DROPS & OTHER THINGS

Rex Denney

During the war years, when we on the Klamath Forest became experts in dropping cargo, an incident occurred involving George James, the forest supervisor.

The KNF had about 70 to 100 lightning fires out of a bad storm. About that time, the Klamath had two C-47 cargo planes with three companies of parachute jumpers to combat possible magnesium fires as a result of the Japanese efforts to start fires in Northern California. We were using a C-47 to deliver food and tools to many small, isolated and inaccessible lightning fires. We were hauling two, two and one half tons of cargo per load which was dropped at many locations. The C-47 crew consisted of a pilot, copilot, two sergeants to handle and push out cargo and one FS officer who supervised the loading and directed the plane to the many drop locations on the entire Klamath Forest.

Naturally, there were no spectators, but on the trip in mind we had George James and two more visitors (FS people) who wanted to see how we delivered cargo to a lot of fires. Well, all went well till the last location (fire) on Black Mountain on the

Salmon River District. We had made two or three drops and had the cargo for the last drop in place by the open door. The drop consisted of four or five heavy boxes of food and two light packages on top. One was tobacco and the other was socks. All went well except the chute (burlap) went over the tail aileron and the package went under the tail aileron. The plane began to lose altitude and the pilot circled back to be over Wooley Creek instead of heading for the Montague Airport. He also announced that there were five chutes, one each for the pilot, copilot, each load sergeant and the fifth one for the forest officer assigned to the plane. George James' jaw dropped several inches on hearing the news. About then the pilot asked me via intercom to come up to the cockpit. When I got there he asked me what the elevation of the Shackleford Creek pass was. I told him 6,400 feet. He said, "Oh boy, we are only 6,200 feet now and losing a little altitude all the time. Where is there a field (airport) we can reach?" Orleans field was too small so it was decided to try to reach the coast and crash land on the beach.

About three or four minutes after this decision was made the shroud lines on the package and chute in the aileron twisted off and the plane shot up very rapidly, and yours truly "sat" down on the floor very hard. The pilot looked down at me and said, "Sorry. Are you OK?" We went on to Medford Army Air Base to get gas and nary a word was spoken till we landed and everyone looked to see what damage had been done to the plane. The inspection showed no damage and only a small amount of paint worn off.

A PARTICULAR FIRE

Ted Niehaus

Editor's Note: A cliché among FS people is that every fire is different. Ted Niehaus has sent in a fire story that for some reason makes us think of the book, "Animal Farm," where all animals are equal but some are "more equal" than others. Ted's fire story is, of course, different from others but also seems "more different." It happened while he was ranger at Challenge on the Plumas NF, about 1944. –E.J.

Sacramento Box Company had just opened a large sawmill at Woodleaf. There was no provision for housing employees and a motley crew found shelter in barns, tents or whatnot. A cluster of one room cabins at Challenge often housed whole families.

One night about 10 p.m. we had a call at the ranger's residence. My wife answered the door as I was busy chasing the cat to put it out for the night. Our visitor waited patiently as I took care of this chore. At the door I found a quiet, polite little man who asked for a little help. When I asked what kind of help, he hemmed and hawed a lot, looked back over his shoulder several times, then said help to get his furniture out of his house. I asked if it had to be done now. He guessed so. When I wanted to know why it couldn't wait until morning he said that his house was on fire.

I pushed the panic button; the crew rolled up to the office. It took the visitor some time to explain where he lived. Finally, it turned out the place was within our view. At the house we found a smoldering mattress, which smelled vaguely of kerosene. The mattress was removed and chopped up in fine fireman style and, since there now seemed no need to remove the furniture, we left.

Back at the station we could now see towering flames at the same location. We rolled again but were able to save only the cabins on either side. I noted a smell of gasoline and my little friend had singed hair and had first degree burns. He blamed his brother-in-law and wife. They had left in a huff and his strong box of personal valuables could not be found in the ashes. He had no doubt the fire had been set to cover the robbery.

I talked to the owner in the morning and advised him that he appeared to have a possible case for arson against the fugitives. He asked that I let him handle it. Well sir, he collected his insurance and shortly rented another cabin to the same couple who were now back and doing fine. That was justice in Challenge in those days. People expected to handle their own problems and did. Why bother the sheriff?

THE FIELD TRIP

Roy Booth

I believe it was Edna Ferber who wrote "Life begins at forty." Without attempting to outdo that illustrious woman, I am about half convinced that for me life begins at sixty, for on the eve of my 59th year, when I was already making definite plans for retiring from my official career with the United States Forest Service which began on May 1, 1907, I received a confidential memorandum from my boss, Bevier Show, that General H. H. Arnold, Chief of the United States Army Air Corps, an old and intimate friend of many of us in the Forest Service, was sorely in need of a little mental relaxation from the arduous duties of his all important job.

That the General's mind in searching for that bit of respite should turn to the quiet grandeur of the Sierras and the finny inhabitants of her lakes and streams, seemed perfectly natural to me, for it had been my privilege and great pleasure to have accompanied him on several such prior excursions, and thus I had known for more than ten years how keenly he enjoyed those kinds of outings.

After a certain amount of preliminary correspondence necessary to the determination of date, route and itinerary of the proposed trip, I was given the responsibility of working out the details of planning the campsites, arranging for the stock and equipment, purchasing food, supplies, etc.

Among the must requirements set up, I was informed that the names of the persons making the trip must be regarded as a military secret and that communication by radio must be assured. Therefore, on August 6, Guy Wood, a Forest Service radio technician and Lt. Woldran, an Army communication man, arrived at Bishop and picked up the food which I had secured for them and proceeded on to Mammoth, preparatory to taking off on an advance trip over the proposed route for the purpose of testing conditions for sending and receiving radio-phone messages between each of our proposed campsites and Bishop. They completed their experimental test contacts with complete success and returned on August 12, suntanned and bewhiskered.

Finally, the date for beginning the trip was set for August 23, and on August 22, I checked in at the Lake Mary Pack Station to be sure that all arrangements were in order and met Messrs. Show and Cronemiller who had arrived by car. Mr. Show returned with me to Bishop where we met Colonel Peterson, General Arnold's aide, and learned that General Marshall, Chief of Staff, was also to be a member of the party and that the four star generals were to arrive via plane from Washington, DC at about 6:00 a.m.; that they would have had their breakfast aboard the plane and be ready for a quick takeoff from the airport via car to Mammoth in order to avoid publicity.

Mr. Show had breakfast with us at 5:30 a.m., and just as we were finishing our breakfast, we heard the drone of the plane's motors. Hurriedly grabbing up our personal effects, etc., we proceeded to the airport where the huge plane had already landed and we were soon on our way, Generals Marshall and Arnold riding with Mr. Show and me. A short stop was made at the Mammoth Ranger Station where we met Supervisor Joe Elliott of the Lassen Forest who was also well acquainted with General Arnold, having been on some of the previous fishing trips with him.

After arriving at the pack station, horses and saddles were chosen or allotted, stirrups adjusted, etc., and finally, we started on ahead of the pack train which was to follow. As we proceeded up the trail, the sound of an airplane was heard overhead and soon excited our curiosity by seeming to continue to circle within a radius of only a few miles of our party. Wisecracks began to be made regarding the aerial umbrella being afforded us by the Army Air Corps, but I think none but the Army personnel themselves knew that they really were a part of our party. There were two planes, an L-5 and a larger and faster plane, a Lockheed. We learned later that the L-5 was being piloted by a woman flyer, a WASP, who was keeping track of our route and progress so that mail and supplies could be dropped from the plane.

We rested nearly an hour at Duck Lake expecting Cronemiller and the "lunch" mule to overtake us, but nothing happened, and since most of us had our fishing tackle on our saddle horses we decided to push on to Purple Lake where we were to camp for the night and do a little fishing. This was done and plenty of fish were caught to assure fish for breakfast. The pack train finally arrived about 6:00 p.m. and a late supper was enjoyed by a tired and hungry bunch, most of whom hadn't eaten since about 5:00 or 5:30 a.m.

Satisfactory radio contact was made with the Bishop airport and immediately after breakfast the plane was on the job. Lt. Waldron set off a couple of smoke bombs that gave up a dense cloud of orange colored smoke to inform the pilot of the exact location of our campsite, and after circling a few times, she swooped down just above the tree tops and dropped the morning papers.

A short trip was taken that day and we camped at Tully's Hole, a beautiful little meadow on the north fork of Fish Creek. After a rest and luncheon, more fishing was enjoyed. I fished with General Arnold and Joe with General Marshall, both generals making satisfactory catches.

Croney was the chef and did a very creditable job. Del Fausett, surmising that the two packers that we started with were not competent to handle so many stock with satisfactory dispatch, made arrangements to send in two more who arrived about noon. They, Don Douglass, a teacher from Whittier, and George Brown, a local from Owens Valley, contributed much to the better enjoyment and quicker getaway time for the rest of the trip.

On the following morning, August 25, the plane came in about 8:30 and the pilot, after circling and dropping several packages, prepared to drop the most important of all, a mail pouch containing the latest communiqués from Washington. In getting it out over the side of the cockpit in position to release at the proper moment, the air resistance tore it from her hand (it probably weighed 25 pounds) a good half mile from where she intended to release it, and it came down on the side of a rugged and timbered mountain rather than in the meadow near camp. About a two hour search was required before it was finally located, an embarrassing incident for the pilot under the circumstances. But all is well that ends well and it was recovered in time for us to proceed on to our next campsite at Quail Pasture on Mono Creek that afternoon.

We found fishing good on Mono Creek and the warmer temperatures and lower altitude helped to create a restful atmosphere, and so the following day was a layover day. The packers shod some horses, some of the party just rested or cleaned up a bit, and some of us rode up Mono Creek a few miles and did some more fishing.

The next day we moved on to a campsite on Hilgard Creek about one and one half miles below Lake Italy, the highest camp of our trip at 10,700 feet. Here we had almost virgin golden trout fishing, with little evidence that anyone but perhaps a sheepman had been there to fish for perhaps a year or two. Limits of six to seven inch golden trout were easily taken and in an effort to get a few big ones to take back, the generals had a rubber boat flown in and dropped via parachute, a striking sight as the bright yellow silk chute came floating down between the rugged granite peaks on either side of the canyon below Lake Italy. Next morning the boat was launched in Lake Italy, but no luck with the "big ones." They were either too scarce or too shy. Two of the packers did catch a half dozen nice steelhead out of a small lake to the south of Italy.

Anyway, we had plenty of fish, all we could eat and all we could legally bring out. The generals made nice catches and even limit catches. Another mail pouch was dropped and the communiqués and maps indicated most satisfactory progress was being made on all war fronts. Our cook was doing an excellent job of feeding us. Intimate candid camera shots were frequently recording the business as well as the pleasurable incidents of the trip. Four star generals were huddling over their mail or listening to the radio reports, displaying their skill with the fly rod, or negotiating rugged and even dangerous sections of the trail. The more domestic phases were not neglected; shaving under difficult conditions in the open, seated on the stream bank bathing their feet, all go to prove that the USA is still a democratic world.

It was the last evening in camp. Some of us were still busy with the camp work, washing dishes, etc., when Mr. Show hailed me up to the campfire to inform me that the generals had just made a very interesting proposal, at least he thought Joe and I would so consider it. In short, the generals were extending an invitation to Joe and me to accompany them to Washington. At first we thought they were kidding, but when convinced of their sincerity, we were (speaking for myself) completely dumbfounded. Even Joe was a little slow with his usual wisecracks as they painted the picture of the fun they and their friends in Washington would have at the expense of a couple of hillbilly forest supervisors when on exhibition as trophies of this High Sierra trip.

We finally went to bed, but I am afraid not to sleep too soundly, for I distinctly recall waking two or three times and usually the beam from Joe's flashlight was visible, and I couldn't be sure whether he was trying to figure out his bank balance or whether he was looking at his hiking boots and clothes trying to determine if they would be appropriate wearing apparel in Washington society. Morning finally came and Joe's saddle horse was on the sick list, not serious, but serious enough in Joe's mind at least that he walked all of the 20 miles or so out over a 12,400 foot pass to the roadhead at the U. S. Vanadium Mill in Pine Creek. The rest of the party started somewhat later and only caught occasional glimpses of him silhouetted against a skyline leading his ailing horse.

Travel that last day was rough and slow. The horses were all a little sore footed from being in the rocks, and this last portion of our trip was via an old sheep route rather than over a regular trail. The terrain was rugged, most of it above timberline, but the views of the vast expanse of jagged peaks will be remembered by all the party as the most inspiring part of the trip and, from my own experience, I am sure that we remember those things longest and with the greatest degree of satisfaction that were most difficult of accomplishment.

The modern industrial development of the U. S. Vanadium Company's mill, tramline, roads and other development, afforded an inspiring and interesting contrast to the absence or void of manmade handiwork in which we had been living for nearly a week, but the pack trip was at an end and we tarried at the roadhead, where our cars were waiting for us, only long enough to find our bags and personal effects. We were then on our way to Bishop where we knew hot baths and clean clothes were waiting, and an informal banquet for members of our party and a few representative local citizens had been arranged at the Golden State. Generals Marshall and Arnold insisted that they were to finance that, and they being "The Generals," that was the way it was.

Arriving at home I informed my wife that I was going to Washington and she was perhaps more bowled over than I was the evening before, when so informed by Mr. Show, but she hurriedly gathered together my clothes while I shaved and bathed and almost breathlessly arrived at the banquet room.

The generals were more than generous in their responses to invitations to briefly comment on matters of international and even worldwide interest, and even gave their permission for flashlight photos to be taken. General Marshall briefly highlighted the pack trip and in commenting on our last day's trip, said that hell was probably paved with such trails. That particular remark sounded like Joe Elliott to me but it could have originated with the chief of staff. In either event, I could have had no comeback.

THE CODE WORD WAS PAPER

Donald Peters

The spring of 1945 on the Fremont National Forest was dry and windy. The early melting of snow at the higher elevations gave fire prevention personnel an eerie feeling of a rough fire season ahead.

District rangers were opening and manning lookouts much earlier than usual. One of these was Bald Mtn. on the Silver Lake District. A man and wife team—Gene and Ruth Groth—had gone on duty on April 8th and were getting acquainted with their various topographical landmarks.

Every ranger on the forest, as well as the staff, were on edge mentally and physically. District rangers were told a secret that was not to be divulged. They had been alerted to the fact that the United States was being bombed with incendiary and anti-personnel bombs. Forest Service administrators weren't given much information by the Army but it was later learned that by April 10, 1945, more than 180 Japanese balloon type bombs had been found in the United States and Canada, scattered in various locations across most of the eleven Western States and Texas, Nebraska, Iowa, Kansas, Michigan, several Canadian provinces and Alaska.

These balloons with their attached anti-personnel and/or incendiary bombs were being sent from Japan on the wings of the jet stream, and having no means of guidance, were dropping all over the Western United States with the enemy's objective of setting forest fires or doing much damage in populated and industrial areas.

The Japanese had to depend on feedback from U.S. or other news media accounts as to the success of their venture. However, U.S. military intelligence, together with news media cooperation, prevented such publicity being dispersed.

The balloons were made of three to four layers of tissue paper cemented together. They were about 33 feet in diameter, inflated with hydrogen to a capacity of 19,000 cubic feet and had a lifting capacity at sea level of 1,000 pounds and at jet stream level of 30,000 feet about 300 pounds. They had an altitude control mechanism operated by means of aneroids and batteries which released small sandbags whenever the balloon reached a preset minimum altitude of around 30,000 feet. It would then rise again to approximately 38,000 feet to be on its way crossing the Pacific Ocean in from 30 to 100 hours.

On April 10, 1945, Bald Mtn. Lookout Groth called his headquarters at Silver Lake Ranger Station to report his observance of something which appeared white and changing configuration as the wind gusted in a group of lodgepole pines on the south slope of Bald Mtn. Much of that high elevation area was still covered by snow, and with that type of background and the object several miles distant, Lookout Groth could not positively determine what was attracting his attention. Security demands were so tight that ranger instructions had been to instruct employees to be on the watch and report anything and everything unusual.

Protective Assistant Guy Martin and I made a hasty trip to the area and by use of a compass cautiously approached the object of concern. We found the balloon partially hanging in small lodgepole pines with four or five kilograms of potentially dangerous incendiary bombs all unexploded. Not knowing anything about the device at that time, both forest officers observed the setup with binoculars from a safe distance while lying flat on the ground.

I then returned to Silver Lake Ranger Station and reported the incident to Supervisor Larry Mays by one word "Papers." Approximate location was also given.

Within a few hours, to the surprise of ranger station employees, several Army personnel arrived. We were told that they had flown from their base at Ft. Lewis, Washington, to the Redmond P-38 fighter base and then driven on to Silver Lake.

The group consisted of an Army major, captain, master sgt. and four corporals and privates. They were escorted to the balloon site and after the captain, a demolition expert, had determined that the balloon was armed only with incendiary bombs, dismantled them, and all the material was taken on to Silver Lake Ranger Station and subsequently to Fort Lewis for further study.

Upon return to the ranger station there was the problem of getting a warm evening meal. There was no restaurant in Silver Lake; consequently, I and wife Ollie had foreseen the need and had food prepared and the table all set. She had given the word for the men to come and eat, but only the major and the captain sat down. She noticed the noncoms had stepped outside, so for a second time, went out and told them the food was on the table. Finally, getting no response she again went outside, confronted the sergeant and wanted to know the reason for their delay at her table. She was then told that noncoms did not eat at the same table with officers.

She immediately returned to the dining room and told the commissioned officers that in the Forest Service there exists an "esprit de corps" that separates no supervisor from the lowest man on the totem pole at mealtime and that she would appreciate all of them eating at one setting. The major apologized for not realizing there was no commercial food service available and then called his men in to share supper with us. That ended a hectic day in the lives of Silver Lake Ranger District personnel.

It was later determined that approximately 9,300 balloons were launched from Japan. Of this number two were found in Mexico, one in Hawaii, 30 in Alaska, 78 in Canada and 75 in the United States.

The only known fatalities from the balloon attack on the United States mainland occurred near Bly, Oregon, in the Gearhart Mountain area on May 5, 1945. At that time a minister's wife and five children, who were on an outing, evidently found a deadly device and one of them must have in some way triggered the mechanism by tugging or pulling on it, thus causing the detonation.

THAR SHE BLOWS—MAYBE!

Harvey Mack

As was the case with most year long fire personnel, Richard Raybould and I usually were assigned to road and trail maintenance as swampers, brushing, installing drainage, etc., during the winter and spring months.

On the Palomar District of the Cleveland Forest, over the years a large boulder had emerged in the middle of a truck trail as a result of repeated grading, traffic and erosion. Forest Engineer Dyke Landweer gave us the assignment of cutting the boulder down to where it would be below the road grade. He suggested that we use explosives to break it up. However, he didn't realize that neither of us had any knowledge or experience in the use of explosives.

Since both of us considered ourselves to be "jacks of all trades," we had no concern about such as assignment. Ranger "Fletch" Hayward briefly explained how to put the blasting caps into the dynamite stick and how to plant them around the boulder. He left it up to us to figure out how to wire and detonate the charges. This was 1946, and we knew of no written instructions, guidelines or safety regulations.

We got a dozen or more sticks of dynamite and blasting caps from the powder magazine and put them into the back of our pickup together, not knowing that they should have been kept separate. When we got to the site, we dug a series of holes around the perimeter of the boulder, inserted the caps into the dynamite sticks, inserted them in the holes and packed the soil around them. We then wired them in series rather than parallel. Someone had told us to use the truck ignition rather than the detonating machine, so we backed the truck up to a safe distance and touched the wires to a spark plug. We heard a muffled explosion and saw a little dust and knew that something was wrong as we had expected a big bang. We went to see what had happened and found that only the first of the charges had exploded, doing virtually no damage to the boulder but leaving a large hole beside it. So, we connected the wires to the next charge and repeated the same process with the same results. After a few more tries we gave up and proceeded to dig up the unexploded sticks of dynamite, remove the caps and put them back in the pickup. After filling in all of the holes, we returned to the ranger station and put the blasting material back in the magazine, not realizing that we unknowingly had violated every procedural and safety regulations outlined in the blaster handbook. We later learned that if the charges had been wired in parallel, we probably would have demolished both the boulder and the road, but someone was guarding over us. I think that they later hauled in dirt and filled over the top of the rock. Since experience is the best teacher, we never tried to use explosives again.

SOME CALIFORNIA FIRES

Gilbert E. Belcher

I was discharged from the Marine Corps, April 17, 1947. FCA Clair Miller hired me as headquarters foreman. I went to guard school at Julian in Southern California. There were many there from all three districts of the Cleveland National Forest. I remember it as a very good week. We were introduced to many new ideas in firefighting and equipment.

There were a few local fires that season and I went to several large fires in San Diego County.

Way in the back yard of the headquarters station was a trash pit approximately 10 feet by 10 feet and walled in concrete. It had a wooden top with a trap door to dump trash into it. Each year the top was removed and the trash burned. We couldn't get it to burn, so we got a can of gasoline and sprinkled it around in the pit. A design for disaster, eh? With three of us standing on the concrete liner and looking down, I lit a piece of paper, which floated down like a feather. It was like looking down the barrel of a cannon. A tower of fire, perhaps 25 feet tall, and taking off all of our facial hair, shot skyward. All manner of trash was airborne and burning. Some landed in the nearby grass. After the three of us tired of trying to stomp it out, we got our fire truck and put out the fire. The fire had made quite a noise and we thought sure the city fire department would roll in, but somehow we were the only ones who knew about it and we didn't tell anyone.

Twenty-five Forest Service personnel from the Cleveland flew to a fire in Lake County. Every one of us traveled to and from the airport in a different type of vehicle, mileage and travel time. We went to Arcadia, then bussed to the Burbank Airport. We flew at night and the plane had a stewardess.

The plane circled the Fresno area, and someone said, "Look, there's a plane above us." ...SILENCE. Then another said, "Plane hell, that's a car down on the road." The plane was in a tight turn with the left wing down. Panic broke out, the plane leveled out and we went north. All of us felt like we had a gallon of urine to pass. We landed at a small strip near Marysville. It was very cold there and there were no steps to get down. It was a Paul Mantz C-47. We did get down and found an empty metal shed to flop down in. Several hours later a bus arrived and took us to the Barton Flat fire camp. We were fresh overhead for the day shift.

I was a crew boss. The crewmen were winos, conscripted off the streets of what is now Old Sacramento. Many had no coats, hats or boots. There were many warming fires. I got a list of names and went from fire to fire pulling the most pitiful of men to a stakeside truck. Some left the truck for the fires and I had to find and capture them again.

The fire had started at Clear Lake and went north over some timbered mountains and into heavy brush. It was the first time I'd been called a "Brush Bunny." All of the Cleveland personnel was assigned to the brush lines, as it was felt that it was the only kind of fire we knew anything about.

This was the first fire that I was on where a helicopter was used. Was I ever impressed. Each night we drew two blankets and slept on the ground. It was the first and last time I had crabs. Everyone in camp had at least one hand in their pants to scratch. If the winos got any revenge for being shanghaied to this fire it was with the crabs. I'm sure gallons of blood was lost.

We went to another fire. We rode in Army trucks. We stopped in some town and cleaned out a drug store of anything that would spell relief for the itching.

The fire camp was an old CCC camp. We flew back to Burbank in a C-47 without a stewardess. I had arrived from Corona in the back of a pickup and went back to Corona in a sedan.

We had all racked up some overtime—manna from heaven. It didn't make a whit to any of us how we traveled. It seems there was a difference in pay for riding in the seat of a pickup and riding in the back. Or if you were the driver, or passenger in the back seat of an auto or an airplane, with or without a stewardess. The front or rear of a stakeside, or a fire truck. Another curve was if you traveled by private vehicle. No two of the 25 traveled in the exact same means of transportation. Only those who have experienced the red tape involved in getting their overtime and travel expenses can understand the maddening frustration we had in this. It was months sorting it out and getting paid.

In early September, Jack Osborn, Bob Bell and I hiked up the Bell Star Ridge to cold trail a small fire. Joe Springer hiked up with a can of beer for each of us and to say goodbye, as he was leaving for forestry college. I asked him why he was going to college, as he was already in the forestry. He said he wanted to be a forester. Dumb me, I didn't know what a forester was. In 1954, years later, I too left on school leave to study forestry.

About a month in the fire season, I went to a fire on the San Bernardino Forest. I hiked a crew of Marines up a fireline to an assignment at the head of the fire, which was south of Big Bear. I thought we took enough rest breaks but the Marines asked me to rest again as they had arrived to the States in April. I asked the name of the ship and it was the U.S.S. General Anderson. I was on that ship also. Several of us sorta knew one another, as we had been in the Island Command Military Police on Guam.

I was on a fire on the east side of the Palomar District. Real brushy and big boulders. I had a crew of Rain Bow prisoners and we ran out of drinking water. I assigned a big husky con to hike back to the paved road and fill several canteens. When he came back, he had the water and a very large watermelon. He had hitchhiked to a store and bought the melon. I asked him why he didn't buy booze and take off for Mexico. He said he didn't have enough money to leave on and that he was a short timer and was never going to jail again.

When serving as a prison chaser in the Marines, if you lost a prisoner, you had to do his time until he was apprehended. I was lucky he came back, as I would have had a lot of explaining to do.

FCA Clair Miller spoke to Joe Sherman, the State Ranger in Orange County CDF, and when the fire season closed at the end of October, Archie Bangs and I became state firefighters at Trabuco CDF Station. Carl Yeager was the foreman, and he had come over from the San Bernardino Forest, so Archie and I were right at home.

I was promoted to fire truck driver in the Riverside Unit. In 1950, I took the state firefighter foremans exam. The exam was right out of the USFS manuals. I passed number one in the state. I retired a state forest ranger the end of 1981.

In 1951, I was a CDF foreman in the Placer County Ranger Unit, District 3. Hugh Maples, another foreman, and I left for Happy Camp to the Pony Peak Fire in a 1949 International pickup. I hadn't been north of Marysville before, so this was a big adventure. It started to get smoky at Yreka.

We left the pavement at the Klamath River and started down the dirt road to Happy Camp. I've been in my share of forest fire smoke before, but I remember this canyon as the thickest to see through and the worst to breathe. We checked in at the fire camp and teamed up with Bill Merle, Associate Ranger in charge of the new Iron Mine Conservation Camp near Auburn. Bill had his whole camp, three or four hand crews, foreman and CDC correctional officers. We were directed to an area to bivouac. We had bed rolls and after a big fire camp meal and a day's travel, we made camp. A small warming fire was started.

Bill Merle had lost an arm in the war. He announced in a loud voice, "Hey, you guys, I can do something you can't do." Clenched in the hook-like hand he had was a large scorpion which he dropped into the fire. My hands disappeared way up into my coat sleeves.

We were up, fed and trucked down the canyon before dawn. It was so smoky all day we weren't sure the sun was up. There was a small army of us. The Forest Service was at the head of the column, with a guide who knew the area. The fire was north of the Klamath River, approximately 5,000 acres of virgin forest and no nearby roads. I carried a misery whip. We hiked, stopped and watched a lot of arm waving at the head of the column, and then hiked again. Hiked 30 to 40 minutes, then the arm waving again. We gained altitude and got into some big timber. It was in the afternoon when a trail was finally found and followed to the fire. We were on the south line in bigger timber than I had ever seen. We started to build line. The duff was four feet deep. It soon looked like World War I trench warfare.

Heard, but unseen aircraft, began to bomb us with camp supplies. Cast iron grills, Coleman lanterns, five gallon cans of water, cases of frozen milk and frozen food came down in 14 foot by 14 foot burlap parachutes. The chute didn't slow them down. It just made them easier to see. About 75 percent hung up in the trees. If a tree had three or more chutes, we sawed it down. It was getting dark and everyone was hungry. The trees were bigger in diameter than the saws were long. I'm sure a lot of timber fallers would roll over in their graves at the beavering we did to get the trees down. The Forest Service boss got all the burlap chutes together, then distributed them so everybody got a little use of them for bedding.

There were at least ten CDF foremen. We slept in our clothes and boots and spent a lot of time scrounging wood to keep the fire going. With no flashlights and competing with all the other crews for wood, it was more work than sleep.

The trench line was extended in both directions until it tied in with other sectors. I don't remember how many days or nights we were there. The food was good, though. When it was time to leave, all the camp gear was piled up and we were told we could have any of it we wanted to pack out. I didn't take anything. We stopped it Yreka CDF headquarters and met Lee Lang, the Associate Ranger. It had felt good to be on a Forest Service fire again.

A STRANGE EXPERIENCE IN A STRANGE PLACE

James R. Pratley

On the Angeles, west of Mt. Gleason and west of the old plantation in 1947, my engineering crew and I experienced a scene straight from the Holy Land. We were cutting a survey line about three feet wide through very dense brush, leaving jagged stumps and rocks, anything but a hiking trail. (Now a part of road 3Nl7.) We were well beyond the end of the existing road, in sweltering heat, with little water and bothered by some of the largest horse flies known to man or beast. All that we heard, other than the sounds of our work, was the buzzing of the flies and cursing as the men were bitten. It's little wonder that horses hated them.

Suddenly, my faithful companion Shorty began to bark and her hackles began to bristle. What on earth did she see and hear that would make her take such a defensive attitude? Though we had not seen bear or puma in this area for several years, it certainly must have been something of that nature. We heard nothing, but decided nonetheless, to rest a moment and find out what her problem was. We began to hear what was obviously bothering our four legged friend, the soft melodic tinkle of a bell coming ever closer. After several moments of this musical sound we saw a small procession emerging from around the last bend. It was something that might have been common over 2,000 years ago. Here came a man in flowing robes leading a docile donkey loaded with all manner of family possessions and followed by a young woman and two small children, also dressed in robes. We were so stunned that we just stood our ground until they reached our position. Here, they had to stop as the brushed line stopped. Had this happened in 1981 instead of 1947, we might have thought it was the result of something that we had smoked.

Unfortunately, my first question to the man was, "Where in the hell do you think you are going?" more out of curiosity than anything else. I was soon sorry that I had asked the question in such a rude fashion as the man then asked in a soft and almost reverent voice, "Where does this trail go?" By then I felt like a rowdy fool for asking him such a question in the first place and by using a word that he had certainly heard but would probably not have used himself.

Obviously, the so-called "trail" ended right where they stood. It was still about a thousand feet to the plantation where we would intercept the road leading to Acton in the desert below. The man then wanted to know what he should do with his caravan and if there might be some other way they could reach their destination without retracing the many steps that would otherwise be required.

Sadly, all that we could tell him was to go back to the end of the road and then follow the firebreak along the top of the ridge to the plantation; then to decide whether to proceed further west or turn north and head for Acton. Seemingly unperturbed by this setback, he advised his family of this turn of events and they set about retracing their steps. Before they left we offered them a drink of our small quantity of water, knowing that they were much more likely to do without than we were. There was no way of knowing where they might spend the night and there was no water within a day's walk of our location at the rate they would be moving. We did not hear or see them again, even though the firebreak that we described was only a hundred feet away. Had they taken that route we would certainly have seen or heard them as they passed by. When they reached the end of the road they must have taken it back a mile and a half to the forest highway and then proceeded north to the community of Acton. A strange experience in a strange place.

SHOW ME TRIPS THAT WILL NEVER BE FORGOTTEN

Harvey Mack

Usually not much happens when you take people on a show me trip other than telling them a lot of history, pointing out points of interest, never seeing the wildlife that you had hoped to see and than bringing a bunch of tired and dirty people back to the ranger station. But, once in awhile, something unusual does happen. Here are a couple of examples:

Shortly after the end of WWII some of the local newspapers were hungry for stories and often turned to the national forests. On one such occasion, I took a reporter on a day long trip over the Trabuco District of the Cleveland Forest. We got off to a good early start but then I noticed the pickup had not been gassed up and I had not picked up our lunches. So, after an appropriate delay, we were on our way. It still was fire season so all of the gates were locked and it became the reporter's duty to open the gates and lock them behind us. His shining shoes and pants slowly gathered dust and scuff marks. It also was evident that he had not planned on doing any work. As we came out on a saddle, he asked that we stop, as he wanted to get some pictures of both sides of the ridge looking into Orange and Riverside counties. I had warned him about snakes and, after going a few feet, we heard the familiar rattle nearby. I told him not to move until I located the snake and there, about 15 feet away was a beautiful specimen coiled, head up high and rattles going. I "casually" picked up a rock about the size of a baseball and threw it underhand as hard as I could at the snake and, much to my surprise, hit it squarely in the head and it dropped, either stunned or dead. Taking no chances, I fully dispatched it with a shovel. He then asked if that was the usual way I killed snakes and I replied boastfully, "All the time." Never did he guess that it was just good luck and never done before. Then he realized that in all his excitement, he had failed to get an action picture of our wildlife. Other than getting stuck in soft sand, the rest of the day went by without incident and no wildlife!

In 1950, members of the Corona City Council requested that we take them on a trip through the ranger district as it was only a matter of time before urbanization would begin to move toward and up against the forest boundary. I borrowed the ranger's 1947 black Pontiac sedan and loaded three of them in for a day long trip. None had ever been on a truck trail before and you could sense some white knuckles, especially since the sedan was relatively "gutless" and wherever I could, I would take a run at the next uphill pull. On one of these spurts there was a loud "bang" followed by the smell of gasoline. We stopped and when I looked at the tank, there was a steady stream of gasoline about the size of a pencil pouring out. Obviously, we were not going any farther and our next worry was getting back down, as we were several miles from the nearest highway. For lack of something better, I took off one of my heavy socks, whittled a stick to a point and drove the sock into the hole. This stopped most of the leak, but we now had less than a quarter of a tank left, a leaking "molotov cocktail" under the tank and several miles of truck trail and highway to get back to town. We found a place to turn around and then started back. Everywhere we could, with some safety in mind, I coasted and used brakes using low gear only on the steepest inclines and taking full advantage of gaining as much speed as possible to make it up the next grade avoiding using engine power, then cresting the grade and repeating the same process. If you think we had "white knuckles going up," you should have seen these fellows coming down. Of course, I was "cool as a cucumber" but sure was glad when we finally got back on the highway and back to the ranger station. Would you believe that we never got a "thank you" letter or a second request to tour the district?

Now they pay good money for a similar ride at an amusement park and this one was free!

WEATHER MODIFICATION

Carl Wilson

Editor's Note: Your editor likes writers to be logical and systematic so is a little annoyed with Carl Wilson. How does one classify the following story? Carl, who was on the Lassen from 1946-1949, gives us a fire story, or so it seems. Then, he switches to an account of weather tampering. He heads the story "Weather Modification," then goes on to say he's not sure that's what actually happened. However, we do like his conclusion—he didn't say, "More research is needed." –E.J.

It was the opening week of the deer season, mid-September, 1947, on the Lassen. I was fire control assistant on the Almanor District. Our headquarters was the Coppervale Ranger Station, which was actually within the newly created Susan River District. (Coppervale no longer exists; there remains just a few concrete foundations. Everything else has been moved!) The Almanor district ranger, to the surprise and dismay of the forest supervisor and his staff, decided to resign and I became acting DR. Then it happened; a small fire, presumably started by a deer hunter was discovered near the mouth of Duffy Creek. Initial attack consisted of Patrolman, Earl Nichols and our "powerful" D-4 tractor. It was too little, too late, and we soon had a big fire roaring up Duffy Creek. When the large fire organization was set up, I was asked to go to the Chester GS to protect the "valuable Almanor District." At Chester, we "girded our loins" and prepared for the worst. Meantime, the disastrous

Peligreen Fire had started on the Mineral District and we were in real trouble.

Shortly after arriving at the Chester GS, I was visited by Jim McDonald, Logging Superintendent for Collins Pine Company. Jim reported that he had just returned from a flight over Butt Mountain in the Collins plane and had created rain and snow by dropping dry ice from the aircraft. He said he had been reading about some of the early experiments in weather modification and decided to give it a try. He also mentioned he had a selfish motive because most of his woods crews were on the Duffy Fire and he wanted them back. He thought that if he could create rain and snow on Butt Mountain by dropping dry ice, it might be possible to do it over the Duffy Fire if we had more dry ice and a suitable plane. This sounded fantastic to me.

I called Bunky Parker on the Duffy Fire and told him about Jim's proposal. He said, "Go ahead, it can't hurt anything." So, Jim and I went to the drug store in Chester and the druggist volunteered 40 pounds of dry ice "in the interest of science." We returned to the Chester airport where he had asked the pilot and copilot of a Trimotor Ford from Mt. Shasta to standby. They had just delivered several sector teams for the fire. The two agreed to cooperate in our crazy test. Jim, the fire crew and I put the dry ice in gunny sacks and pounded it up into powder, just like we used to do with ice for making ice cream. We then returned the dry ice to the cartons. Our instructions to the pilot and copilot were: "Go as high as you can into those cumulus clouds over Lake Almanor. Then, dump the stuff into the clouds," which were then moving east toward Duffy Creek. The Trimotor took off and disappeared into the clouds and presumably did what we had asked.

"Did it work?" You ask with bated breath. Because of my research background and the inherent difficulty of proving the effectiveness of weather modification efforts, I must say, "It appeared to." There was some good news and some bad news!

During the afternoon we heard reports from the Duffy Fire headquarters that rain and snow had fallen on the fire. Spread of the main fire was slowed but the backfires were also dampened. Even worse, the crews were upset at having to eat ice cream for dinner while sitting in the snow. The bad news was that lightning started at least one fire near Westwood and another on Stover Mountain that afternoon. The first was easily suppressed by crews from the Duffy Fire. But, I spent most of that night helping fell a snag on Stover, and then we found that water in our hose line had turned to ice! Probably no one can say for sure that our dry ice actually triggered the rain and snowfall. But, the Lassen records do not show any other precipitation elsewhere on the forest on that date.

Ten years later at the PSW Station, in co-op with the California Division of Forestry, we were trying to prove that seeding with 50 plus ground based silver iodide generators could increase precipitation and/or reduce lightning fires. After three frustrating years, we found that our experiment could arouse the ire of locals. One group said we had created a drought by our seeding. Another reported that the seeding had caused "unseasonal rains" which ruined their down hay. Yet another group said that seeding had caused defoliation of the oak trees. This event occurred during the oak leaf moth infestation! In any case, our early opportunistic experiment in weather modification in 1947, did emphasize that weather modification may produce a mixture of good and bad results. Maybe it does not pay to mess around with Mother Nature.

THE WILD BLUE YONDER

Bruce Barron

Whenever I climb into the cockpit of an airplane, I'm always reminded of the pilots motto "thou shalt fly with caution lest mother earth rise up and smite thee!"

It was in the late 1940s when this admonition kept racing through my mind. I had agreed to participate in some aerobatic maneuvers with Frank Embree, Ranger on the Magalia District of the Lassen Forest. He had recently acquired a Stearman Biplane, the type coveted by agricultural crop dusters due to its exceptional maneuverability.

Frank had barnstormed his way up through that historical "Paul Bunyan Country" to join the Lassen Forest personnel at Susanville in a party celebrating the close of fire season. It was a hearty, hardy party that night and as we got to the Susanville airport the next morning, we were both suffering the effects of overindulgence.

Frank was a rugged square-jawed ranger that exuded that macho appearance of a typical movie star ace, especially when he donned his military style leather jacket, helmet and goggles! I was supplied with a similar mode of attire. The adrenaline began to flow when he trussed me up into a bulky parachute and explained how it operated and how to extricate myself from the cockpit should the occasion demand.

After hand spinning the propeller to get it started, the plane backfired a few times, blew out a cloud of oily smoke and settled down to a noisy roar. I climbed into the open air cockpit and gave the thumbs up signal that I was ready for take off. We immediately zoomed down the runway in a cloud of dust and hurtled our way "up into the wild blue yonder!"

As we started climbing higher and higher to gain a safe altitude for aerobatics, mother nature started to play her hand. The lofty mountains surrounding the Susanville area have a habit of creating waves of cold air flowing downward, while warmer air from the valley starts rising. This action creates mild turbulence which under normal circumstances is tolerable to airplane pilots and passengers. But in this particular occasion the unstable air caused the contents of the two overindulged stomachs to oscillate in a very disturbing manner. Frank's usual tan complexion was starting to turn a greenish yellow to match that of mine. We exchanged glances and I was more than happy to see his thumb point down toward the ground, at which time I nodded in positive agreement.

I was disappointed over losing the opportunity to do some "stunt flying." It was like having your pistol all cocked and loaded and having to let the hammer down again without firing a shot. It was best that we canceled the flight when we did, as I would have hated to spend the rest of the morning cleaning out the cockpits of a thoroughly messed up airplane.

STRIVE AND SUCCEED ONCE MORE

Everett Jensen

In 1945, with the end of WWII in sight, leaders in Region 5 were beginning to think about filling vacant spots in the organization. There were two open spots in the lands division of the regional office, in the land exchange and land uses section. Millard Barnum, head of the section, needed someone to do the legwork. Barney, as he was widely and affectionately known, was a planner. Out in the field I had made a few ripples by talking up multiple use planning, so I was detailed to the division for three months, a size-up situation. I passed inspection and was offered a place on the lands team.

Barney's section had previously included two men in support, John Edwards and Fred Grover. Edwards was on leave of absence to a wartime project and Grover was long gone to the Washington office. During their tenure they had compiled a valuable appraisal tool, a so-called "soil chart" which I inherited. It was a forester's approach to valuation of timber land and was based on years of experience in negotiations involving privately owned timber land, usually, but not always, cut over. Prices ranged from a basic 35 cents per acre for rocky or brush covered land to $5 per acre for fully stocked stands of young pine on the best timber growing sites.

I understood the soil chart and gratefully used it or taught it to field foresters when their assignments included land exchange.

In 1948, Mr. B. F. Manbey, Regional Chief of Lands, U.S. Park Service called on Barnum informally to ask for assistance. The Park Service wished to acquire two parcels of land, totaling 260 acres located in the southwestern part of Mt. Lassen National Park. Manbey had a problem. On the one hand, purchase funds were limited. On the other hand, his Washington office required an "outside," disinterested appraisal to guide negotiations and purchase. Could a sister agency provide services of an appraiser, thereby reducing costs and leaving more for actual purchase?

Barnum was generous by nature. He was also a shrewd operator. Some time down the road our region might need a friend in the other camp. I was sitting in a nearby chair and began to feel a bit uneasy. I knew Barnum wouldn't undertake the job, his talents lay in timber valuation. I was the only candidate in sight, and it was obvious from Manbey's remarks that soil charts wouldn't do the job. We were talking about an alien world known as real estate. This specialty had never been discussed in any training program I had experienced.

I came to attention when I heard Barnum agree to cooperate. Mr. Jensen would be available, there was only a need to check schedules and arrange a date. For me, it was a case of sink or swim, to join the club composed of those who are asked to do the unexpected. After the first mild shock, I realized it wasn't a life or death matter. It would be a learning experience. It might even be fun.

I was driven to Lassen National Park by one of the park's winter sports experts who briefed me on winter use on the south side of the park.

It turned out the Supan people owned two parcels which had been labeled Supan Sulphur Works with 100 acres and Tract No. 2 with 160 acres. The park highway gave access to both. It was easy to see how the first parcel got it's name. There were numerous sulphur deposits, pot holes with boiling water and mineralized springs. Supan Tract No. 2 was located about a mile and a half farther up the highway. Here a clear stream flowed into Little Hot Springs Valley. A geothermal area lay in the path of the flow. High acid content foreclosed any possibility for "mineral baths."

Supan Sulphur Works supported a modest commercial development consisting of a lodge and inn, four cabins and a curio shop. Supan Tract No. 2 was undeveloped, but I believed it to be suitable for a single recreational home site. I learned that subdivision had been considered at an earlier period, then abandoned. Kepler Johnson, engineer on the regional engineering staff, calculated the value of improvements on the Sulphur Works Tract.

Subsequent negotiations with the owner failed to reach agreement and the Park Service moved to condemn the property. Time had gone by. It was now 1950. Franklin Dill, special attorney with the Department of Justice in San

Francisco, was assigned the case. The Forest Service library had a book titled, "The Volcanic Activity and Hot Springs of Lassen National Park." Bill borrowed it for possible guidance. I had looked through it and found no reference to the Supan tracts.

Time went on. It was now 1951 and updated appraisals were requested and prepared. On June 20, 1952, Dill notified the Park Service that a jury trial date had been set for September 30. I was expected to show at the trial which would take place in Sacramento.

I drove to Sacramento at the appointed time and located Dill in an anteroom at the courthouse. He was accommodated by two men. There were introductions and I learned they were professional appraisers retained by the Justice Department. They sat apart and conversed in low tones, comparing their findings I supposed. I wasn't sure that I belonged here.

The trial got underway. The first expert witness to testify for the Supans waived his arms and spoke dramatically about the marvels of nature, meaning the hot springs. Impossible to place a value on the marvels, but if one must, then the properties were worth $300,000. The second expert seemed more restrained. His figure was a moderate $70,000.

Now it was the government's turn. My two colleagues testified and I wondered if they had looked over each other's shoulders. The two opinions coincided at $30,000. Dill let it rest at that point. I felt relieved and approved his strategy. I felt no need to wait for the jury's verdict and headed back to San Francisco.

I have a letter in my files dated October 29, 1952, from Franklin Dill. It was politely complimentary but it also included a set of figures that surprised me—the jury award to the Supans. The results are tabulated below, together with my valuations of the parcels.

Parcel	Jury Award	Jensen Report
Supan Sulphur Works	$ 39,950	$ 43,080
Supan No. 2	9,000	4,750
Totals:	$ 48,950	$ 47,835

TESTING HELICOPTERS

Charles Arment

In 1948, I was district ranger at Callahan on the Scott District of the Shasta NF. We were selected to do some experimental flying in mountainous terrain. The contractor was Armstrong and Flint Helicopters and they had a Bell Ranger of that period. Our mission was to test the helicopter in high altitude landings and takeoffs. These early helicopters had wheels, as compared to today's skids. The first area we selected was the high country in back of Bolivar Lookout. We landed nicely on an open ridge that had a slope to it but not enough to be of concern about the rotor blades digging in. The problem was that the helicopter started to roll and Freddy, the pilot, yelled at me to jump out and get something under the wheel. I did so, with alacrity, you can bet because it was a long walk back to Callahan. After getting a fair sized rock in front of the wheel and getting back in the copter, I asked Freddy why he didn't put on the brakes. Guess what, the copter didn't have brakes. What a time to find out! In spite of that situation we made several more test landings under the same circumstances and returned to Callahan.

Helicopters in 1948 were made to land on airports or helipads that were nice and flat. Brakes were ordered immediately. In the meantime, we continued to test fly for performance at high altitudes and in mountainous terrain.

After the testing was completed, the helicopter proved to be a valuable asset in our fire control program.

SEEDING THE CLOUDS

Bruce Barron

I first developed an avid interest in aviation during the late 1930s. It all started when a friend took me for a ride in his three cylinder Curtis Wright Jr. "pusher" type aircraft. It was called a "pusher" because the engine was mounted high above the wing and faced the rear of the plane. It was a fun airplane, since it had open cockpits and no front engine to inhibit visibility or to blow a lot of "prop-wash" back in your face. I became shamefully persistent in taking advantage of his generosity and flew many exciting hours chasing coyotes and jackrabbits around the open prairies in the western edge of the San Joaquin Valley near the little oil producing boomtown of Coalinga, California.

In later years, after joining the Forest Service, I was still quick to seize every opportunity "to break those earthbound ties and soar with the eagles."

In the late 1940s, Northern California was besieged by a plague of campaign fires that stretched all available manpower, equipment and fire camp services to the limit. I was service chief on that nasty conflagration on the Lassen Forest called the Duffy Fire. The firefighters were bitterly complaining about the poor quality of our fire camp food. To help placate complaints, I stopped by a local dairy in Susanville and bought a large quantity of bulk ice cream, which was stored in big insulated containers copiously packed with dry ice.

By coincidence, our Forest Service Pilot, Cal Ferris had just arrived in fire camp with a plane load of overhead fire staff. Cal and I were soon discussing flying experiences and as we gazed toward the western skies, several huge cumulus clouds had started to develop and were drifting our way.

Having read recent articles in the papers about "seeding clouds" to induce rain, I suggested to Cal that maybe he and I should take that dry ice off the ice cream containers and seed the clouds when they arrived over the fire area. He was quick to see the potential and I was eager for the chance to do some impromptu flying. Needless to say I was keenly disappointed when I was aced out of the flight by a staff member whose prestige placed him on a higher echelon of the totem pole.

The cloud seeding episode was a fabulous success. It precipitated a torrential mixture of rain, sleet, hail and snow. It brought that stubborn fire under control in a matter of a few minutes.

Firefighters were jubilant and Cal Ferris was an instant hero. The story appeared in the October 1951, issue of the

Reader's Digest. The only glory I gained from this historical event was in the satisfaction of knowing that I had supplied the impetus through my futile attempt to finagle a little flying time!!

HAPPY HUNTING

Bob Gray

For about three years just after World War II, deer hunting was fabulous on the McCloud District of the Shasta Forest. At Bartle we would validate between 150 and 200 bucks each year. Day and night, hunters would stop by with their trophies. The area, then as now, was a three point or better zone, so we saw some beautiful large bucks. In those days we weighed the deer, measured the antler diameter and spread width, reported the physical condition of the deer, spotted on a map the location of the kill and listened to many incredible hunter tales.

Most hunters overestimated the weight of their deer, especially when they had to pack or drag it a long distance. "Come on," they would say after we had weighed a deer in at 130 pounds, "He's got to be 175 pounds." Usually we could convince them our scales were accurate by weighing the hunter himself. Every time there was a lady in a hunting party, the tags indicated she had killed the deer though the male members usually would have a mental lapse, saying something like, "I shot him just as he cleared the brush," then pausing when he realized what he said, then adding, "Of course, the little woman was shooting too, so we couldn't tell exactly who hit him."

A few forked horn bucks were brought in, the hunter not knowing he was in a three point district We would have to confiscate the deer until we got hold of a game warden. One man brought in a 210 pound forked horn. He expected us to validate it because of prominent eye guards on the antlers. I would have done so except that Don Chipman, the game warden was right there. Any hunter would have shot that buck without counting points because of the size of the deer and horns. The warden told the unhappy hunter to report to the judge at McCloud the next day, which he did. Don had obviously not told the judge because he did not show up, thus, a compromise settlement was made. The hunter lost his deer but didn't get arrested. Don later told me that he would have given the deer to him too, "If all you guys hadn't been around!"

Another large forked horn was brought in by a man with the location of kill shown as Four Corners. There are two Four Corners locations, one in the three point area and the other in the forked horn area. This deer was similar to the other with huge antlers and eye guards. I validated it, assuming that the Four Corners on the tag was the one in the forked horn district, believing that justice was served in this case.

It was in one of these years that a four point doe was killed near Medicine Lake. I didn't validate it, but the hunter stopped by Bartle and showed her to us. She was a medium doe, with delicate doe features, and a small rack on her head.

Several times hunters stopped by to tell me that an illegal buck or doe had been shot and was lying in a certain location, usually alongside a road. Our crew would go and pick up the deer and use it at our camp. Probably a mistake had been made by the hunter reporting the deer but we never pressed the issue. Better to salvage the meat than let it rot in the woods.

Our saltlick was about 200 feet back of the Bartle Station and for a couple of weeks two big bucks had been coming to the lick at dusk. An idea occurred to me that I would like to see how close the deer would come to the salt while I was there.

The lick consisted of a salt block on a stump with a single rail pole fence, about 30 inches high. It would keep cattle out and allow deer to either jump over or go under.

About a half hour before dark I sat down in the closure with the stump to my back and the salt block behind my head, facing the direction the deer always came from. The weather was still and warm, and a few flies and gnats were buzzing around making a nuisance of themselves.

Right on schedule the two bucks showed up with darkness rapidly approaching. They both got to the fence and stopped to check their surroundings, while I sat stonestill looking right at them. Deer have sensitive ears, but like most animals, don't see too well unless something is moving, so they hadn't seen me yet.

Their noses too, are sensitive, but the still air hadn't carried my scent to them. The larger buck, a four pointer, jumped the fence and stopped about four feet from me, and the other one lowered his antlers, and crawled under the fence but stayed behind the larger one. Old "four point" stood like a statue, with his nose toward me, while "young buck" fidgeted in the background. My nose started itching and the sweat was beginning to run down my forehead, but I did not dare move for fear of frightening the bucks away. That beautiful wild creature only four feet from me was breathtaking and a little scary.

Suddenly that buck took two steps toward me and was literally nose to nose with me. Wild thoughts raced through my mind. Is he going to ram a horn into my face, or cut me with his front feet, or what? It had reached a stalemate between us. I was scared and I didn't know what he was thinking. Surely he had something on his mind. We made our move together, both bucks whirled and cleared the fence as I rolled around the stump. You can't imagine how good it feels to scratch your nose when you haven't been able to for a while!

THE ELLIS QUARTET

Phil Lord

It was a hot summer afternoon and a cavalcade of Forest Service vehicles was en route from the McCloud Flats to Alturas. The reason for this expedition was that there was a new face or two in the RO office of range and wildlife management and these new faces came from regions where range management was a much more important activity than it was in R5. So, it had been decided that to put that activity in what they felt was a proper perspective, a "show me" trip was in order. Seemingly, it was believed that range conditions in R5 were not as good as desired. A "show me" trip might point this up to the grazing staff men and also indicate to some supervisors that the range management activity should have a little more of a "place in the sun." Hence, there were both

Model T Ford on Klamath River Hiway - c1923. Large decal on windshield. Metal shield on radiator.

Eben Cady, wireless operator, working Heliograph at Orleans Ranger Station, Klamath NF, California - 1921.

supervisors and staff men on this trip. There had already been such an expedition in the southern part of the region and now we northern people were getting the treatment.

It was a very well planned, organized and managed trip, including the assignment of certain people to certain vehicles. It came to pass that assigned to one sedan were Guerdon Ellis, Bill Peterson, Al Crebbin and Phil Lord. All in all, some people might have considered this a rather motley group but as Crebbin remarked to me, "who were we to argue about it." As is usual, or used to be anyway, with Forest Service cavalcades of this type, they were just like a bunch of cattle strung out on the trail. The eager and ambitious were soon in the lead. The more conservative in the swing, and the drag was, in this case, the sedan containing Ellis, Peterson, Crebbin and Lord. Knowing that it would be too late when we got to Alturas for any official duty (except "children's hour"), we drove along at a sedate pace, discussing many things, although I don't think plant succession on the McCloud Flats or range conditions in the region were included. Finally, I don't remember who, Ellis, I think, said, "Poor Charlie Rector, he is up there on the Modoc with nothing but grazing troubles and I understand those permittees are really giving him a hard way to go. I think we ought to try to cheer him up some way." This was all too true. There had been a grazing survey made on the entire Modoc Forest and the final figures were not encouraging. "Adjustments" were indicated. The hassle about these adjustments had been going on for sometime. Charlie Rector, from out of region, was the third supervisor to make a run at the problem and there were no signs of any peaceful settlements or agreements. The local newspaper had taken a side against the Forest Service and was printing some very nasty editorials. Meeting after meeting was held with no results except everyone getting angry.

So after Ellis said, "We ought to try to cheer Rector up," I said, "Well, we can try but what in the world will we do?" Ellis answered, "There are four of us here, enough for a quartet; let's sing him a song at children's hour." So, we all said, "That's a fine idea. What will we sing?" Guerdon's reply was, "Don't worry, we'll think up something." And we did, after many false trials and a short stop in Fall River Mills to obtain some throat medicine. By the time we got to Alturas we had a beautiful, haunting, melodic song adapted to the tune of Beautiful, Beautiful, Texas. It was all about the Modoc and Charlie Rector. At the appropriate time we delivered the song and I will say that we were all in fine voice. As the last beautiful notes died away we waited for the thunderous applause we felt we deserved. There was a small, very small smattering of applause, mixed with a few cat calls and hisses from some of the rougher element at the bar. Charlie Rector rather than being cheered up just looked down his nose, and our leader, who was with Rector, had a terribly shocked look on his face.

We later learned that with the situation what it was on the Modoc Forest, nothing, but nothing, could have cheered up Charlie; that, actually, he was on the verge of a nervous breakdown. Also, our leader looked so shocked because he realized that we had traveled all that way from the McCloud Flats and hadn't used all the time discussing range management, plant succession and the few mangy Stipa plants we had seen on the McCloud Flats. After everything was practically over, a stove up, weather beaten, grazing staff man (probably from the Mendocino) was heard to mutter into a glass he was sipping from, "Well, what could they expect, putting those four in the same car." So ended Guerdon's quartet. Who knows, under different circumstances, we might have gone on to fame and fortune.

SOARING WITH THE EAGLES

Bruce Barron

It was in late summer of 1949, and the embers of that raging Peligreen Fire on the Mineral District of the Lassen Forest were still turning cold.

Fortunately our Forest Service Pilot, Cal Ferris, had just stopped by the Lassen Forest headquarters at Susanville. I explained to him that as service chief on that fire I was very much concerned over having lost a complete 200 man fire camp outfit somewhere in that Peligreen area. Apparently, when the off-forest personnel were released, whoever was in charge had failed to take responsibility for the disbandment of this mobile "spike camp," or notifying us as to its final location.

Cal volunteered to fly over the area in the Forest Service's newly acquired low-wing airplane. With luck, we immediately located the missing camp perched on the top of the Mill Creek Rim along the Lassen Pioneer Immigrant Trail. The camp was later retrieved with considerable difficulty due to the inaccessibility of that steep rocky terrain.

In many places, the unimproved serpentine road into the Peligreen area clung precariously to the edge of huge lava rimrock precipices. In mute testimony as to the hazardous conditions in the darkness of night, a Greyhound bus driver delivered a load of firefighters. On his way out after sunup the next morning, he gazed down into the awesome bowels of one of those bottomless canyons and was seized by a severe case of acrophobia. He immediately abandoned the bus and a relief driver had to be rushed in to move it out of the way so traffic could resume to the fire camp.

While we were flying in that area, Cal decided to show me where a WWII P-38 airplane had crashed down in the bottom of Antelope Creek. Those who know Antelope Creek are aware that its confines are deeply precipitous like a miniature "Grand Canyon." As we proceeded up the gaping maw of this canyon, Cal was intently peering down over his shoulder searching for the derelict remains of that old military pursuit plane. I was nervously looking forward at the narrowing canyon walls. All of a sudden there was a bend in the canyon and a solid wall of rock loomed directly ahead. It was too late to warn Cal who was thoroughly engrossed in his backward gaze. I pulled back on the wheel, kicked the right rudder pedal and we zoomed over the edge of the canyon wall with only a few feet to spare. Cal seemed nonplused that we had almost added another aircraft to the one at the bottom of that gorge. His unperturbed comment was, "Thanks old buddy, I owe you one!" (Luckily our aircraft was equipped with dual controls.)

As we proceeded back over the Cascade Mountains toward Susanville, Cal took the scenic route over Brokeoff and snow covered Lassen Peak. Just as we approached Brokeoff, we saw a big eagle soaring directly above us. We were amazed when it folded its wings, and like a kamikaze pilot, started a screaming dive toward our airplane. Again, lady luck was with us (and the eagle). As it zoomed on by, it passed between the propeller, the engine cowling and the wing. Six inches in any direction, and that eagle would have been killed. There was a strong possibility that we and the airplane would also have suffered some grave consequences.

As we were about to terminate our trip back to Susanville, we witnessed a huge pall of smoke rising from an out of control fire at a local sawmill. We made some low passes over the fire and were able to get some good detailed photographs which were utilized and much appreciated by the local newspapers covering the event.

Being confined in a small aircraft on long flights is often boring, however, this particular flight generated a profusion of sweat on palms, brows and under the armpits!

THE FIFTIES

POWER SAW ON A STICK

M. M. "Red" Nelson

Early in 1950, I was transferred from forest supervisor, Umpqua NF, R6, to be assistant chief of division of fire control in the WO. Carl Gustafson was the chief of the division, which at that time was very small. It consisted only of Gus, Clare Funk, two secretaries and myself. We had three rooms in the USDA South Building. Gus had one, the women one and Clare and I shared a double desk in the other. Clare Funk's job was staffman for equipment development and for air operations. The division of fire control was the leader in the FS for equipment development (as well as use of aircraft), and had some field activities specifically for fire equipment development. Example was the Arcadia Fire Equipment Development Center started by Funk (when in R5) and then headed by Gene Silva. (Later the development centers were broadened to include development of all types of equipment used in all FS fields.)

I recall one day very clearly in the spring of 1950, when Gustafson came into our room, sat down and talked about what was needed most for fire control. He told Funk he should give high priority to the development of a power operated hand firefighting tool for cutting brush and reproduction in building a fireline.

In those days, when need for new equipment was recognized, the policy was: First, survey commercial markets to find such equipment or something that might be adapted; second, if not available see if a commercial company could be found that had an interest in development of the equipment wanted; and third, if such private development could not be stimulated, assign it to a Forest Service equipment development center. Clare Funk followed that policy for development of a "power saw on a stick." Funk found that nothing of the type wanted had been developed but he did find a company in Richmond, Indiana, that was interested in development of such a tool. Clare worked closely with that company. The first development was with a small chain saw for cutting on one end of the "stick" which was powered by a small gas motor on the other end. It was not long before it was found that a circle saw performed better than a chain saw. Several makes became available. They became more practical as better and smaller motors were developed. This led to adaptions of the tool for such as cultivation, and then into the modern "weed eater," using a plastic string as the cutting edge. I claim that Carl Gustafson's directive to Clare Funk to develop a "power saw on a stick" made him the "father" of all such modern day power tools now in common usage.

CUT AND QUARTERED

Harvey Mack

In the spring of 1950, a small military reconnaissance plane was flying from March Air Base to a base in Orange County. Low clouds enveloped the top of Sierra Peak on the Cleveland Forest and the pilot misjudged his altitude and crashed into a ravine a short distance from the closed Sierra Peak Guard Station. Fortunately, he was not seriously injured and managed to hike up to the station, broke a window to get inside, rang the old magneto phone and got the ranger station. I called March Field immediately and they arranged to send a rescue/salvage team to the ranger station for me to escort to the crash scene. Since the pilot stayed by the phone, it was possible to connect him with the military people and he assured them that he was not in need of any medical help and that he would stretch out and rest on the bed until they arrived.

About an hour later a Jeep and a flat bed truck arrived at the ranger station and we proceeded up Skyline Drive to the crash site. The wreckage was well entangled in heavy brush and all of their attempts to pull it out with the truck wench failed and they were in a quandary as to what to do. I hiked down to the plane, looked it over and came up with an idea. I then asked the sergeant if he cared what it might be like when they got it out and he felt that anything was OK. So, I got out my large bolt cutters and proceeded to cut off both wings and the engine. This made it easy to hook onto and drag out each part. But when the sergeant saw what I had done, he got real concerned about what would happen when he got back to the base and they saw what remained of the plane.

We never heard anything further about the incident and received no kind of appreciation or recognition for the "ingenuity" used by the troops in the recovery of the aircraft. As far as we were concerned it was just another day's work that added a little excitement away from the office.

SAGA OF THE CORK BOOTS

Bob Bjornsen

It all started on a hot July afternoon in 1950, when a message reached me: "Report to the ranger station in Bly (Oregon) for a fire detail in Region 5." A memorable experience I'll never forget, especially the ruination of a pair of forty dollar Curren cork boots.

Arriving at the RS from the timber sale where I had been woods scaling, I barely had time to gather my fire pack and board the waiting black (not green) FS panel truck and roar off for Klamath Falls, cork boots and all. There the Fremont NF group was last of a R-6 overhead contingent to board a United DC3 for San Francisco.

Eventually the venerable Douglas aircraft landed at SFO and taxied up to the old terminal building—remember the one with the tile roof and Spanish architecture? Stepping on the waiting room tile floor, my cork boots started a 15 foot skid with me frantically waving arms and legs trying to catch my balance. That caused some titters from the waiting passengers and laughs from my compadres.

Soon we boarded the Doug again for the last leg to Monterey Airport. There, after an interminable wait, we boarded taxis (yes taxis) for Big Sur and a winding road to the top of the ridge where a heliport was located. Someone remarked if we'd included a boat in our journey it would have been one for the record books.

None of our hosts bothered to ask if we were tired, or even realized most of us had put in a day in the field before answering the fire call. It was a "get your a__ into that copter and out on the line" kind of greeting.

It was the Indian Creek Fire, Los Padres NF, Monterey Division. It was my first of quite a few R-5 fires I went to over the next 30 years. I can say I left a lot of sweat and tears on many chains of California fireline.

Now the memorable part. The copter was a Bell G1, or some such early designation for a primitive craft that had no bubble, only a narrow windshield. Freddie Bowen was piloting it. I believe he became a legend over the years for his fire aviation exploits.

I was third of the single trip passengers to board the copter. We each had a pulaski and shovel. Our instructions were to clear a helispot in the brush for bringing in recon scouts and overhead. Once a crew arrived (by shanks mare) we were to take charge as crew bosses and start building line.

I had trepidations about my first copter trip after watching it struggle to get aloft with each of the first two passengers. When my turn came Freddie informed me he would make two passes; the first to drop my tools and the second to drop me off (literally, as I found out). He added that he couldn't hover but would slow down for my exit from the craft. The first pass I dropped my tools at the proposed helispot with no thought about landing on them when I exited. On the second pass I stepped on the skid awaiting Freddie's near hover pause. As we came into the spot he signaled for my departure.

Only then did I realize that the brush was 10 feet high and his "near hover" was in my frightened state about 30 mph, although it was probably a lot less.

Finally, responding to Freddie's frantic nodding, as he fought the controls to keep the machine aloft, I bailed out, ripping my shirt and pants in a crash landing. It was only then that I realized I had taken the canvas seat cushion with me in a death grip as I left the copter.

Some years later I was amused to see how helitack crews suited up in protective gear similar to that of smokejumpers to make their exits from copters. Perhaps we were not the first but surely we were among the pioneers in barely clothed helijumping into California brushfields.

We put in a full shift, building and losing line. Then a night shift patrolling to pick up hotspots. A relief crew came in the early morning on day two. The last thing I remember was lying down on the fireline exhausted to the bone and awakening about 10:00 a.m. to see crewmen stepping over my body as they went up the line.

It wasn't until I got home that I discovered a big crack across the sole of one of my ruined Currens.

SMOKEY BEAR IS SAVED

Jay H. Cravens

1950 was a bad fire year on the Coconino National Forest. The dry season was prolonged and we had many man-caused fires before the lightning season started. I had become an experienced sector boss by this time. The first large fire that season started shortly after I arrived on the forest. Fire danger was rated extreme. A Santa Fe Railroad welding crew was replacing standard rails with welded continuous rails three miles west of Flagstaff. The right-of-way was filled with highly inflammable vegetation and partially decayed ties that had been thrown aside with other debris the railroad crews had failed to remove or dispose of properly. This problem existed all across the forest, from the eastern to the western boundary, some 35 miles. Repeated reminders and pleas from the Forest Service to the Santa Fe to remove the debris in accordance with their 19th

Century right-of-way easement agreement went unanswered. Sometimes they would make a token start at clearing and then a train wreck somewhere along the line would occur and all the men would be pulled off the clean-up job. They always conveniently forgot to return and finish the job. The welding crew started a fire on the north side of the double tracks and burned over 23,000 acres before we stopped it. I was a sector boss and my crews were on the head of this dangerous fire. The Santa Fe acknowledged blame. However, months later the Department of Justice neglected to meet a deadline for filing a judgment and the Santa Fe had a free ride.

The fire did not cost them a penny, while it had cost the government thousands of dollars in fire suppression costs, fence replacement, reseeding and reforestation. After the fire, Santa Fe cleanup efforts were still periodic or nonexistent

That same spring I was dispatched to a large forest fire on the Lincoln National Forest in New Mexico along with Harlow Yaeger and four other Coconino crew bosses. Harlow supervised project crews on the Elden District and had visited us at the Elden Guard Station. While he never talked about it, Harlow had been a prisoner of war when he was captured by the Japanese in the Philippines during World War II. He was an excellent person to work with and get to know. He taught me a great deal. En route to New Mexico we saw a series of large fires burning on the Sitgreaves and Apache National Forests. Our fire on the Lincoln National Forest was known as the Capitan Gap Fire. A very experienced forest officer by the name of Dean Earl was fire boss. Harlow and I were assigned separate sectors that burned hot all night and crowned and ran during the day. The temperature was very high; it was extremely dry and windy. The fire camp and our food was constantly buffeted by gale force winds. Wind blown debris and sand were in everything we ate. It was a very dangerous fire.

Harlow was in charge of a group of soldiers from Ft. Bliss and was assigned to the southwest sector running up on to Capitan Mountain. His company of soldiers were endangered by a crown fire and almost trapped before Harlow led them to safety on a rock slide. When the fire had passed around them, one of the soldiers heard mournful crying and whimpering sounds. He found a tiny black bear cub clinging to the trunk of a burned tree. The cub's paws were badly burned and one of the soldiers carried him back to the fire camp. I was there when they arrived and the bear was handed to me. It was crying, whimpering and licking its burned paws. I recall it crying as it chewed my leather gloves. I did not know at the time this 10 inch ball of black fur I held in my hands was to become the original Smokey Bear. For years afterwards, my daughters Melissa and Cindy asked me to tell them the true story of Smokey the Bear. Many people think that I found him. I did not find him, but I held the little creature for a short time before he was taken away to become that famous symbol for fire prevention.

THE BREATH OF SATAN

Ralph Bangsberg

As fire control officer on the Shasta NF on the day of September 2, 1950, the papers on my desk seemed unimportant. Leaving my desk to go to the dispatcher's office, I listen to the radio. All is well. Labor supply was becoming critical because of the approach of the Labor Day holiday. Lumber mills would be shot down, road contracts closed and people in general were traveling on extended vacations before the opening of schools. Even our own supply of trained firefighters had suffered losses as some returned to colleges. You are as ready as you can possibly be you keep telling yourself. Why then, those little feelings of fear and despair, bring out the sweat in the palms of your hands. Forget it.

Having been raised in the mountains of the Kern River Canyon in central California, I had fought fire in my teens. During the depression, as a forester in Northern Wisconsin, I had fought it. Later, back in California on my own ranger districts, I had helped teach it and during World War II had been on a flying fire squad that went all over the state, down to the Mexican border. I thought I knew demon fire.

Suddenly a brisk wind flows up from the south, loaded with heat and haze coming from the great Sacramento Valley miles below. Any moisture in the air or in the forest fuels is quickly sopped up by this gigantic dry sponge of hot air. The distant mountains, especially mighty Mt. Shasta, that stood out crystal clear this morning have disappeared, and closer ranges are indistinct. I find myself fighting off a strange nervousness. The day wears on slowly.

It is nearing 3 p.m. suntime, and I return to the dispatcher's office. The radio comes alive with three lookouts competing for airtime all at once: "Reporting heavy, black smoke along Highway 99 a mile north of Mt. Shasta!" They give the usual azimuth readings for locations estimate of size, then silence. As the dispatcher starts ordering men, I head for my car, a green Chevy sedan. In seconds I am off with red lights and siren; the same old lump in my gut, dry throat, heart pounding. The car radio is squawking as orders go through for tankers. Fire trucks are calling for directions, as into this comes the lookouts reporting a tremendous spread, followed by the supervisor's voice: "What's going on?" In minutes I am there. My God, what is the matter with that fire? It has gone crazy and nothing man-made will make a dent in it. The lump, dry throat and all are gone. When you meet the enemy face to face you muster up everything to do battle. My mind races. The fire is already a quarter of a mile long, roaring ahead in a ball of greasy red flames, lapping up pine and fir like little puffs of cotton and rolling all into frightening billows of black smoke some 2,000 feet in the air. The lid is off, but it isn't hopeless, as right in its path is the huge, volcanic cinder cone known as Black Butte. It is solid rock, so nothing there to burn. Already traffic and sightseers are jamming the nearby highway. The approach of a bulldozer is signaled by its high pitched squealing and can be heard over the general roar of sirens in the distance, the Southern Pacific chugging upgrade and above, a commercial passenger plane circling. Boy! What a place to observe from! Wish I were up there! Above this cacophony of sound I hear my car number on the radio: "Car 22, return to headquarters." This is frustrating but back I go. I learn that another fire has broken out, this one seventy miles east along the McCloud River railroad tracks. I am to proceed to this area. What a spot! The devil himself would have trouble getting to it, miles from any road.

Off I go. A fire truck with half a dozen men and the ranger of the district in his pickup had earlier been dispatched to this fire, but no word had been received from them since. The lookouts report: "She is rolling." Time: 4 p.m. No knots in the stomach, just a tough fight ahead now. As I ride along, I am keeping track of the fire near Mt. Shasta. It had not stopped at the big cinder cone Black Butte, but went right up the side! What was burning—pure rock? The lookout might have to be evacuated with such thick smoke. Also, one of the flanks had broken out and was headed for the town.

I am alerted by another report. A third fire has started south, down toward Redding. I am hoping it is on state protected land as we have all we can handle. That thought is soon swept away by the fact that I can hardly see where I am going. I am still over 30 miles from my destination, air so thick with smoke I can see only a few hundred feet ahead. Hurrying now is not only dangerous, but foolish. The giant fir and pine hang in the air like toys. I try to pass some tourists on the narrow, dust choked road. What possesses them, I wonder, to hang around! Ashes begin to slither down the windshield like raindrops. Take it easy, I say to myself. At last I can see the fire. What I saw when it first became visible made me wish I had taken up anything but fire control. There it was in all its demoniacle splendor! Two thousand acres more or less, miles from the nearest road in a brush covered lava bed that would rival a moonscape. Angry black clouds are rolling thousands of feet up where they develop their own cumulus or thundercloud. Somewhere out there were seven men with shovels and one fire truck, good only for transportation. Some odds! As the choking dust settles, I pull up behind a pickup. It is the ranger's. I begin to relax and reflect a bit. This had been an endless 70 miles, and it may be home for a week or more. Where is the crew?

Now to get what it takes to stop a fire of this magnitude and apply those forces at the right place and the right time. I reached for my canteen—phooey! Dishwater! I had forgotten temperatures were well over a hundred degrees. Sweat was the order of the day. I try radio contact; no luck. Time passes slowly.

It is about 6 p.m. and still no word from the ranger. He and his men have buried themselves in an impossible task. It will be getting dark and we need a fire camp, tankers and God knows what. We need them now! If the fire is not stopped tonight, it will come roaring off the lavas tomorrow and head into virgin timber. Out comes the trusty notebook and pencil. We'll need at least seven dozers, 20 fire trucks, 1,000 men, fire camps, fire bosses, first aid facilities, food and so on and on. I rush to get this over the air to the dispatcher. He is efficient, but polite: "The Black Butte Fire is threatening the town of Mt. Shasta and all requests for manpower and equipment must first be directed to that fire. Anything not needed will be sent your way. You will maintain radio silence until further notified." "Good God," I mutter. Later, as I prepare to wait, I stare into the smoke and gathering darkness and I am aware that something is beginning to materialize out there. It is the crew! Seven sad and grimy men straggle in, minus one fire truck. They are elated to see me but I realize I have only bad news to give them. I wait to hear their story. They had had about the same luck as a mosquito would have fighting an elephant. They had almost corralled the fire until it got on both sides of the railroad tracks. The rest of the time had been spent trying to get around it safely. With these erratic winds, you find yourself being chased at any moment through a land of sharp lava ridges, thorny brush patches and sandy flats. They had taken a short cut and the truck would arrive later. This was all laced with pungent remarks, while washing down C-rations with warm water.

Telling them about Black Butte's priority during this holiday weekend didn't improve their morale. They knew they would have to take potluck and like it. Night work would now be difficult. We needed a road and the crew must get some sleep as they would be needed at daylight. If we are to contain this sleeping giant, the ranger and I must make some plans. The time of darkness at a fire passes frightfully fast. We are working now against time, trying to hold it back. We start making plans. On a controlled fire during the night is the time you hope you have done your work well. I keep thinking of that magic hour of 10 a.m. when any known or unknown fire suddenly springs to life. Tomorrow we either break the back of this monster by that hour, or we better damn well get out of its way and try again; there's no compromise.

As the night hurries on toward midnight, we think we hear a familiar sounds Can it be—yes, they are coming! First two dozers on the backs of trucks fully fueled with two tough loggers to run them. Then, 40 or so men arrive in a clatter, with shouts of, "When do we eat?" The kitchen is going up, lighted by long strings of lights fed by a portable power plant giving sort of a carnival effect in the night. Men grunt as they carry tons of food. Some dig latrines. There is joking, shouting, pounding of stakes, all spiced with profanity. From out of seeming chaos soon comes a well organized fire camp. By 2 a.m. we have only one chance; we can put the fresh crew and dozers on the fire and try to turn it back on itself into the lavas. It is quite a sight to see two men leading the giant dozers into the smokey darkness, followed by a string of men each equipped with headlight and tools. If it were not such a grim business, it could be likened to a fairy tale with the gnomes going to work. I looked up to see them disappear in the distance in an eerie, red glow. They spent the remainder of the night trying to reach the fire because of bad terrain. Scouts and guides became lost and rimrocked in the lava reefs and deep fissures. They had to stop and save their strength for the next morning. All men and equipment were pulled back to the road about a mile away from the fire, there to make a stand where the lavas give way to virgin timber. And there I hoped we might be able to backfire.

The ranger and I spent some uneasy hours until near dawn. Our own small crew awoke with hearty appetites and enjoyed steak for breakfast. I always marvel at the amounts of food consumed in a fire camp. During these hectic hours of responsibility, I lose my appetite but consume gallons of fruit juices and milk. As we near the magic hour of 10 a.m., we observe the telltale signs of increased fire activity. Right on time, the fire moves out! First, the dense smoke that has clung to the ground all night lifts and thins. It is almost clear. A sudden breeze draws in toward the fire and now a solid wall of flame starts advancing, not too violently, just steadily. Then the anticipated happens. With a roar that shakes the ground, the fire leaps into the tops of the tall fir and pine. It takes but a moment to calculate the odds; out with the men, out with the tankers and dozers. Is it too late? The dozers manage to escape, but the tankers are soon blocked by falling timber and so am I! A quick decision is made for the ranger to drive the tankers off the narrow dirt road to a wider logging railroad grade. Here, their water can be utilized to keep the men and machines cool by spraying them and hopefully keep them safe.

Being plans chief, I must return to try to get in touch via radio. I rush as fast as sanity will allow back down the road. The fire winds have increased to almost hurricane velocity. Straight ahead is a large burning log across my path and another soon to fall. Should I turn around and try to reach the tankers, or get out and run for it? But if I leave my car, we will have no radio contact. I start backing, straining to see behind me. Storms of ashes, pine needles, twigs and large branches are flying in all directions. Now I am suddenly blocked as a tree starts falling behind me! I've got to go ahead and try to find a spot. My God, it looks like the main fire is coming up the road towards me and the heat in the car is getting unbearable. I can't sit in this damned oven and be roasted like a turkey! Deciding to make a run for it, I am through the car door and back again in seconds. The air outside is superheated and violently strong. I can't even breathe! In the instant I was out and the door was open, the car seat had ignited in two places. Where was my canteen? I grabbed it and drenched the seat. I was beginning to tire and streams of sweat were running down me. I needed my salt tablets but there was no time now to look for them. I must rev that car back, back, in and out of burned over spots—and time ceased to be.

I was aware of the pitiful wild things running for their lives, fur aflame. There was the smell of burning paint on the car. Would the gas tank explode? The tires go flat? Radiator dry up? At times like these the subconscious seems to take over. I revved that Chevy from low to second and then reverse. At one time it grew very dark and looked like a tunnel ahead. Finally it happened; the engine died. I slumped over the wheel, too exhausted to care. I don't know how long I lay there. Something kept saying, take slow, deep breaths, relax. My heart was pounding, beating hard and fast. My head throbbed. When I tried to raise it, my vision was blurry. Somehow, I groped for my salt pills and found them.

The violent winds had died. Everything was strangely still. When I looked out the window the thought came—this is what hades is like. I was alive, but I had felt the breath of satan!

Long after dark, when it had cooled, I started down the road in ankle deep ashes. What a depressing sight—the tall trees blackened, still smoking. Bits of fiery bark falling down from above like glowing, miniature waterfalls. I was getting uneasy now as to how the crew and tanker had fared; would they be there? Thank the Lord, they were there! Of course, shocked at my sudden appearance. I learned it had taken all of their water to survive. There would be no let up in this fight now, day or night, and by tomorrow we could have a city on our hands. We all settled down, more than ready for sleep.

Suddenly, all hell broke loose! There was a dull pitched roar; something flew in the air spreading an orange-red ball of flame. Another and another! We knew then that the fuel barrels stored along the railroad grade had become superheated and were blowing up. Being far enough away and tired enough, nothing could worry us now. With that climax, I fell asleep to be awakened near dawn. Something kept hopping over my feet. It was a little, homeless rabbit.

That morning, there was a welcome sound in the air. Bulldozers were coming up the road, clearing logs and debris as they came. When I walked that road again, I gave a loving pat to my paint blistered Chevy, She was given water—she started! (She continued to run the remainder of the fire season and I was reluctant to part with her when she was replaced.) In the next few days we would have 12,000 acres of fire to control instead of 2,000, but happily, the men and equipment were arriving and our troubles should be over. But would the summer ever end?

I was to learn later that somewhere along in this deadly game of my battle with the unleashed forces of nature, I had radioed in that I was trapped; I do not remember. Also, at that same time, my wife, who was at a movie had become so restless and worried about me she had left the theater, coming straight to the dispatcher's office to ask about the fire. They hurriedly told her everything was fine and got her out of there. She had never felt this way before when I had been away on fires. I like to believe her prayers helped me,

The local paper later reported: "Forest Service overhead were flown from the Wash-Ore Region to fight the Porcupine Fire. Six of the 18 Forests of California have major fires on them. All Forests in the Northern California Zone have been stripped of men. At the present time this is the worst fire season since 1924. The Weather Bureau reported 101 degrees September 1 and 2, the highest recorded for September since the year 1896." The other fire, Black Butte, had been stopped on the outskirt of Mt. Shasta at 1,350 acres.

MUD NEEDED

Don Bauer

The military discovered that it was not well equipped or trained to fight a mountain war in cold country because most of the officers and noncoms of the early 1950s had gotten their experience in the Pacific. These were marines, on coral atolls and flat, warm, balmy tropical climates. When they got in steep, mountainous, rocky snow country in Korea, they didn't know what they were doing. A fellow named General Bowser, a two star marine general from Pendleton, came up one day with one of his staff people. Just dropped in at the San Jacinto Ranger Station and described his problem. He had to train marines to win the war in Korea, and he needed training areas in rough, mountinous terrain, for bulldozer operators especially, and whatever you would call the construction people. We were the nearest mountain area that had trees and snow. Our policy in the manual was that we would cooperate with the Secretary of Defense with a memorandum of understanding.

We sat and talked and I asked them what they would do. Well, they were just going to chew everything up. So I explained that we have some guidelines, but would be glad to work with them. We had a lot of dead end roads that the CCC program had closed when World War II started, so I said that if those would fit into their program we would be glad to work with them and get them engineered so that they could get them tied in. He said that was just what they wanted to do. Then he said, "I need mud. Lots of mud, and ice. The worst possible things." That's pretty tough, but the thought came to me. Lake Fulmor had been in operation for a few years and we knew there were a bunch of beer cans and crap on the bottom and I told them we could drain Lake Fulmor and they could train to their little heart's content and skipload. Any silt or sediment they took out of there would increase the capacity. And after they were through we could fill the lake up again and you wouldn't see any damage. They could also maybe improve the underwater configuration; we had no beach area or anything like that.

When we went out there the lake had been drained and there was a little skin of ice where the puddles were and it was a blustery, bitter, wintery day. Within 20 minutes they had three dozers stuck. They wanted to build a bridge, too. That meant cutting timber. They wanted to cut all the pine trees down. I said no to that, but they could cut the live oak along the old road. It was mainly a firewood species. We marked some pine for stringers, but I made them make all their decking and the trestle sections out of live oak. They wanted hardwood experience, too. They had some hardwood in Korea, so they built a bridge. It was very popular. It helped me, because people could then walk around the lake.

BRASS HATS

M. M. "Red" Nelson

In 1951, as a matter of my training in the Forest Service, Chief Lyle Watts assigned me the job of working with Assistant Chief Earl Loveridge in making a general inspection of the California Region and the California Experimental Research Station (now the Southwest Station).

One of the things you quickly learned from Mr. Loveridge was that you paid no difference in work on a weekend than you did during the regular workweek. Thus, it was a Sunday that we found ourselves busy with inspection work on the Lassen National Forest. There was, however, one difference on that Sunday.

Bunky Parker, the forest supervisor and his wife, along with the wives of the staff personnel, had arranged a delightful Sunday picnic with families and with us inspectors as special guests. As we prepared to get in our cars to continue with the inspection, the small daughter of the ranger (age about seven or eight) came over to where we were loading and very politely asked one question. It was, "May I please see your brass hats?"

HUNTING ON THE MIGHTY COC

Jay H. Cravens

In May 1951, I was offered a promotion to district ranger and moved to the 425,000 acre Long Valley Ranger District on the Coconino National Forest—the "Mighty Coc" as it was called.

Long Valley was the place I was able to fulfill my wartime dreams of real hunting. I harvested elk, deer and antelope each year. I acquired two .30-06 rifles through the National Rifle Association and the Director of Civilian Marksmanship. Both were bolt action models. One was a Springfield and the other an Enfield. I traded the Springfield for a K-4 scope, had the "ears" ground down and barrel shortened on the Enfield so I could carry it in a scabbard when I was on horseback. Later I had it bedded in glass on a sporter stock to ensure more accuracy.

I enjoyed tracking and hunting elk late in the season, preferably during a snow storm. One season I was given the assignment of escorting one of the Outdoor Life editors, Charles Elliott, on an elk hunt in the Mogollon Rim area. Unfortunately, my elk hunting permit was in a different part of the district. I acted as a guide for almost a week while he observed our management programs. He visited with hunters and took many photographs. The elk he finally shot was a huge bull, but it had antlers on only one side. Nevertheless, he was proud of it. After we delivered it to a processing plant for cutting, wrapping, freezing and transport to Atlanta, Charlie went on to the North Kaibab for a deer hunt. I was then free to go hunting. My crew was sympathetic and took over while I went hunting on the last day of the season.

I asked Alvin Teague and Ken Cook where to hunt. They both suggested I go down and have a cup of coffee at daylight with Irvin Walker on the "poor farm." That's what Irvin called his log cabin headquarters. It was snowing and had accumulated about a foot of fresh snow. I visited with Irvin about his cattle and plans for the winter, had coffee and then asked him where I could find a bull elk. He suggested I cross the fence out his back door and walk along the fence until I found fresh tracks. It was just breaking daylight and off I went. The snow was in my face as I hunted up wind. I had gone about 300 yards when I came across the tracks of a large bull elk. The tracks were so fresh they were literally "smoking," and heading into the wind in the direction I was going. I went another 500 yards and through the snow I saw a man and woman coming towards me. They were back tracking the elk. They asked if I had seen it, I crossed my fingers and said "no" and wished them luck in tracking it down. I was not about to give them a lesson in tracking on that last day of the season. Within another 200 yards I saw where these two people had crawled through the fence and started following the tracks. Snow fall was increasing and daylight had arrived. In a short distance the elk circled to the west and climbed a low hill. The tracks disappeared into a young pine thicket. I hunkered down and scoped the area and finally located the elk and dropped it with one shot at less than 100 yards away. I never shot a deer or elk that was more than 100 yards away. The two hunters I had met earlier came running up to me and asked, "How did that elk get over here?" Then I gave them a tracking lesson. They went away bewildered and scratching their heads.

One deer hunt was made with Roy Barnes, our timber stand improvement foreman. We hunted on the adjacent Mormon Lake District where the ponderosa pine merged with the pinyon-juniper type. We attempted to keep one another in sight in dense juniper. I jumped two large mule deer with racks like an elk and took a shot. The brush was dense but I had the cross hairs behind one deer's left front shoulder and fired. The other one ran towards Roy. Mine never appeared to slow down and I found no sign of blood. About then, another large buck came busting out of the junipers. I sighted, fired and the deer tumbled head over heels. I put my deer tag on an antler and then field dressed it. Still no sign of Roy. I got to thinking how could I have missed that first deer and I started circling around for sign. I found it folded up and partially obscured under a juniper not too far from where I had first sighted it. I field dressed it but had no tag, in the land of many game rangers! Still no Roy! I spent a couple of frantic hours locating him and led him back to tag the deer. He wanted the meat and was glad to have that chore done. That relieved my "guilty conscious." All this brought to mind the transgressions of Ranger Vince's assistant on the Kaibab.

During dove, duck and goose seasons, I had good jump shooting on livestock ponds. Once in two shots, I dropped more than my legal limit of ducks and had to call on Bill Phillips to come to my rescue. I still remember his "yuk, yuk, yuk." Duck hunting moved to the lower country in the Verde Valley when the ponds above 7,000 feet froze. In that lower area there was good Gambel quail hunting. Earlier I commented on my success in squirrel hunting. In the ponderosa pine forest the turkey season was also good. We used a light center fire turkey load. Shotguns were not permitted. I brought down my first turkey with a shot that connected at the hip joint. None of that 25 pound gobbler was damaged. The shock of the bullet dropped him on the spot. My next turkey was a young six pound hen that I spotted at a livestock tank. I jumped out of the pickup truck, crossed into the woods, loading a shell as I went. Unfortunately, I had a 220 grain "Silver Tip" elk load in my jacket pocket and loaded it instead of a 110 grain turkey load. I hit a small hen dead center and feathers, blood and guts flew in every direction. I took home a small wing and drumstick from that mistake.

Elk hunts took place in late November, usually after we had burned the accumulated logging slash piles. We finished burning one Friday in a light rain. Elk hunters were coming into the area while we were burning in advance of the opening of the season the next day. Many came from Phoenix and were not well dressed for the high Mogollon Rim country. From the rim the ridges extend north to lower elevations. The only way out was back to the rim road. On Saturday a light rain turned to heavy snow on the rim. By Sunday night there was a good 30 inches of snow in the area. The weather cleared and the temperature plunged to below zero. We had moved to Flagstaff for the winter. Sunday night I received a call from the sheriff's office. We had over 1,000 hunters stranded in the rim country and down under the rim on the Tonto. Monday we started a full scale rescue operation, involving bulldozers, snow plows, 4-wheeled equipment, airplanes and rescuers on skis and snowshoes. The sheriff was in charge and his efforts were coordinated with the Forest Service, State Game Commission and Army. Our headquarters was on the rim at General Springs cabin. A *Life* magazine photographer and writer recorded our rescue efforts. Within a week we had most everyone located and out. Hunters tried to return as fast as we could extract the stranded ones. It was a real mess! Five hunters perished, four in the rim country and one on the Tonto. All who stayed with their camps were safe. Those five had tried to walk out or were unable to find their camps and never made it.

THE FEATHER FALLS TRAIL

Bill Tikkala & Lyle Klubben

FROM BILL

A new crop of junior foresters helped build a portion of the California Riding and Hiking Trail around Feather Falls on the La Porte District, Plumas NF, back in the winter of 1952. We called our tent village, Max Peterson's "rag camp." Max was then assistant forest engineer and in charge of the project. We (you and I) progressed from trail grubbing to powder monkey to tree falling, only to drop a snag on the chain saw, the only firewood cutting tool we had in camp.

Then, there was Johnny, the tough as nails trail superintendent and the overweight cook from Portola Valley who had to ride a horse into camp (the rest of us walked), and got knocked off by a tree limb. Then, there was a heavy snow that collapsed one of our tents, causing some mad scrambling by its occupants in the middle of the night. It seemed like everyone was waiting for a call from their home districts telling them that timber sales operations had started and they wouldn't have to return to the "rag camp."

FROM LYLE

Yes, Bill, the Feather Falls Trail job remains a vivid memory after all these years, though many of the names of those in the motley crew (JFs, scalers, fire people and others who also needed winter financing) have left me. While Murphy's Law (anything bad that can happen will happen) was in full swing in the "rag camp" many of the memories are good and in the end we had built a fine piece of trail.

I must add some meat to the chain saw saga. As I recall, we had planned to fell the tree up the hill until the boss, Johnny came along and forcefully convinced us that it would fall downhill. I think that was wishful thinking on his part, since that would take the tree off the trail right-of-way and make further cutting unnecessary. Regardless, I was the lookout up the hill from the tree, which was a limby snag about 150 feet tall, and you Bill, ran the saw. I still have a clear recollection of the tree trying to fall uphill, teetering and binding the saw so you couldn't remove it, doing a full 130 degree turn over my head and eventually falling downhill, sort of jumping off the stump and landing on the saw. You had wisely departed the scene once you were able to decide which way to run. Johnny arrived back at the site about that time and said, "Just the way we wanted it," and you replied, "Is that the way you wanted the saw?" No answer from Johnny!

Speaking of Johnny, I heard later that he had a reputation as a tough boss, as you implied earlier. I was told that during his CCC foreman days he had three crews at all times—one coming (hired), one going (fired) and one working (presumably temporarily). Be that as it may, I liked him and was especially appreciative of the fact that he reported to my Ranger Ben Beard, that I was a good worker. Somehow, such praise from people like Johnny carried more weight with old-time rangers than similar praise from others.

Other remembrances:

> Bob Powers spilling all our syrup on his pants one cold morning. We soon found out that pancakes are not very tasty without syrup. Doesn't do much for pants, either.
>
> The powder monkey we worked for who had a gimpy leg and who calmly limped down the trail after we had detonated a blast, with his shovel protecting the back of his head as rocks and other debris rained around him. Apparently, he didn't want to go very far because walking was hard for him. I don't recall that he was ever hit by anything, but then, I was hiding where I couldn't see much.
>
> Looking up from my bunk one night and seeing our tent pole on fire. These were wooden poles cut from local trees (we didn't pack in proper tent poles), and someone had fastened the stovepipe to a pine pole which had eventually dried out enough to burn. No harm done—just a rather normal occurrence in "Max's rag camp."

NON-TRESPASSERS WILL BE VIOLATED

Jack A. Godden

I moved on to Kentucky from Tennessee. I may have been threatened a few times by folks with guns in Tennessee but never shot at. Folks down there called me "the skinny kid with glasses. Don't shoot at him because he's ok." (Melvin, my helper and native had made sure the natives knew I never passed any words to any law about what I found in the woods. This was whiskey making country!)

The Red River Ranger District on the Cumberland NF (now Daniel Boone) anchored the north end of the forest; some 135 miles long, 25 miles wide, about 135,000 acres, scattered over nine counties. Two GDAs were the backbone of the district. We "professionals" came and went but they knew and had the respect of the locals and were the successes for the Forest Service image in the area. This story relates to one of the GDAs, Clarence Henson.

We had a Sunday afternoon fire down below Frenchburg, Kentucky. Clarence Henson and a warden crew were on the fire, and I joined Clarence to do some investigative work on who started this "smoking game" fire. It was up the hollow from a farmhouse and the likely first witness was a young 14 year old boy in bib overalls and no shoes. We had permission to go through the barnyard to the fire, so we asked the youngster if he would mind answering a few questions. He agreed and in the middle of the county road, Clarence and I interrogated him. He didn't supply us with anything positive so we told him we were going to talk some more with the warden on the fire.

Clarence and I proceeded over the front lawn to the barnyard gate some 150-200 feet away from the back of the house. As I followed Clarence through the gate there were shots from a 22 caliber rifle. One or two bullets hit the side of the barn. I turned to see an elderly woman on the back porch unloading a Winchester pump 22 in our direction. Clarence's first response was: "I'll split her head wide open with this pulaski (he was carrying) if she comes much closer with that rifle!" Mine: "If she gets much closer, you won't have to worry." About as quickly as she pumped three to four rounds at us, the shooting stopped; the teenager had taken the gun from the woman. He hollered to us asking us if we were OK. We answered and hoped the excitement was over. We asked if we could get our crews out the same way they went in to the fire and he gave us his OK. There was no further interrogation of the boy or his family that afternoon.

The next day, Monday, Clarence and I visited the county courthouse to talk with the DA about the family and the shooting. He thought us lucky to be alive; the boy's grandmother and his father were mentally ill and he was taking care of both of them. "She could have killed you both. We would have to send her to the state mental hospital, split the family up and it would cost the state!"

TALES ABOUT A COW, A CALF AND A WEASEL

Bob Gray

Rangers get called on to do everything. Like rescuing a new calf in the deep snow. "Just call the Forest Service, they'll do it," and they usually do.

It was late fall and three feet of snow had covered the McCloud area when Paul Friday, the fire control assistant, got word of a cow and her newborn calf floundering around in the snow about four or five miles out of McCloud. The "weasel" wasn't an animal, it was an Army surplus, open air, track laying vehicle which the Forest Service had acquired. Paul and I took off from the ranger station headed toward Mud Creek in the weasel. Sure enough we found a furrow in the deep soft snow, obviously made by a cow and calf. We followed the tracks in erratic circles until we found the mama and her baby under a pine tree. If there was anything that cow didn't want it was anything to do with us. We figured on herding her and the calf back to McCloud in the tracks we'd made coming out. Cooperate? Not that old heifer! She glared at us, shook her horns at us and refused to do anything but circle those trees with us in hot pursuit in the weasel. The calf watched in curious amazement. What kind of a world have I been born into? he must have been thinking.

After failing miserably to herd that old gal in any direction, I told Paul to let me out and I'd lasso the cow, then we could lead her home and the calf would follow. Well, I ran at her with my lariat swinging over my head. No cow is supposed to react as she did when the noose settled around her horns. Instead of pulling away from me, she charged. Now we were in three feet of snow, and I wasn't prepared for a charge. As I back pedaled in my tracks, she came on like a locomotive, hit me in the chest with her head, knocking me backwards about five feet on my back. Fortunately for me, she also stumbled and fell, giving me a chance to scramble to my feet. What was Paul doing all this time? He was laughing so hard that tears were rolling down his cheeks! I think he did manage to ask me if I was hurt before continuing with his uncontrolled laughter. When I finished brushing the snow out of my collar, I climbed into the weasel, still holding onto the rope. Since I had no injuries, no thanks to Paul, the situation began to appear funny to me too.

Old bossy wasn't about to be led by that weasel. I guess she thought we were going to take her away from her baby so she laid down. Well, we thought, if we tie the rope to the hitch she'll soon get up and walk. She didn't. After the rope stretched, her neck started stretching too. Did you know that a cow's neck can stretch about twice its length without any bad effects? We dragged her through the snow but she would not get up. The calf stayed under the tree. We simultaneously thought, if we load the calf in the weasel, that old biddy will follow us. The calf sat in the back seat and we finally got the cow on her feet An hour or two later we got to the barn in McCloud with our livestock. All of us totally tuckered out except the calf who was probably beginning to think that this is a pretty interesting world.

The rope was still around the horns of the cow when we released her into the barn, and it was too dark to try to get it off then. I'll get it off in the morning, I thought, and I did, but not before she had her last fling at getting even for the injustices of the day before. When I went into the barn I took the end of the rope and snubbed her to a post so I could get the rope off without her getting me. She pushed her head against my ring finger and mashed my ring so that I couldn't get it off. I finally succeeded in getting it off by straightening it with a pair of pliers.

The owner of the cow and calf came and picked them up without a thank you or any comment.

THE MONROVIA PEAK FIRE

Carl Wilson

Editor's Note: The year 1953 was a disastrous fire year on the Angeles NF. Carl Wilson, Arroyo Seco Ranger at the time, gives us a firsthand account of major outbreaks with emphasis on the Monrovia Peak Fire. Somehow, there is always a touch of Hollywood about Southern California affairs, even including its fires. In this instance we have Angeles Supervisor Bill Mendenhall telling his ranger: "Carl, either hide that fire or put it out. The Rose Bowl show must go on." –E.J.

It was years ago, but it seems like last week when the historic Monrovia Peak Fire roared down Big Santa Anita Canyon toward the cities of Monrovia, Arcadia and Sierra Madre. The disaster started during extreme Santa Anas, near Newcomb's Saddle in the West Fork of the San Gabriel in late afternoon on December 27, 1953. Before the nightmare was over, more than 14,000 acres of precious watershed were blackened; 30 special use cabins were destroyed; more than 2,000 firefighters had Christmas vacations demolished but no lives were lost.

The 1953 fire season on the Angeles Forest had started out with a bang during the first week of May when a 100 acre fire burned out the Crescent watershed within the City of Monrovia. By mid-June, we were again alerted to the potential for a bad year, when a 300 acre fire burned over Gold Canyon near the C. B. DeMille Ranch.

The July disasters came next. On July 2, the Sulfur Fire started near a small campground on the Valyermo District, and before it was over, nearly 40,000 acres of the Valyermo were charred. Even before the Sulfur Fire was controlled, the Maddock took off on the Arroyo Seco near Fish Canyon and headed north. I rolled for Duarte, and we tried to gain first night control. By the second day, we had a line around it, with the help of high-powered overhead from the RO, including Norm Farrell and Paul Stathem.

The Angeles had a brief breathing spell until the first week of September, when the historic Fish Fork Fire started near Big Pines. Unfortunately, our one and only contract helicopter had been wrecked on a small fire on the Baldy District. Hence, no one could reach the Fish Fork Fire in its early stages. Before it was finally controlled, more than 20,000 acres of very valuable watershed in the San Gabriel were denuded.

Most of us thought that the fire season was over, but December came along with its strong Santa Anas, and the Barrett Fire took off in Barrett Canyon near Mt. Baldy Village. The fire threatened dozens of homes in the canyon and rolled downhill under the influence of cold, very dry winds. The Barrett Fire burned into the San Bernardino Forest in the lower portions of Cucamonga Canyon. Within several days this new disaster was controlled and everyone was convinced that the 1953 fire season had to be over. The Arcadia Warehouse was trying to dig itself out from under the piles of dull tools, dirty blankets and cooking outfits.

Then, along came December 27, just two days after Christmas. My parents had come down from Idaho to view the Rose Parade (with tickets for the stands) and enjoy New Years with us. We had gone to dinner in LA and had just finished when I noticed the dust filled air and wind whipped trees. Just about the time I said, "Oh! Oh!" FCA Loy Winters called and said, "We have a fire and it looks bad!" We left the dinner table in a hurry and headed for the Arroyo Seco RS. I drove to Red Box after checking with Angeles FCO Harry Grace and to my dismay, found that it was hard to stand erect in the vicious Santa Ana. My windshield was so pitted that it had to be replaced after the fire.

By now, I had word from Harry and Dispatcher Virgil Shoemaker that there was no point in being at Red Box. The fire was roaring down canyon toward the three cities of Monrovia, Arcadia and Sierra Madre. I returned to the front country and met Harry at the dispatcher's office in Arcadia. This became the fire headquarters and we had clearance to use Arcadia City Park for our fire camp.

Selmo Lewis, then DR on the Baldy District, was designated zone fire boss on the West Fork side of the fire; Lynn Biddison (my assistant DR), his plans chief and I became zone boss in the front. Harry Grace served as overall fire boss. As it turned out, Selmo didn't quite make it to the Monrovia Peak Fire because his "controlled" Barrett Fire broke loose. Lynn quickly received a "fireline promotion" to zone fire boss.

At about this time we began to worry about the people who lived in cabins in the Big Santa Anita Canyon and Bill Adams and his pack outfit at Chantry Flat. Supervisor Bill Mendenhall and I drove to Chantry and met Bill Adams, who had just come out of the canyon after alerting weekenders and yearlong residents. He said one old-timer refused to leave. Then we began to worry but the old man arrived at the trailhead after deciding the fire was moving faster than he liked.

Adams decided to abandon Chantry, but to leave his horses and mules there. There just wasn't enough time to trailer them out. The forest supervisor and I decided that it would take about one or two hours before the fire hit Chantry. We returned to the fire headquarters and I began to estimate the number of tractors and men needed to keep the fire from crossing the main divide between Big and Little Santa Anita. (It had all been pre-attack planned.) Unfortunately, by the time the calculations were finished, the entire canyon had blown up and the fire had crossed over Chantry Flat and was moving downhill toward the three cities. Meanwhile, thanks to Harry and Virgil, LA County and the local fire departments were heavily involved in trying to save structures.

During the next three days we were able to build control lines on the North (West Fork) side, east sector, and to prevent loss of any structures in the front country. We now had more than 2,000 people committed to the suppression effort and dozens of fire trucks (now fire engines). Nolan O'Neil said he had about five acres of fire trucks of assorted colors at Mt. Wilson helping him protect the observatory, radio and TV stations. The Santa Ana had tapered off, and it seemed that control was imminent. Then, the wind picked up and our weakest section was the chunk of country between the Mt. Wilson Toll Road and Sierra Madre. Harry Grace told me, rather directly, on the morning of December 31, "Get out there" and expedite the backfiring. Our burning was going well until late afternoon when we reached a small saddle above the monastery in Sierra Madre. As our firing reached the saddle, the fire spotted through and we nearly lost several Mexican Nationals, who could not understand our panic stricken Spanish! Finally, we reached the monastery and I was greeted by Regional Forester Hendee and Bill Mendenhall, who said, "We got it made."

I went back to Arcadia and Harry told me to go home and spend some time with my folks, since it was New Year's Eve. That I did, but didn't even have time to change clothes before a call from Virgil Shoemaker told me the fire was loose again. It was threatening to move uphill toward Mt. Wilson. The headline in the *Pasadena Star News* said "Wilson Threatened" but only Bill Mendenhall and I knew that he had told me we had to get that fire under control before the Rose Parade and the Rose, Bowl Game—or else!

In the meantime, Lynn Biddison was doing a fine job of controlling the fire in the West Fork-Monrovia Peak zone.

We stewed a lot during New Year's Eve and made plans for taking care of the two major slopovers below the Mt. Wilson Toll Road early the next morning. We were fortunate that Thelmas Biddison, Lynn's and Don's father, was available. He, with his special crews, was able to control the west side of the Monrovia Peak Fire before noon on January 1, 1954. The Rose Parade and Rose Bowl Game went on despite the smoke in the background. Prior to control, our huge fire had been the "best show in town" for the Los Angeles Basin. The threat to the TV broadcast of the parade and game was gone and Wilson was no longer threatened. We officially called the fire under control on January 2, but were still mopping up when the rains came on January 10. The rains fell in Arcadia, but several inches of snow fell on the Spring Camp fire camp, where Biddison and his bivouac were located.

After our holiday disaster was controlled on January 2, 1954, Hatch Graham, then the Santa Anita ranger, went into the canyon to survey the damages. First, he found that the horses and mules at Chantry Flat had survived, primarily because the area was "open and park like." Secondly, he discovered that 33 special use cabins in the canyon had been destroyed. On the bright side, about 100 had survived the fire. We reported that most of the "survivors" were the result of adherence to the 30 plus feet clearance requirement, which we had insisted on. He said some cabins burned because of wood piled against exterior walls.

When the rains and snow fell, the loose soil on the denuded slopes and in the canyons began to move downhill. Chantry Flat was soon closed, and mud was moving downhill into homes in Arcadia and Sierra Madre, all of which had been spared by the fire. It was soon apparent that emergency flood control measures were going to be needed. A public meeting was held in Arcadia at the insistence of local citizens and politicians. This was the initial session of what later became known as the very influential "Los Angeles County Watershed Commission."

The snow in the high country also had an effect on the dissolution of Lynn Biddison's fire camp. Vehicles without chains had trouble evacuating the area. Lynn's exotic WKFL (Fountain of the World) robed men and women firefighters had to walk to Red Box on their callused but very cold bare feet.

Meantime, we had more than 50 compensation cases—many of them caused by poison oak, but one serious case of pneumonia as a result of the cold weather.

One individual made the mistake of volunteering his services while on leave from the National Park Service. Jim Ruppelt, who retired in 1979, was hired as a scout on a fire time slip and then stayed with the USFS for the next 26 years.

SQUATTING AWHILE ON THE "UNCLE"

Jack A. Godden

One of Harry S. Truman's last strokes of his many pens in 1952, was to sign over some Tennessee Valley Authority land to the Forest Service in exchange for the national forest land flooded by construction of the TVA reservoirs in the late 1930s and '40s. By this executive decision some 20,000 acres along the Hiwassee River combined with lands from the older Tellico and

Packing fire equipment at Orleans Ranger Station, Klamath NF, California - 1927.

Fire Guard Station, Klamath NF, California - 1923.

Ocoee Ranger Districts created the Hiwassee District on the Cherokee National Forest. I arrived there as a junior forester in 1954. I was the third forester to serve under the second Ranger, Alden Norris Quam, since the district was established.

Our land status charts had been quickly amended and copies of the legal descriptions provided to us. Only this "recently acquired" land had never been surveyed, marked or posted on the ground after purchase. There had been no surveys of what TVA acquired. The owner(s) settled and exchanged titles and papers. Our job was to go find it. Its condition on the ground is what this story is about.

It is believed the residents remaining in the few home sites along the river were given a reasonable time to vacate their homes and land. They did, but an empty house with nobody in it was too much of a temptation to poor folk without a roof over their heads. What Ranger Quam found were the remnants, both of homes and people living in them. The task given to him by Forest Supervisor Bill Paddock was to clear both off the land. It was easier said than done.

The ranger gathered his intelligence usually from the Webbs, the local store owners (his wife, the postmaster, he, one of our fire wardens) who the locals owed money to. Over a Coca Cola or two you could usually find out who and how many were living in the wood framed structures that most folk would call "shacks." By the time I arrived in October '54, Norris had successfully moved four or five families by legal eviction notices. Very little salvageable lumber or glass windows and hardware were usually left, just the shells waiting to fall. Even the rats had moved.

One house with four children and a man and the mother of the children remained. They had wintered over in '54-'55. My first scene of poverty was this family. The children were barefoot, in rags, and the youngest, diaperless, her legs wet and cold from urine. I was standing there cold in my Army O.D's, with lined field jacket and hood, wool sweater and shirt, in their front room. Their only heat came from a four plate stove in the kitchen. The father was out "with the boys." He had no vehicle so he probably was picked up and now sucking on a mason jar of "shine."

A six months notice had been served for the family to move by April 1, 1955. As winter subsided the ranger found a job for the man of the house on one of the county road crews. He even found him transportation to the pickup site. He worked one day, didn't show on the second and quit on the third day of his starting week. The mother and her kids were still no better off. The father seemed worthless. Ranger Quam verbally repeated the emphatic order to the father: "Out by April 1!" They moved. Where they went we never knew.

Within a few weeks after their departure my task was to clean up the homesite. The "lazy cuss" had torn the siding off the front of the house and burned it for fuel. It left a strange sight—the standing two by fours supporting the roof were coated with layers of old newspapers. These had been glued to the inside walls to cut the winter drafts from the outside. How they obtained all this newspaper was a mystery to me as Webb was probably the only person in the valley that subscribed to a newspaper. There were only a few porch boards remaining to the front opening. The two doors and any hardware were gone. I was surprised he hadn't burned the place down for the nails. With my one helper, we soaked the structure with coal oil and torched it off. There was no forest fire threat that day and within a couple of hours we banked the few burning timbers for the night. Within a week I was back supervising a contracted dozer burying the site, its trash and others along the river road. Following this cleanup I laid out the planting sites for machine and hand planting for the following spring. (The tall grasses along this bottom land were "chigger" heaven as my body that summer could testify to.)

The spring of 1956 saw the major planting of all the once tillable bottom land along the river with loblolly pine. Several troops of Boy Scouts from Chattanooga assisted in the hand planting; most was machine planted with excellent survival. On my last personal visit to the district in the mid-80s the ranger told me these pines had approached 100 feet in height, some were 18-20 inches in diameter. I hope the children I met and their children and those of other families we evicted from their "squatter" homes grew and were cared for as well as these planted pines.

A MIGHTY FINE PILOT

Ralph Bangsberg

There hadn't been too much flying lately in the Shasta-Trinity National Forest. The late deer season was upon us, with hundreds of hunters jamming every campsite and expanding elsewhere into dry and dangerous areas. As fire control officer on the Shasta-T, of three million or so acres, I thought, wouldn't it be nice to reach this many hunters with a fire prevention message! I knew the helicopter pilot would welcome a little flight time. We could take along a megaphone to call down to the hunters in the most remote places. It would take days to cover this same ground using ordinary transportation.

Let's go! We load on drinking water, good maps, megaphone, shovel and axe, message tubes, in case we want to get a message to ground as we could not land in most places; also a basket stretcher. We were now a floating fire prevention unit, an emergency rescue vehicle and a flying fire crew. We really felt we were ready for anything!

The pilot fires up the helicopter. We strap ourselves in while the noise of the motor drowns out all attempts at conversation. The machine begins to shake and jump as though it were anxious to get into the air. The pilot nods at me. I return the nod. In an instant the ground pulls away from us as we soon soar high above the tree tops. The motor responds with an easy flow of power as we head for the start of our patrol area.

I had chosen one of the most remote areas north of the great Shasta Lake, known as the Clear Creek area. Roads were few and navigable only by 4-wheel drive vehicles. In all, a very rough, steep canyon area with no open areas, heavily timbered. Yet, somehow the hunters were there! It takes only one careless person and the whole wilderness can explode in fire. We had flown perhaps an hour, contacting dozens of campers. It would have been better if we could have stopped to chat, but no way? I was feeling mighty good about the work we had done and leaned over the pilot to shout, "Good show!" when a terrible clatter came from the motor, a sound much like that of someone throwing a handful of rocks on a tin roof. To say I was startled is to put it mildly. Just then the noise came again. I looked at the pilot. Talk was out of the question. He was pointing to the gauge that shows the cylinder head temperature. It was rising like a hot air balloon! Somehow, he got it across to me that we had about one minute—60 seconds of flying time, as we had lost both of our fan belts! These, I learned later, cool the cylinders, and we were done for if we didn't land. Land? Where? In the steep canyon, or a tree top? Suddenly I remembered a narrow logging road and a small spot at the end of it. It just might be big enough. All I could do was point the pilot in the right direction. At about the same time, I grabbed the radio mike and gave the usual cry for help: "May Day! May Day! This is helicopter number one having engine failure. Will attempt a landing at the end of the Clear Creek logging road." No more time for messages.

In a few seconds we saw the narrow little logging road and the tiny opening at the end. Before I knew it, branches and leaves were flying as the chopper cut into the oaks. Smoke and heat were pouring out of the engine. We landed with a bounce! The pilot fell out and rolled on one side and I on the other. I realized I had a shovel in my hand! We wouldn't go near that hot vehicle, expecting that at any moment it would explode in flames as the two saddle tanks of fuel were dangerously close to the motor. Now, we couldn't use the radio, so spent a good long time wondering if anyone got our message. I only know I was blessing a mighty fine pilot for a remarkable landing.

The helicopter gradually cooled down and we were overjoyed to finally hear vehicles grinding up the long grade. When they arrived, they were thankful also, to see us unhurt. The mechanic went to work and by evening they were able to fly it back. I had had enough and rode home on terra firma.

CATCH THE THIEVES

Maren C. Scholberg

The Christmas season was not always a time of good cheer and relaxation for my husband Carl. He dreaded the knavery of some of the Christmas tree thieves and in the 1950s there were a good many of them trying to steal trees from government land. We were on the Sierraville District of the Tahoe National Forest when this incident took place.

Some of the Forest Service employees had discovered a large number of Christmas trees stashed in the woods, quite hidden from public view. They knew the trees had been cut from Forest Service land and that some night soon the cutters would be coming in with a large truck to haul them out. So, they planned an elaborate scheme to catch the thieves in the act.

There was a Forest Service telephone line that crossed the road which led to the stashed trees. Red Giffen, the assistant ranger, devised a plan whereby a small telephone wire was strung across the Forest Service road so that it was about six inches above the roadbed. It was connected to the Forest Service phone line which ran to our house. If a vehicle was driven over the road it would break the line and by listening in on the Forest Service phone in our house we could tell when that happened for there would be a constant buzzing on the phone.

We had planned a station potluck at our house for the evening and everyone was gathered for an enjoyable time with the added excitement of knowing that the men were going out on what could be a dangerous mission to catch the Christmas tree thieves. The men took turns listening in on the phone every 10 to 15 minutes to see if the buzzer had been set off. Nothing happened for over an hour and we wondered if this was not going to be the night they would pick up the trees. About two hours into the festivities Red picked up the phone, listened briefly and called out, "Let's go! The buzzer is ringing!"

Five men dashed off in two pickups and raced out of town. Those of us left at our house had a hard time waiting for their return. We were anxious to know just how they had cornered the thieves, how many of them there were and what the culprits had to say about being caught red-handed!

Well, five men returned looking a little sheepish. It seems that a porcupine had walked across the road and had broken the line which set off the buzzer! All their ingenuity and well laid plans were to no avail. No culprits were caught.

Those Christmas trees are still piled there and were never moved. And although the men had a good idea who the cutters were they never had a chance to "get the goods" on them.

EARLY DAYS OF AIR TANKERS

Carl Wilson

Although the dispatch of air tankers on California's wildfires is almost automatic in the 1990s, a few of us old-timers can recall when the dropping of water or fire retardants was just a dream. Some scientists said it couldn't be done. Yet the concept has been in the back of our minds for at least four decades. I recall in early 1954, after the notorious Monrovia Peak Fire on my Arroyo Seco District on the Angeles, that a gentleman came in while we were mopping up and asked, "Isn't there some way that you can drop a fire retardant like sodium bicarbonate on fires, as bombers drop bombs in support of ground troops." He went on to point out that, "You are using the same primitive hand tools to put out these fires that you were using nearly a half century ago." Somewhat defensively, I replied that some physicists at the University of California at Davis had proved that liquids wouldn't get to the ground in any significant amount if dropped from a fast moving airplane. Later that month, I heard from my L. A. County friends that they had seen an operation at Rosamond Dry Lake where a DC-7 made four free fall water drops over the flat surface of the lake, and water did reach the ground. This had been triggered by actions of the Douglas Aircraft folks who earlier in 1953, had dropped ballast water from another DC-7, which resulted in a strip of water 200 feet wide and almost a mile long on the airport.

This was so exciting that Operation Firestop, a giant cooperative R&D venture in 1954, "tested the feasibility of holding small fires in check by dropping water or chemicals on them in free fall." Some of the drop tests were made on Camp

Pendleton, the site of Firestop. I was fortunate to be host for the first such drop on a wildfire. At that time, I was responsible for a large portion of the Jamison Fire on the Cleveland, and the Firestop people asked if they could make a test drop just below the Main Divide truck trail. I gave them an OK, but they said, "You must remove your firefighters from the line for safety reasons while we are dropping." So, a modified TBM made two drops on September 1, 1954. Sad to say, our line was lost when the wind changed and the fire roared downhill. This calamitous event demonstrated the truth of the statement that aerial attack can supplement or support, but not replace ground firefighting forces.

As one might guess, my feelings about tactical air support were largely negative until I transferred to the Experiment Station in Berkeley in early 1956. Then, I found to my surprise and later pleasure that as project leader of fire control research, I was responsible for R&D work with aircraft which would be dropping water on fires.

So, there I was. I learned that Joe Ely on the Mendocino had been working with the operators of Stearman crop duster planes, and during 1955, had shown that the planes could carry and cascade as much as 120 gallons of water at a time. Trials on several wildfires were so promising that in 1956, thanks to "Red" Nelson in the RO and Joe Ely and Bob Dasmann on the Mendocino, a fleet of seven Stearman and N3N planes became part of California's firefighting force. During that year, these planes successfully fought 25 fires from the Oregon border to the Mexican border; they cascaded 100 to 150 gallon loads of sodium calcium borate and plain water. More than 1,000 separate drops were made which led to the first major publicity for the air tankers. Besides, it was the first large scale use of sodium calcium borate from air tankers, and the press quickly coined the term, "borate bombers." The name seems to have stuck for many years regardless of whether the substance dropped was borate, bentonite, ammonium phosphate, or ammonium sulfate.

And, as they say, "the rest is history." It is a rare wildfire in California that doesn't have at least one air tanker sent to it.

EARLY AIR TANKER TESTS IN CALIFORNIA

Jack Horton

A research project was conducted in 1954, to develop new wildfire suppression techniques and to test equipment needed to utilize these techniques. "Operation Firestop" involved numerous federal, state and local wildland fire protection agencies in California, as well as several support agencies and organizations and the University of California. The U.S. Forest Service, California Division of Forestry and the Los Angeles County and City fire departments were among the key agencies participating. Field tests were conducted at the Camp Pendleton Marine Corps base in Southern California.

Use of aircraft in wildfire suppression operations was an area of study which received considerable attention during "Firestop." Aircraft had long been used for detection of wildfires, reconnaissance and delivery of smokejumpers and their gear. "Firestop" evaluated use of helicopters to drop water on fires, lay fire hose, ignite backfires and deliver personnel and equipment to the fireline. Fixed wing aircraft was also evaluated for water dropping. One of the first opportunities for a fixed wing aircraft to make trial water drops on a California wildfire occurred in late 1954, on the Jameson. This fire occurred on the Trabuco District of the Cleveland National Forest. This was a large fire which burned in heavy brush on very steep terrain. I was assigned to the fire as a helitack crew member. This helitack crew was assembled at the fire from members of our sector team, as full-time helitack crews did not exist at that time. We operated from a ridge top helispot near Los Pinos Peak and were flown to hot spots on the fireline as the need arose.

Our ridge top position afforded a good vantage point from which to watch the converted World War II torpedo bomber make its drops. The plane approached Los Pinos Peak to make its first water drop on the fire. The doors to the bomb bay opened and a 600 gallon bag of water was released. The bag ruptured and the water was discharged over the fire. Subsequent radio traffic concerning the effectiveness of the drop elicited a response that the aircraft appeared to fan the fire. The plane returned to base for another baggie and flew back to Los Pinos for another drop. This drop did not appear to be particularly effective either.

In 1956, several small air tankers and a torpedo bomber were used to drop fire retardant and water near Mt. McKinley on the San Bernardino Forest. This fire consumed more than 10,000 acres of vegetation and threatened several mountain communities. Small Stearman fixed wing aircraft were also employed extensively to drop fire retardant (borate) on the Inaja Fire on the Cleveland Forest in 1956. The planes provided much needed support to ground forces and helped to save several structures. This fire burned nearly 45,000 acres of watershed and killed 11 firefighters. Aircraft also made retardant drops on several other 1956 fires. The era of the air tanker had arrived.

TREE MARKING TALES

Jack A. Godden

The formative years of developing timber volumes in standing trees was behind me when I went to work in 1954, as a junior forester at $3,410 per annum on the Hiwassee District of the Cherokee National Forest. I had learned of the Doyle, the International one fourth inch, and Scribner log rules in college. In Region 8, I was to use the Scribner tree/log scale. I had good teachers in Ranger Alden Norris Quam and Forestry Technician Hilman Hargis. I had to dust off my forest dendrology book and learn to identify the numerous commercial tree species growing in the Southern Appalachians. We had 23 commercial species on one sale, including holly and persimmon, which was sold in four foot length wood. I became the third qualified person on the district to mark, measure, estimate and record standing trees for sale and cutting.

The earlier method of marking trees was blazing the tree at dbh (diameter breast height, 4.5 feet from the ground) and at the stump, then branding the blazes with a "US" marking axe, after which its measurement and species were recorded. In

1954, the marking hammers and axes were left in the office. We now had paint guns with nozzles. A squirt of paint on the butt log, about eye level and on the stump were the markings on a tree for the logger to cut.

"Southern Glow" and later "Nelson" were the registered tree paints of those days.

They were expensive paints in the eyes of purchasing. Tom Hunt, the administrative officer, approved the purchase of "traffic yellow" paint for tree marking use. It was delivered to us in five gallon cans. The paint guns only held a quart with about a one and one half inch opening for filling. If it wasn't the refill problems, there were the mixing problems and difficulty in getting it to flow through the nozzle.

I had several heated telephone discussions with Tom and Paul Russell in the supervisor's office in Cleveland, Tennessee. Paul saying, because of costs, we couldn't afford the registered brands. Tom saying "paint is paint," and "traffic yellow should last on trees as well as it does on roads." The weather got cooler and the paint just wouldn't flow through the paint gun nozzles. Adding kerosene to thin the paint was thought to be a cure. Only to me as "inspector," I found the diluted paint faded too quickly on the marked trees. Worse yet, was seeing our marking crew stir a five gallon can of traffic yellow in the woods, transfer it to gallon and then quart cans of the marking guns. Time was wasted and there was enough paint left at the mixing site to provide evidence of the waste and our frustrations.

I wrote up a time analysis on our efforts, the waste, the quality, etc. It was good enough to allow us to even buy "Nelson" paint that was to become the preferred marking paint for many of the forests in the eastern and southern regions.

A tree marker would always remember the ground. It was flat, steep, rolling, covered with grass, brush, laurel, rhododendron, vines, shale, rock, leaf or needle covered, dry, wet with rain or snow, marshy, boggy, burned over, ice and wind damaged, a few snakes—are some at the descriptive elements. Wrapping a diameter tape around a 24 inch tree on a 60 percent slope and getting the merchantable height was difficult. Crawling through laurel and rhododendron, face to face with a white faced hornets nest or a black snake raiding a bird's nest are experiences I can relate to.

Today's environmentalist use the expression "Go hug a tree." I guess I'd qualify as a true lover because I hugged many a tree in my earlier years. What was I hugging? Pine trees wore easy, the bark plates usually flat, easy to fix the hook of the diameter tape and usually "clean" of vines and ivy. Hardwoods seem to attract more vines, thicker, hairy poison ivy, four and six legged lizards and insects. Pines held their needles and with their particular bark were quickly discernible. Hardwoods lost their leaves in the fall and you had to learn to identify bark and its inner colors to distinguish tree species which meant difference in stumpage prices, e.g. black vs. southern red oak.

The following are some stories of tree marking on the Cherokee National Forest:

Within a few weeks on the job, Ranger Quam, Hilman Hargis and I were out marking a unit heavy with old growth shortleaf pine. It was a dry ridge top overlooking the Hiwassee River. I had learned to "cover" my line working between Quam and Hargis as I was the new man. Hargis and I would call out our tally; Quam recorded it. He had the easier ridge top line and could narrow his strip as he saw fit to keep up with the tally and his forward progress to "lead" me a little on our marking strip.

After calling out a tally, I saw Norris jumping up and down in one spot. Surely he wasn't stomping out a yellowjacket's nest! After a dozen or so hops he hollered down to come up and see what he stomped. Hargis and I climbed up the slope to him. "I want to show you a copperhead." I obliged as it was probably part of my training program. Norris had "tenderized" with his caulk boots the poor fellow who was just trying to get a little more sunshine before fall weather set in.

Norris admitted it had surprised him which prompted the "snake hop." Usually we left both copperheads and rattlesnakes alone or got out of their way if they weren't challenging us. They belonged; we were just passing through.

In the spring of 1956, a tornado took a swath across the Hiwassee River and laid low everything in its path. There were no residents in this area so there was not much alarm. It did blow down a few million feet of some prime timber for which

Ranger Quam was given sales authority to expedite salvage sales before the "bugs" got to work on it. Two stories relate to handling this blowdown with a little personal twist.

A patch of blowdown southwest of Wolf Creek, Hargis and I classed as a "piece of cake." It was road accessible and fairly level; an open stand of oak before the tornado. We marked and sold it quickly. My task was finding the more difficult.

The old logging up Wolf Creek had been blocked by uprooted large white pine and hemlocks. The best access was up the creek, in and out of the water to determine the extent and size of the blowdown area. (Aerial reconnaissance was not in vogue in 1956.) I worked my way up stream for about a half a mile, finding only a narrow swath of blowdown, maybe a quarter mile wide running west to east across the creek. I decided it was a good day's work and began my walk out. Coming down stream I found a shallow pool teeming with trout which was dammed by a large pine and hemlock at opposite ends. What better way to relieve their overpopulation of the pool by catching a few native style. About ten were enough of a challenge and a fish dinner. I gutted them and placed them in the bottom of my aluminum lunch pail. Back on the road I put the pail in my open quarter ton Jeep for the trip back home.

Above Reliance I met up with the one conservation officer for the area. We stopped to shoot the breeze and got into a conversation about the blowdown on Wolf Creek. He was concerned about the fish trapped by the trees in the creek. "If our sale is timely enough, the stream course should be open before summer," I said. I couldn't assure him of the amount of logging damage in getting the tops and logs out of the creek bottom. Within four feet of his nose were a few fish that didn't survive the salvage operation before it even started! (They made a good dinner that night. A few of them did survive the salvage operations.)

After the marking crew got into the lower area of Wolf Creek, Ranger Quam thought it best if I walked the drainage to make sure we didn't have any other large areas of blowdown left. He dropped me off on the upper road. The two mile hike out was reasonable and I would meet up with Hargis's crew for a ride home. With a light knapsack, lunch, map, aerial photos, compass and a lightweight marking axe, I took off. The first mile and a half was easy. Any blowdown had to be below me.

About lunch time I stood at the junction of two streams that formed the creek. With the exception of a fluttering leaf or two and the trickling waters, all was quiet. This serenity was suddenly broken by the sound of splashing water that came from around the bend in the creek below me. Within tenths of a second there appeared a black bear on all fours coming my way, running for what appeared its life. My only defense was a marking axe I carried in my hand. There was no tree big enough to climb. In the next few tenths of a second the bear went by me on the left fork of the creek. I swear it was close enough to touch (probably 20 feet away). The next second he/she found itself about 60 feet from me; she (I'll call her a lady) stood up on her back feet, sniffed the air for what seemed an eternity, then tore up the slope through the rhododendron. I survived my first "close encounter" with an animal of size and proportion.

There was no sudden dampness in my pants but my knees and legs were so wobbly I had to sit down and gather my composure. After a few minutes I gathered my gear and hurriedly walked downstream another 300-400 feet where I found a gravel bar in the shade for a hopeful, quiet lunch. After meeting up with Hargis, I had a greater confidence. I figured the crew must have kicked the bear out of her bedroom and she vacated her premises for a few hours. I just happened to be on her way out!

FOREST SERVICE WOMEN—THEY ALSO "HIRED OUT TO BE TOUGH"

Lyle and Barbara Klubben

There is an old expression that we used to throw at each other when things got hairy: "You hired out to be tough!"—the implication being that we FS people could triumph over all adversities. Most often it was applied to the men but it has often been proved that the women also "hired out to be tough."

It was 1955, and I was assistant ranger of the Mineral District, Lassen National Forest. We were living at the Mineral Ranger Station in an old (condemned) but very comfortable house. Our family at that time consisted of myself, my wife Barbara and our children; Kathy, age five and Keith age two. The Mineral Ranger Station was one of the summer-winter stations and it was getting on towards the time when we would all be moving to Red Bluff for the winter—usually early in October—and the station would be closed.

At the time, Barbara was about eight months pregnant. The doctor's prognostication was that the baby would be born in early October. As a precaution we went to Red Bluff and rented a small apartment as a place to hang our hats in case the doctor was wrong and the baby decided to arrive early. In view of what followed it turned out to be a wise move.

About the fourth or fifth of September I was called to a fire in the Sequoia NF. My first impulse was to ask to be excused, but several things caused us to decide that I should go:

First was the doctor's estimate as to the time of the baby's arrival.

Second was that Ranger Wes Brokeshire had begged me off several fires during the summer because I was involved with some important property line surveys with the Diamond Match Company.

Third was that the station was heavily manned as the fire season was still in full swing, which meant that there was a lot of help there.

Fourth was that I needed and wanted the fire experience.

Fifth was that we needed the money that would result from overtime pay. Maybe the fifth reason was really the second reason but we relatively new forestry recruits knew better than to admit that we considered money more important than professional commitment. However, this all happened before the days of health insurance—we had to pay full price for babies. At any rate we decided that I should go to the fire.

The fire was the usual; hard work, smoke, dust, sweat and long hours. About the fourth or fifth day, I was called to the headquarters tent and told that I was needed at home—my wife was in the hospital. I was not told why she was in the hospital, so I had to hope it was because the baby had come and not something more serious (not that having a baby was not serious enough). Arrangements had been made for a plane to fly me home, so I quickly packed my things and I was driven to Bakersfield. I had not had a bath since I left home and my clothes were dirty but those things seemed trivial at the time The plane was waiting, so we took off immediately and in about three hours we landed in Chester, about 25 miles from Mineral. Ivan Bakerville was waiting to pick me up to take me to Mineral. He told me that the baby had been born in the Red Bluff hospital and the kids were at a baby-sitters, so we agreed that the pilot should fly me to Red Bluff. About 45 minutes later we landed at the Red Bluff airport where I was able to get a ride to the hospital

When I got the hospital, Barbara was awake and perky and the baby was fine. Now my only pressing problem was to rescue the other two kids from the baby-sitters—on a farm outside of town—where they had been deposited after being rousted from their beds late the evening before. Suffice to say, I rescued two very happy youngsters in an hour or so, after I had a bath and a change of clothes. The shirt I had worn on the flight home was grimy and torn, so I had stopped at a store and bought a new one. I think I picked up the first one I saw in my size and had to endure tack from Barbara because the color was not a good choice but it was clean.

The most important and most impressive part of the story belongs to Barbara. She was left at Mineral with Kathy and Keith. At that time of the year there was a sizable crew at the station but it turned out that there had been an inordinate number of late season fires. Thus, the people were constantly answering fire calls, some of them off-forest and often the only ones left at the station were Barbara and the kids and the office clerk and his wife. The clerk, a man named Babcock, adamantly refused to leave the station in any situation and Mrs. Babcock did not drive. Babcock, in my view, was overly impressed with his importance. He was relatively new to the job and had not yet learned that the FS put medical emergencies above everything else.

At any rate, Barbara was faced with a situation where she might be in very dire straits, some forty miles from the nearest hospital and no sure way to get there except to drive the car herself. Thus, as the week wore on she became increasingly nervous, perhaps the experience of two earlier pregnancies gave her ominous signals. So on September 8th (Kathy's fifth birthday), she and the kids packed the essentials into the car took off down the hill and arrived at the apartment in Red Bluff in the evening. Some time later the signals of impending birth could not be ignored. Barbara had made arrangements with the baby-sitter earlier. She piled the kids (sleepy, confused and somewhat scared) into the car. She drove to the baby-sitters, parked them there and drove herself to the hospital, arriving there shortly before midnight.

Nancy was born at about 3:00 a.m. on September 9th. She was perfectly healthy and it had been a relatively easy birth. When Barbara got to the hospital she called Elsie Bakerville, wife of the aforementioned FCO Ivan Bakerville, and told her what was happening. Elsie came to the hospital at around midnight to see how she could help. She was the one who set the wheels in motion to get me home. Another example of how the FS takes care of its own.

This account would not be complete without recognition of the part Kathy and Keith played in the affair. Imagine at their tender age (and they were virtually hillbilly kids) being moved from their home at Mineral while their father was absent to a new place in Red Bluff. They were then almost immediately packed off to a baby-sitter (and being left there for an indefinite time) who was a complete stranger while their mother went off to a hospital. That was a lot for them to handle. No wonder they were extra happy to see their dad after what must have been a long night and day. Incidentally, they spotted me the moment I drove into the yard. The baby-sitter told me they behaved very well. I am happy to report that now, over 40 years later, they show no signs of trauma from that experience. Maybe they too, without even knowing it, had "hired out to be tough."

SURPLUS PROPERTY MILITARY STYLE

Harvey Mack

Military surplus property listings were a "gold mine" for meeting forest needs. Due to our relative remoteness, we were not in a position to actually check things out and had to rely entirely on the printed description. Harold Stratton, who was staff for engineering and fire, really worked this source, as there were so many things he could use. In 1955, he saw a listing for 20 foot three quarters inch chains with hooks on one end and rings on the other. These looked like the perfect thing for tow chains for all of the equipment on the forest, so he ordered about 50 which would give us a few to spare.

About a month later, the warehouseman called the SO and said that a military semi had just arrived with a large crate on board—what was he to do with it? He did not have a crane or any means for lifting it off. There was a dozer in the yard, so they started it up, raised the blade high, and shoved the crate off onto the ground. On impact, the crate burst open and out spilled the chains—three quarters inch but that was the diameter of the link stock, not the size of the links. They were cargo chains used to unload ship holds! One man could hardly move one let alone pick one up!

Eventually, some uses were found such as using one as a drag behind a tractor for clearing brush, or suspending between two posts as a rubbing chain for livestock, but not for putting in the tool box of a pickup or tanker for towing! At least not on the Modoc.

CREATIVE INGENUITY

Chuck Smay

Early in my career I learned how resourceful employees in the Forest Service were in their determination to get a job done. The source for such determination must come from a combination of living in remote locations, where supplies are not readily available; or, a depression era background of limited funds. It was reinforced by the fact that it took so long to get approval and funds to replace anything in the government. I do not think Rube Goldberg ever worked for the Forest Service, but he would have been proud of the accomplishments of dedicated district employees. However the trait got started, many employees had a fierce pride in their ability to create something out of nothing. Every ranger station had a well stocked "bone yard" that was the source of the needed parts for virtually any project. The untrained eye would view it as a pile of junk that no one had bothered to clean up. Once in a while, a new ranger, not understanding the nature of such technical matters, would order the cleanup of the junk pile. If some fast talking could not redirect the devastating order, it might be necessary to relocate the "treasures" until the threat to these valuable assets passed.

Sometimes an entire project camp could be constructed from such a collection of abandoned pieces and parts. My first home at Bearskin Meadows on the Hume Lake District of the Sequoia National Forest was such a place. The camp was located at the 6,000 foot elevation along Bear Creek, a tributary to Ten Mile Creek that flowed into Hume Lake some three miles to the north. I arrived there one day late in May of 1955, as a brand new member of a five person blister rust project crew. While the camp was in a beautiful location, it was a proud collection of the recycling process well before the term was politically correct. The tents that covered the wood platforms were hand-me-down from the abolished Bureau of Entomology and Plant Quarantine. Plumbing was the area where creativity was often the most evident. The shower house represented this skill at its height. Heating hot water was accomplished by a series of creative "S" shaped pipe coiled inside a 55 gallon drum, cut open at one end and water heated by a wood fire built in the drum. This connected to a series of every imaginable size and type of coupling, union, tee, reducer, nipple and valve. Naturally the pipes and fittings were of random assorted sizes. Every possible part available was displayed in this plumbing system. Somehow it worked with one minor precaution necessary to those using it. Whoever was first in the shower had to stand back when it was turned on, since the initial delivery was pure steam that would clean your skin to a level well beyond what most users desired!

The center of the camp was the cookhouse. If there was an original piece of lumber anywhere in the building, I never located it. The building had originally been somewhere else, dismantled, and moved to this new location of service. The exterior paint decor was a combination of standard red oxide with dark forest green roof and trim. Sometimes the various shades of the touch up paint even matched previous paintings. The inside of the building was a combination of originally white and battleship gray. All of the colors were readily available on the surplus market. The idea there is no building as permanent as a temporary Forest Service building, could have originated with buildings such as this one. The building was about 16 feet wide by 40 feet long with a dining area in the front, a kitchen in the middle and the cooks room and storage in the back. Naturally, a crew person never ventured beyond the front dining area unless specifically given permission. Although the code of conduct in the Forest Service dining area was not as strict as in a logging camp, it was still an area where you had best be on good behavior.

Camp operated from mid-May until late October. The big wood stove in the back corner of the dining area took the chill off the cool mornings in early and late summer. It was a good thing the hand-me-down stove was too big for the building, since there were an ample number of air leaks around every window and door. The open rafters above exposed the single layer of corrugated roofing covering the building. The metal roofing gave ample evidence of past life as there were rows and random nail holes in numerous places in the roof that let in small beams of sunlight or moisture as seasonally available. The floor gave evidence of the short supply of framing material. It seems that the only support for the four by eight foot sheets of battleship gray plywood flooring were around the edges, so there was a three to four inch sag in the center of each section of the plywood floor. The summer afternoon thunderstorms would often provide enough rainfall, combined with the leaking roof, to fill the sag area with small puddles of rainwater. The cook kept asking Bill Prinz, the foreman of the camp, for some roof patching to deal with this rather unwelcome intrusion into his kitchen. It seems that there was just no money to fix the roof, given the great cost of a can of tar. After all, it had been this way ever since the temporary camp was there, so why fix it now! After several storms and the extra effort to mop things up, the camp cook demonstrated for all of us that he too was a creative Forest Service employee who could fix things with nothing. When we returned from the field one day, after a rainstorm, he proudly showed us his solution to the leaking roof and how he used the sagging floor as a part of the solution. It seems he discovered a brace and bit (forerunner of an electric drill) in the toolbox and used it to drill a one and one half inch hole in the floor at each puddle location. Instant solution; water in, water out, no mopping. Eat your heart out Rube Goldberg! It stayed that way until they tore the building down several years later!

ACCIDENT PRONE TERRITORY

Scollay Parker

One day during my tenure at Orleans (Six Rivers National Forest) Don Marriott, who was, at that time, suppression crew foreman, radioed that he and another driver had collided on the Bark Shanty Road. You always dislike to hear about an accident as worst of all there may be injuries, damaged equipment and the record goes to hell. So it was with some trepidation that I grabbed my camera and headed out to investigate.

Upon arriving at the scene I saw that a private sedan going uphill and the government vehicle coming downhill had collided on a switchback. To my relief, I observed that our pickup was as far to the right as possible. Jasper Reese, who had a homestead off the Bark Shanty Road, was the other driver and he told me that his brakes had failed. Don had seen Jasper coming and almost had his truck in reverse when he was hit.

Jasper's passengers were "Tee Bar" Annie and one of his brothers, Bert or Bill; I never could tell one from the other. Annie was wailing a bit, as she had impacted the dash board pretty hard with one of her breasts. There were no other injuries and not much damage to the vehicles.

Jasper was sober, and as far as I knew he didn't drink. With Bert and Bill it was a different story, rarely sober.

In summing up the accident with Jasper, I mentioned that he admitted to being on the wrong side of the road. With that, his brother who had obviously had too much to drink said in a loud voice, "What do you mean on the wrong side of the road? That's the side we always drive on in this country." And, I will have to admit, many times it was and probably still is.

FIRE AND WILDLIFE

Dick Worthington

Around Labor Day of 1955, northern California and southern Oregon experienced a series of the most severe lightning storms ever recorded in that area. Many major fires were started, including the Haystack Fire on the Oak Knoll Ranger District of the Klamath National Forest which was eventually to exceed 150,000 acres. With the normal air flow in the area the Rogue River Valley quickly filled with smoke. Not only were the airports closed, most of the state and Forest Service lookouts could not see.

On the Rogue River National Forest we had a lot of fires but because of the smoke no one really knew how many, how big or where they all were. All of us were pretty busy for more than a week on a minimum sleep and maximum effort basis. I remember getting home for the first time in a week after it all started. It was for the first full night's sleep I had for the period, only to be rousted out about 4 a.m. to take a logging crew into Crater Lake Park to take over one of their fires.

The fire, about a mile from the highway, was smoldering in very heavy duff. No water was anywhere near, so it was a dry mopup show, just the kind of operation a logging crew absolutely hates! I remember about 10 or 10:30 a.m., as we were eating lunch, a yearling bear showed up. We threw him some sandwiches and he thought we were great. Then we noticed that his paws were burnt and he was obviously in pain because he'd lick them frequently.

One of the crew hiked out to their crew wagon and picked up an old end hook rope. Then, much to my consternation, they lassoed the bear and tied him between a couple of small hemlocks. After using all the burn ointment in our first aid kits, the next problem was to let him loose. Well, we got the rope off of him, but I'm quite sure the bear had fewer cuts, scratches and bruises than most of us. The last we saw of him he was just getting into high gear. His hind legs were reaching up about his ears and his front paws were just visible under his rump.

I was always real partial to young bears after that, for this one kept a logging crew working on a fire they didn't like by providing an hour or so diversion.

THE TASTY TAIL ROTOR

George N. Newhall

It was in September 1955, on the Los Padres National Forest. I was the forest engineer. My job on the catastrophic Refugio Fire was intelligence officer. One duty was to have a map of the whole fire ready for the plans chief at 3:00 p.m., when he would assign night crews to control the last two hot areas of the 60 mile perimeter of the fire.

One of the best helicopter pilots, Freddie Bowen, was at the Los Prietos Ranger Station with his copter, so I asked if he could fly me around the fire to finish my map before noon. "OK," he said. His cruising speed was about 60 mph and we completed the mapping flight about noon. He set the copter down in the small yard used as the heliport in front of the mess hall. We went to lunch with other fire headquarter's people.

When we came out from lunch Freddie exclaimed something like, "What in the hell has happened to my tail rotor?" One of the men said, "It looks like maybe one of those mules got hungry." The excellent Los Padres pack string was quartered at Los Prietos, and of course, the mules were usually in the corral or coming and going on the fire job. But sure enough, there was a gap (a piece missing from the rear edge of the tail rotor), about four inches to six inches and about four feet above ground. There were mule hoof prints visible in the ground nearby.

The fire boss asked Freddie if he could fly the copter into Santa Barbara airport and order a new tail rotor. Freddie said no, it would not be safe to fly in its present condition. So, they went into the office and phoned for a new rotor to be flown from the factory to the Santa Barbara Airport and Freddie Bowen and his helper got the pickup and towed the copter away. The mule was never apprehended.

AN EXPLOSIVE SITUATION

Bruce Barron

It was in September of 1955, when the Los Padres Forest experienced an explosive 75,000 acre record setting fire in the Refugio (pronounced rey-fu-hio) Pass area north of Santa Barbara.

Lynne Correll (R-5 Chief of Personnel Management) and I were serving as alternate public information officers. We were flying in a big Coast Guard helicopter above a burned over area east of the big petroleum storage tanks at Goleta, when we noticed what appeared to be some small abandoned acetylene welding tanks. It aroused our curiosity as to why these tanks would be lying on the top of this ridge where there were absolutely no signs of roads, trails or habitation.

The mystery may have been solved, when 25 years later an article appeared in a Dept. of the Army publication relating the story of a Japanese submarine, that on February of 1942, had lobbed over two dozen projectiles at the nearby oceanside petroleum installations. It was said "the projectiles went wild in their trajectories, and did little or no damage to their intended targets." However, during the Refugio Fire, several loud explosions were heard that may have been attributed to some of these bombs that were detonated by the heat of the fire.

Another unusual event occurred on this fire when two big transit mix concrete trucks were pouring concrete on a construction job in the Refugio Pass area. They were suddenly engulfed in a firestorm. Unable to outrun the fire, the ingenious truck drivers crawled back inside of the big concrete mixers and utilized their extra water supplies (normally used to thin the concrete and wash out the mixer) to wet down and cool the trucks as the fire roared over the top of them.

Also, a freak incident occurred when a weary firefighter worked his way down a fiercely burning canyon to rest in a small clearing. While relaxing, he lit a cigarette and was killed instantly by an explosion from an accumulation of volatile wood vapors that had drifted down from the fire.

In the late 1950s, a similar phenomenon occurred on the Rattlesnake Fire on the Mendocino Forest when volatile wood vapors exploded and killed around a dozen volunteer firefighters who were members of the New World Mission religious organization.

It may be prophetic that one interpretation of the Spanish word refugio is "bomb shelter!"

A DAY OF REDISCOVERY

James R. Pratley

South of Oogaromtok Lake and west, across the Klamath River from the Irving Creek Ranch, near what is now forest road 13Nl3, my survey crew on reconnaissance for this road, came across a piece of ancient history, and later in the day, a piece of more recent history, two very diverse impacts of man on the land. The year was 1956, and before any roads existed south of Dillon Mountain.

Having marveled at the size of the bull frogs in the lake when laying out forest road 13N35, we were intrigued by the rugged nature of man who before the turn of the century, and for many centuries before, managed to live off the land. The frogs were introduced into the lake by gold miners who wanted to supplement their diet of fish and venison.

Our first discovery on that spectacular day were two communal pits dug into a small flat of land. Both pits were in the form of a dixie cup with the small diameter in the ground and the larger diameter at the ground surface. The banks were sloped inward from the outer rim and ended in a rounded bottom about three feet in diameter, complete with a small fire ring of stones from the river, leaving enough room for the feet of the people who would occupy this haven from the wind. The larger pit was about six feet deep, obviously for adults and the smaller pit about four feet deep for the children of either of the two Indian tribes that migrated up and down the Klamath River long before white men discovered the region. It was evidently a camping spot, close enough to the river of life to be convenient, as remnants of an old trail connected the two.

As far as we could determine, and without benefit of any written history in this regard, we could only assume that the pits were used for sleeping purposes by the Indians who leaned against the sloped pit walls, feet to the fire ring. Not only were the people out of the wind by being below the ground surface, but they all benefited from the warmth of the fire and the communal body warmth. The pit walls as we saw them, were as smooth on that day of rediscovery as they had been during the years of use over the past several centuries. We now understand, that these communal pits were located a days journey apart where better flat country campsites were not available such as at Happy Camp, at the confluence of the Klamath River and Indian Creek.

The second discovery on that amazing day was the remnants of an old mining cabin that had all but collapsed. This cabin was somewhat south of the pits, and it too, had been connected to the river by a trail, much of which was still quite evident. But perhaps the most startling find was the existence of a small cemetery behind the cabin, enclosed by a corral type fence of cedar about eight feet square, with corner posts beautifully carved, including a rounded ball at the top of each post. It was obvious that one of the occupants had been there longer than the other, as only one of the grave sites had collapsed, as if the coffin had rotted away. My only question on that day and to this day has been, "Who buried the last survivor?" of what was perhaps a man and wife who lived out their days in almost total isolation, for at that time the only means of transportation were primitive trails along both sides of the river and all personal gear and provisions were brought in by backpack or at best on a mule or a horse.

On our way back to Happy Camp that night, a trip that took over four hours by the time that we could hike back to the pickup and drive the Klamath River Road, we had the opportunity to marvel at the stamina of both men and woman who not only achieved this most difficult life style, but who very likely took it as a natural way of life, hardships and all. No communications, no news less than a month old and certainly no help in the event of sickness or accident. Can you imagine people of today who would even venture into such country, let alone scratch out a living under such conditions?

But, who buried the last survivor? We have to remember that people in those days were much more concerned with each other out of necessity and probably considered it a badge of honor to help out in times of need, even if it took several days out of their difficult life.

TALES FROM THE TAHOE

John Ed Tonnesen

THE GOLD MINES

I worked on the Downieville District my first summer in 1956. Jack Woolfolk was the district ranger and George McCluskey was the assistant district ranger. I had just come

2,800 miles from New England and didn't know much about gold mines. I learned a lot during the two and a half years we worked on surface rights. The first summer was the most memorable. I soon found out that most of the claims hadn't been mined for years. A few had cabins or houses that were occupied seasonally or permanently as either recreation or permanent residences. The Forest Service would struggle with these for decades.

There were a very few that were actually being mined. These miners either lived on the claim year round or seasonally. These were the interesting ones! I especially recall three of the miners that were working their claims that summer. All were solitary men who lived in rustic cabins under primitive conditions. I realize now that they had the "gold fever." They lived miles from civilization and each other. Each man had a different way of working his mine. One fellow shoveled gravel from cuts into a series of sluice boxes. Another had an adit (a horizontal opening in the side of a hill or mountain) with tracks and an ore car he pushed in and out of the mine. The third man wouldn't tell me where he mined, but he was anxious to show his success. After asking me to step out of his cabin, he came out with a one pound coffee can full of placer gold! One nugget was nearly as big as my Ronson cigarette lighter. He agreed to let me photograph the nugget next to the lighter and I still have the picture.

The old famous Sixteen to One Mine was still operating in the little town of Allegheny. The foreman took me down through the many levels on the elevator. I was told that in the old days, when winter storms made travel difficult and dangerous, the miners from Allegheny would ride down the elevator to a bottom level, walk north until they met other mining shafts and come up near Goodyear's Bar. There they would go to the Saturday night dance and return the way they came on Sunday morning.

SIERRA BUTTES LOOKOUT

One day, while working near the Sierra Buttes, we decided to visit the lookout. Sierra Buttes is a rugged, scenic peak at the top of the Sierras. In fact you could see it from the old forest supervisor's office on Banner Mt. Road. The lookout had a commanding view of the countryside. The lookout's name was Jerry Little and he was quite a character. I later bought a .30-.30 cal. Marlin carbine from him for $15. It had an octagonal barrel with buckhorn sights and weighed a ton. I killed my first deer with it.

Anyway, after that long, bumpy ride to the lookout, I asked Jerry where the latrine was and he pointed towards the edge of the cliff. There I found a small outhouse that was somehow fastened to the side of the rock so it was hanging out in space. When you were in the little house and looked down, it was several hundred feet to the first rocks. Needless to say, it was a breezy outhouse. Years later, I asked someone from the district if that outhouse was still there. They looked at me with a blank look, so I assumed it had been replaced with more modern equipment.

THE GREEN HORNET

During some surface rights examination work in 1958, we stayed in barracks at the different ranger districts. One such barracks was at White Cloud on Highway 20, west of Nevada City. The Nevada City district ranger was Frank Sodolski at the time. There was a cookhouse and tent platforms that housed project crews. We were all available for fire duty, if needed. The fire crew had a 4X4 Ford truck with a 300 gallon pumper unit that went on the back of the truck. All the Forest Service vehicles were painted a dark, forest green in the 1950s. This truck had a four-wheel drive unit made by Marmon-Harrington. The cab was so high off the ground that a stepladder would sure help you get into the cab. The cab held two and the rest of the crew had to sit in a jump seat in the back. There was a sign on the dash that sternly warned the driver not to exceed 45 miles per hour. As you approached that speed the front end howled and roared as if it would blow up. I don't know where they got the nickname for the beast, but everyone called it the "Green Hornet."

You could hit 45 mph if you were on the straight away, going downhill, however, you could trot alongside the fire truck going uphill. Several times I ended up on the truck when it was dispatched to a fire. Off we would go, roaring up Highway 20 grades at 25 mph with the red lights flashing and siren wailing. After awhile, we agreed to shut off the siren and lights since we couldn't pass anyone and vehicles would be backed up behind the lumbering "Green Hornet."

AN INTRODUCTION TO CALIFORNIA WILDFIRES

David M. Waite

The Malibu Fire of December 1956, was my introduction to Southern California brush and structure fire suppression. My previous experiences in wildland firefighting ranged from the Appalachian Mountains and Piedmont foothills to the Dismal Swamp, all during my ten years with the Virginia Conservation Department.

I had been accustomed to working with small hand crews, fire plows and lots of backfiring. Most of the firefighters were semi-volunteer local people, paid a small hourly wage for time spent on fires. Many were high school age boys and they all went home at the end of their shift. Fires rarely burned more than a day or two and fire camps were mostly set up for training exercises not really used on actual fires. I had much to learn about large fire management.

My Forest Service career started December 19, 1956, when I reported to Supervisor Bill Mendenhall, Angeles National Forest, in Pasadena. He explained to me that my title was fire prevention officer but the job was really public relations and for my orientation I would tour each of the ranger districts.

On December 26, I started my tour of the Mt. Baldy District with a pleasant hour or more of listening to Ranger Anselmo Lewis expound the advantages of living in Southern California. Later I accompanied ADR Cliff Stevens to Crystal Lake

Recreation Area and on to South Hawkins Lookout for an overview of much of the Mt. Baldy District. When we arrived at the lookout, we could see a large cloud of smoke far to the southwest. The observer told us Los Angeles County and City Fire Departments were fighting a fire in the Malibu Mountains. Cliff and I decided to eat our lunch at the lookout and while we were there the Malibu smoke increased considerably.

As the day wore on and Cliff showed me more of the district, we kept getting radio reports of the fire and requests for assistance. As I left the district office in Glendora, Cliff jokingly said, "I'll see you on the fire in Malibu." He was more correct than he knew for early the next morning, Forest Dispatcher Virgil Shoemaker called me at home and told me to be ready to go to the Malibu Fire as soon as I could get to the Arcadia dispatcher's office. Soon after I arrived there, about 7 a.m., I joined a carload of various personnel assigned to the fire. It was very strange to be en route to a wildland fire and at the same time be entangled in a major city's commuter traffic, however, this was a unique fire situation as I was about to learn.

After about an hour we were at the Malibu area and were directed to fire headquarters at Webster School. I was supposed to meet Dick Johnson from the San Bernardino National Forest to get on-the-job training as fire information officer. I never did find Dick but was greeted by my newly found friend and colleague, Ranger Anselmo Lewis, and he soon "shanghaied" me into his service organization. Anselmo had arrived at camp about 6 a.m., was very short of help and he was building up his "staff" any way he could.

The fire was in the second full day and there was a massive build-up of firefighting resources. My first task was to inventory all the fire engines in camp. I had never seen so much equipment and was sure we had all the fire trucks in Southern California. I was wrong, for they continued to pour into camp. At one time in this build-up we had 140 trucks inventoried but as the fire spread to new areas, most of them were dispatched by mid-afternoon.

The fire was driven by a northeast wind, known locally as "Santa Ana's" or devil winds. By the third night there were reports that the fire had jumped the Pacific Coast Highway and was burning dried drift wood and even seaweed on the ocean beach.

There were several memorable events during the five days I was on the fire but there were two I recall most clearly. The first occurred on the fourth day of the fire. Fire Boss (pre IC times) Roland Piercey, Deputy Chief, Los Angeles Fire Department, requested 200 line personnel. Anselmo called Dispatcher Virg Shoemaker and started to place the order but Virg stopped him in midsentence with a reply that all available resources that we could get in less than 24 to 36 hours were exhausted. However, the Governor had declared the fire a "disaster" and that meant we could call on the military. The Camp Pendleton crews were already trained and could be in camp within eight hours. Anselmo ordered 10 crews of 20 marines each, with supervision. They were in camp on time and went through the chow lines. They were fine looking crews, dressed in field fatigues and obviously in superb physical condition. Our manpower problems were solved, or at least Anselmo was congratulating himself for getting out of a jam. The next morning at 6 a.m. another detachment of marines, 200 strong, came into camp ready to take over the lines from their comrades. This process continued each shift. Anselmo had more marines than Camp Pendleton. Well, not quite. The post commander was very proud of his trained firefighters and had 40 crews of 20 men each and he wanted all of them to get "hot" line experience. Anselmo solved the problem of too many marines by working each crew one shift, feeding them and sending them back to their base.

The second event that left a lasting impression occurred about 3:00 p.m. on the fifth day of the fire. As I mentioned before, this fire was declared a "disaster" and of course all the top officials in the state were duty bound to make an appearance at fire headquarters. Virg Shoemaker alerted the camp that the Regional Forester Charlie Connaughton and State Forester Swede Nelson, were going to give us the benefit of their presence. They were to arrive by helicopter and wanted to land in the staging area which happened to be adjacent to the kitchen and eating area. Our camp officer was "Nappy" (Napoleon) Martin, Construction Foreman, Angeles engineering department. Nappy overheard the radio conversation between

Anselmo and the helicopter pilot and when Nappy understood what was being planned he became violently opposed to the landing of any d___ helicopter near his kitchen. After a few more exchanges between the pilot and Nappy, they landed in the regular heliport some 500 yards away. When Charlie Connaughton came into camp he was more than a little upset about the last minute change in landing location, that is until he met Nappy Martin and was given a logical explanation. Charlie immediately changed his attitude and was man enough to congratulate Nappy for having the fortitude to stand up for what he knew to be the right decision.

After I went home on New Year's Eve, as the fire had been contained, I had time to reflect on these two and other events that had been so vastly different from my past experiences. I did learn much that was to be useful to me during the nearly 30 years I spent on the Angeles and Los Padres National Forests. I learned from the two events related here, the value of clear, concise orders for resources. Also, that when you make a proper decision, stick by it and don't waver. Even high ranking officers will support you.

A LAND OWNERSHIP PLAN AT LAKE TAHOE

Dick Carson

Editor's Note: Dick Carson has sent us a story of the time when urban development escalated at Lake Tahoe South and Truckee River Valley. Mixed ownership, national forest and private, created problems for forest managers and developers alike. Eldorado County Supervisors responded by hiring a professional planner to bring order to the area. At a meeting with the planner and Forest Service people (Ralph Bangsberg, Dick Carson, Everett Jensen) present, a plan simple in outline was agreed upon. The basic concept was a green belt on the slopes surrounding the valley floor. The Forest Service would acquire private lands on the slopes and use valley floor tracts as exchange base. Dick Carson proved to be an apt pupil and was a key actor in simplifying the ownership pattern. The low key approach described by Carson was lost in the fire and smoke. –E.J.

The mid-1950s saw the start of a boom in the subdivision of real estate at South Lake Tahoe. The Forest Service, which managed most of the lands surrounding the lake, found itself under attack on three fronts:

1. National Forest land within and adjacent to the urban area was under intense "special use" pressure for everything from schools to dumps.
2. Subdivisions were being built right up to national forest boundaries, creating management, access and fire problems.
3. Private land parcels one-to-three miles inside the "solid green" were being proposed for subdivision, which if developed, would greatly impact all adjacent lands, with a good chance they would lose their "national forest" character.

The first two problems were political and one could only strive for the best political solution. That's another way of saying that any solution would not be a "win" for the Forest Service. The third problem had a resource solution; acquire the isolated private lands.

Land acquisition here was old stuff to the Forest Service. Exchanges had been completed for several thousand acres at Lake Tahoe twenty years before from the Eldorado Wood and Flume Co., and the Carson-Tahoe Lumber and Flume Company. At that time, however, timberland was $3 per acre, lake front land (the Nevada 4-H Camp) was $10 per acre. A deal for the Nevada Tahoe Basin lands had been rejected by the WO as "not being national forest in character." So now the Forest Service was faced with paying subdivision land value for privately owned parcels.

Bolan Peak Lookout, Siskiyou NF, Oregon – 1925.

There were a couple of important details:

1. There was no money for land purchase.

2. There was barely enough administrative money to pay Everett Jensen's salary in the only land adjustment position in the RO, let alone any funds for a project.

The three problems mentioned above were identified and with a natural treasure at stake (Lake Tahoe), land exchange there needed priority. Barney (Millard Barnum, ARF, Lands and Recreation) was easily convinced. Barney, an old-time deal maker, loved acquisitions and was still savoring his "coup" of the Pope-Baldwin exchange a few years earlier. Convincing Eldorado Forest Supervisor Guerdon Ellis to make it a priority for his Lake Valley District, Ranger Al Mullen, was another matter. However, Jens had an ally in the Fire & Lands Staff Officer, Ralph Bangsberg. (They gave Ralph the lands responsibility since fire was not a big job on the "asbestos" Eldorado Forest.)

Well, Guerdon was pessimistic about the success of the project. He knew that Jens was the only one working on land exchanges (planning, appraisals, title, etc.) and there were other regional priorities. He also knew the chances of negotiating successfully with several different land owners was almost nil. Still, it couldn't do any harm and if Ralph and Jens wanted to "go play with their toys," it was OK with him.

The first task was to identify which national forest parcels would be given up as exchange base. Lands adjacent to the existing private land "core" were chosen and a line drawn that would be a future boundary between subdivisions and solid national forest lands. This was done in coordination with El Dorado County, since county approval was also integral to success. The line roughly followed Pioneer Trail, then a narrow dirt road. While actually using Pioneer Trail as a boundary line might have been preferable, two reasons precluded it:

1. Roads invariably change location over time. (In fact, Pioneer Trail was relocated and widened about ten years later.)

2. It would have necessitated land exchange surveys, then done only by the BLM and next to impossible to obtain.

Ralph, in October 1959, tapped the Lake Valley District Acting FCA, me, to implement the project with guidance from Jens. (Ralph was finding himself swamped with SMUD's Upper American River Project in full swing and a very "un-asbestos forest" with the Ice House, Camp 7 and Bear River project fires.) I had started the year before cruising an initial exchange with Gary Plisco under the direction of timber beast Bill Charlton, Ass't District Ranger. Soon Jens had me gathering sales and other market data.

There were nine parcels of private land that had to be acquired under the plan. I immediately started work contacting the owners and trying to convince them of the benefits of a land exchange. However, the two year processing time was a hard sell. Even then it took a lot of effort on Jens' part and great support from Bill Jones at the Washington office. (At that time all exchanges had to be approved by the WO.)

Amazingly enough, all private land owners were successfully negotiated with, and exchanges completed by 1966. In addition, several other key Tahoe Basin parcels at Rubicon, Glen Alpine, Camp Richardson, Big Meadows, Meiss Meadows and Angora Ridge were acquired in eight other exchanges. In the middle sixties, some purchase money became available from the new Land and Water Conservation Fund. Consequently, five other properties were purchased:

1. A lot adjacent to Pope Beach;
2. The Valhalla, Pope and McGonical lake front property at Camp Richardson; and
3. Upper Cascade Lake.

BEAR SLEDS FURNACE CAPER

Bob Bjornsen

Remember the old wood burning furnaces? These were in vogue before heating oil models became available, particularly in rural areas. Such was the case in the old Bear Sleds RD office at Wallowa, Wallowa-Whitman Forest.

The furnace was located in the office basement adjacent to a two cord wood storage area. Living next door in the ranger's house, I would come over about an hour before opening time to load the temperamental wood eating monster in an effort to cut the chill when we went to work. This was often an exercise in futility because either a smoldering low heat fire resulted or a roaring inferno that drove upstairs occupants to open windows in a self defeating exercise.

Compounding the problem was a lone register near the second story stairs that was supposed to heat the entire building. It intermittently belched smoke that failed to go up the main flue. If the heat didn't drive one out, the smoke would.

Having long since given up trying to repair the infernal apparatus, and having no success in persuading the SO to give us funds for a new oil burner, we decided to give the forest supervisor a full smoke treatment on his next winter visit.

On the appointed day, Supervisor Hal Coons was seated by the open door to my office in proximity to the register. By rearranged signal, District FMO Bill Maxwell stoked the furnace with wet rags and green alder wood to ensure a smoky fire. You guessed it! The old furnace never broke stride as it burned that fuel with a degree of efficient combustion never before achieved and nary a smidgen of smoke to sally forth from the register.

Meanwhile, Bill was frantically stuffing more wet fuel into the firebox ignoring the fact the heat was driving us to work in our shirtsleeves with the windows open. Hal's comment about a "damn good furnace you have there," shot down our last hopes for its replacement.

OUR MISTAKE JACK—THAT'S AN ACTIVE PIPELINE

Jack A. Godden

There are only a few people in the Forest Service that had the experience of working in an active oil field on national forest land. I did on the Morehead District of the old Cumberland National Forest (now Daniel Boone). I quickly found out what a "B.S." pit was—never attempt to walk across a snow capped pond (and pit) in the winter. They were the only level places in eastern Kentucky. Some were dug, most were dams across a gully containing a few feet of oil over water. They caught the overflow from drill sites, wells or leaky transmission lines.

You had to be observant working in these oil fields. Not all old drill holes or wells were capped and "sucker" lines running through the woods were an above ground hazard. These rods were supported by notched posts, sometimes five to eight feet above the ground, greased to allow a one half inch or three quarters inch steel rod to move back and forth in a near straight line from the pump house to the well. They provided the power for the lift pump at the well. The rod line sometimes reached out 300-400 feet from the pump house which housed a one lung engine (most powered by natural gas at the site). Anyone in the area knew they were pumping crude by their sound. One revolution of a many-yoked axle in the pump house moved the rods four to six feet back and forth on the posts. The rods provided their own groans as they slid back and forth, working deeper into the notched posts.

There was plenty more steel or iron in the woods. There were the distribution, transmission, vacuum and pipe lines, ranging from two inches to twelve inches in diameter. The distribution lines were usually from the well to a storage tank, the transmission lines of four to six inches from tanks to bigger tanks. If truckers didn't haul it out by tankers, a pipeline transported the crude from the storage fields to the refinery. Around the Indian Creek area of the district, they hauled out the crude; on the lower half of the district, north of Beattyville on the Kentucky River, Ashland Oil had a pipeline transporting its oil to their refinery in Ashland, Kentucky.

In 1957, I laid out a timber sale on the upper reaches of Miller Creek in an inactive field, but with remnants of all kinds of pipe. There were what appeared to be two pipelines, one on each side of the creek following an old road. One 12 inch or maybe 16 inch line crossed the road that we planned on improving as a temporary road for hauling out the timber. With maps and land status I visited the Ashland Oil field office in Oil, Kentucky. There I talked with one of the company officials who checked their maps. No, that visible length of pipe was a piece of abandoned "vacuum" line. They agreed to have a "pipe monkey" out there when our timber sale operator was ready to pioneer in the road. They even offered the old piece of pipe for culvert material. This and a release to the Forest Service was all put in writing, in the briefest of terms. It looked official to me and agreeable to Stan Ockers, lands staff in the supervisor's office in Winchester, Kentucky. The sale area was then marked, advertised and sold. The operator paid his deposit and wanted to go to work on Monday of the next week.

He met me on time with a dozer and operator that next Monday. Then came the Ashland Oil official I had met earlier in his pickup truck. He had no time for any "Good Mornings." In his excited voice: "There's a problem. That old vacuum line is full of oil!" "What do you mean?" was my reply. "Our records weren't posted when we had a break on the line across the river and we switched over to the old vacuum as a temporary a year ago. We've had more time than necessary to make this change back to the pipeline, but..." There his sentence ended by the timber sale operator butting in: "What the hell next?"

I was no law school graduate but I had a contract in hand for an operator to improve a road bed as his haul road, a release from an outstanding third party interest and an oil company with rights to transport oil and a dilemma. I couldn't remember any cases like this in my business law course in college but nobody's temper was aroused yet. Common sense prevailed. The representative from Ashland Oil asked me for the sale area map. Over the hood of my truck we pointed to the "problem" area on the map which was within a couple hundred yards from where we had parked. We all walked to where the now active pipeline stuck out of a berm and exposed where we had planned to blade the surface. There was no way around unless we built up the grade which the oil company wasn't about to let happen. The oil company's rep thought 300-400 feet of road built below the pipe and burying the exposed vacuum line would be satisfactory to them and would allow the logger to skid the few logs across the covered line with little possibility of damage. "Who's going to pay for this?" was the question in my mind and the timber sale operator's.

"I'll have a dozer and crew in here Wednesday to build the relocation and place a culvert. You mark and bill us for the timber cut and damage in the road relocation," was the answer from the company's representative. We drew up a one page agreement on a lined tablet that he signed, the timber sale operator witnessed and I signed. Even before our morning cup of coffee the crisis had been solved. The sale operator, not to lose a day's work, walked his dozer through the woods to a section of the road he could work on that day and the next. On Wednesday, I returned to meet their foreman, dozer operator and a helper. I marked and flagged the temporary relocation and estimated the timber for billing the oil company.

Ashland Oil continued to pump oil through the old vacuum until after the sale was completed. We became good friends and to show their spirit of friendship they helped me out in determining who owned some of the transmission lines and miles of pipe laying in the woods. This was to become my assignment while still on the Morehead District. This special project proved a financial gain to the "Uncle," as some of the earlier leases held by the landowners and third party interests expired in the 1960s. The royalties, sometimes as little as 1/32, 1/64, 1/128, 1/256 ownership of minerals, gas and oil then were paid to "Uncle" on the crude from the active wells and fields. I left in 1958, but helped establish higher grade level positions for those who replaced me.

REGION FIVE BACKFIRE EXPERTS

Lyle Klubben

Sometime during my service on the Modoc NF (1957–62), a call came from the regional dispatcher asking the Modoc to send a team of "Backfire Experts" to the Fremont NF to help control a fire. The Fremont adjoins the Modoc on the north but is in Oregon, hence in Region 6. The fire was located some 40 miles west of Lakeview, Oregon, near the California-Oregon line.

Apparently, R-6 had no backfire experts, at least in southern Oregon, so looked to R-5 for help. That made good sense because of the large amount of backfiring done in R-5 during the average year. What didn't make much sense was to look to the Modoc for such experts. At that time, probably about 1961, the Modoc had experienced one campaign fire in the past ten years and that one was controlled at about 1,500 acres and required no backfiring. Otherwise, the most recent large fires had occurred in 1950 and '51. Obviously, we Modocers had been to off forest fires where we learned something about backfiring but usually as drip torch or fusee carriers as per instructions from others, presumably real backfire experts. Seldom did we participate in planning the strategy.

Thus, I still harbor a strong suspicion that proximity rather than expertise was the main criterion for the decision to send backfire experts from the Modoc to unsuspecting R-6.

Be that as it may, three of us—Hugh McLean, Big Valley DR; Ben Casad, Devils Garden ADR; and yours truly, South Fork DR—were quickly designated as "Backfire Experts" and dispatched. We left Alturas about dinner time and arrived at the fire some two hours later where the line boss and others were eagerly awaiting us. En route to the fire the three of us had indulged in some "gallows humor" about our qualifications as backfire experts, or more accurately, our lack of same. We decided to put the best face on it, act confident, and at all costs, avoid R-5 arrogance. This last was logical since, as Winston Churchill might have put it, we had very little to be arrogant about.

At the point where the R-6ers felt backfiring was needed, the head of the fire was moving slowly down a wide, rather shallow canyon in a mixed timber-brush-grass cover. It was burning vigorously in dry fuels. Fortunately, winds were light but were predicted to increase the next dav. The line boss felt that it was very important to control the fire at its present location or it would progress into an area where heavy fuels, rough terrain and lack of roads, coupled with high winds, could cause major problems.

Upon arriving at the fireline, which had been constructed across the canyon some distance from the fire, we inspected it with the line boss, made a few suggestions for minor changes and decided it was adequate. The line was about one half mile long with the most critical part in the middle.

We three "experts" spread ourselves about 100 yards apart among the crew members. We all smoked the filthy weed at that time, so we simply held cigarettes between our fingers and when the convection currents from the fire began to pull the cigarette smoke toward the fire, we ordered the backfiring to begin. The currents from the main fire pulled the backfire toward it until they met in one big pyrotechnic show. In what seemed like only a few minutes but was probably considerably more, the flashy fuels were consumed and the fire died down. Not a single ember made its way across the control line. It was a textbook backfire. The fire was stopped!

After accepting congratulations from our hosts, we climbed back into our pickup and headed for home. Of course, we stopped at the first tavern we saw for a congratulatory beer or two, well deserved since we had upheld R-5s backfire reputation. We were home shortly after midnight.

Several times since, the thought has crossed my mind that maybe we really were experts but didn't know it until that night, but other thoughts intrude, such as, maybe we just hit a situation that was simple enough for our skill level or maybe we were just lucky. As far as I am concerned, I still hold the "Backfire Expert" classification though fortunately, in my opinion, I was never tested again.

LONG TOOTH

Scollay C. Parker

While district ranger at Orleans in 1958, we had a clearing project around Fish Lake prepatory to raising the water level. Besides our regular crew, we needed some extra hands. One of those hired was Jack Leslie who had migrated up river from Pecwan. Jack was a good worker and even supplied us with smoked eel he had obtained from Pecwan friends and which we barbecued over the slash fire for lunch.

After a month or two, Jack developed a legitimate sickness and couldn't come to work. Unfortunately he doctored himself with the traditional Klamath River cure—alcohol, which takes a long time to work its magic.

In those times we had no clerical help in the winter and it was up to the district ranger to fill out the time reports. A while after Jack was incapacitated, I received a memo from Ralph Accardi, the forest administrative officer, that said I had over-credited Jack's sick leave by $7.50 and if Jack couldn't come up with it, he would look to me for reimbursement. This ticked me off a bit as my job description didn't include accounting.

Armed with the memo, I stopped by Jack's cabin and explained that he owed the government. He protested that he had no money and I was sure he was telling the truth. Knowing that Ralph "Long Tooth" Accardi would soon be following up, I told Jack that I would be satisfied for the moment if he would give me one of his guns to keep until his monetary situation changed. This was agreeable to him, but only one of his old guns. I returned to the station and put an old worn out rifle in my closet. "Long Tooth's" memo I taped up on the wall next to my desk as a reminder to not be sucked out of position again.

That summer a jubilant Jack Leslie came into the office waving a government check he had earned on a lightning fire up at Happy Camp. He said, "Let's go to the post office and cash the check and pay you off." We did, and I handed Jack his gun.

"Long Tooth's" memo was still on the wall that fall when Accardi made a visit to the district. He seemed a little embarrassed by it and suggested I take it down. "No, Ralph" I replied, "With it close by as a reminder, I'm not likely to make any more time report errors."

That winter, Ralph transferred to the regional office and I presented the memo to him at his going away party.

GRABBING LEATHER

Bob Bjornsen

It was one of those bad yellowjacket summers on the Wallowa-Whitman. The critters were everywhere in the woods—neither man nor beast were safe from their stings.

Tom Griffith, Baker Assistant Ranger, and I were working our way horseback up Indian Creek below Anthony Lakes, when his horse stepped on a jacket's nest and all hell broke loose!

We were on a sidling hill when they started their rampage. Tom frantically tried to stay in the saddle as the "Speed" horse turned every way but loose to avoid the stings. Have you ever tried to stay with a bucking horse, head pointed downhill while swatting yellowjackets?

"Turn his head uphill," I cried, sitting calmly astride my "Smokey" horse as if such a feat were possible. About that time the yellowjackets hit me and I was grabbing leather too. Miraculously, we both stayed aboard our horses but not because of our skill at bronco busting.

We got out of there fast and found a safe spot where we rubbed our horses down, all the while dabbing ourselves and the horses with spirits of ammonia, a great remedy for neutralizing bee stings.

Our travails weren't over yet. As we reached a meadow near the ridge, it started to rain hard. The meadow soon became a sheet of water and we found ourselves in the middle feeling our way towards solid ground. Next "Smokey" went down in a bog to his hocks, pitching me into the mud and slime. I hollered to Tom to stop on solid ground until we could find our way out.

Next thing we knew, "Smokey" was on his back, legs in the air, gasping for breath and not responding to our trying to right him. This called for quick action because a horse doesn't last long with his innards pressing against his lungs. Somewhere I'd heard that a stick thrust in a horse's anus would cause a violent response—I guess it would with humans too!

I quickly found a smooth stick, and with Tom holding his tail aside, did the job. That horse did a 180 degree flip and in a split second landed on solid ground with all four feet planted! The rest of the tale was anticlimatic as we made our way safely to solid ground.

CLASS "A" CORNERS, METES AND BOUNDS AND ALVA TOO

Jack A. Godden

Most foresters who worked on Forest Service land "northwest of the Ohio River, above the mouth of the Kentucky" never knew of lands where "metes and bounds" was the method of land survey. There were no sections, ranges or townships, just a tract of land with bearings and distances between corners usually closing upon its starting point. It all "fit" together with the adjoining owner's land in earlier years when recorded in their deeds following the first formal surveys. Walking the "lines" was a neighborly thing to do; it sometimes ended in differences of opinion as to where the line went or the corner laid after it rotted away or was disturbed by encroaching uses of the "other party."

As land was acquired on the eastern forests of the thirteen colonies, land surveyors adopted a standard for marking key corners of FS land with a class "A" corner. Not many of these ever became hot items for souvenir collectors. They were a cylindrical piece of concrete, about 24 inches tall with a three inch brass cap on top, tapering to four inches at the bottom. Upon setting this piece of concrete in its measured location, the brass cap was stamped and dated with the tract number.

On hilltop land belonging to Alva Shahan above St. George, West Virginia, was an interior tract of land that justified placement of a class "A" corner.

Alva had more than his share of local color and legends. Somebody told me he was gassed in WWI, but he looked too young for that. He did walk with a limp. His oversized coveralls allowed for plenty of ventilation. He was married, had a clutch of kids and even cared for a mentally retarded senior who was a possible relative. All seven or more lived in a two story frame house overlooking his and "Uncle's" land. He grazed a few cows under permit and he was always delinquent in paying his $15 grazing annual fee. He went into the charcoal wood cutting business when Kingsford's opened their St. George plant and offered $8 per delivered cord of hardwoods. Alva got his truck running by placing a Ford engine in a Dodge (or vice versa). It ran good downhill to St. George but on the return trip needed his team of horses to pull the truck up home. His kids would meet him at the foot of the hill and run home to fetch the team from mom who was usually working the fields. He never did perfect the right gear ratio for that transmission and truck.

I made a real friend of Alva's one spring day by saving his upper barn from an unneighborly "set" grass fire. He was "always owing to ya, ranger," but that didn't mean paying his bills any earlier.

The supervisor's office gave me a few extra dollars that spring for surveying and posting lands. The tract around Alva's had a high priority and after only one day's work the crew leader reported the class "A" corner bordering Alva's property missing. The witness trees were still standing, just no corner physically present. The crew was leery of getting into any long discussions with Alva, so I set out the next morning to find Alva and question him on the whereabouts of the corner.

When I drove up in his front yard there wasn't much question as to where it was. It was supporting the corner of his front porch. Alva appeared in the front doorway and quieted his hounds. He had to yell at his "ward" who sat on the porch humming some tune known only to him, to shut up. Then I moved closer to the porch steps, pointed to the corner post and said, "How in the world did that concrete post of ours get there?"

Alva stuttered, looking for words or another lie, but surprised me with, "Well ya know Ranger Godden, that post just fell out of the ground down there by the edge of the road and my porch post just rotted away, so I knew you wouldn't mind if I just borrowed it until I could cut a new oak post."

I threatened him a $300 fine. He said I might own his farm if he had to pay that kind of money. Next was the possibility of taking him before the county judge. Neither the fine nor any judge would change Alva's way of life. He knew I was mad at him and after I scolded him not to touch any more corners, he replied, "I was hoping you'd come and visit with me but sorry I made you mad, ranger. You're too good not to be called my friend." He and I found a length of white oak in his cordwood. He got his saw and together we sawed out a length for the porch corner post. I squared it up with my axe and it substituted for "my" class A corner. Together we worked up a little sweat but we both got what we wanted and a renewed friendship.

I left the district in 1960, and on a return assistance trip to the Monongahela and the district I asked about Alva. I was sorry to hear "they mistook him for a squirrel and shot him." Alva didn't deserve this. He was a simple man that never hurt anybody but I guess he occupied the wrong place at the wrong time.

WHEN GOVERNMENT MONEY WAS NOT GOOD

James R. Pratley

During the fire season of 1958, I was directed as a staff officer to take all of my engineering personnel from the Klamath to several fires in Southern California. We flew in two separate flights out of Montague, picking up additional manpower on the flight south. Though the first flight left somewhat earlier than the second, both groups were to rendezvous at the Burbank Airport for transport to our first assignment. However, due to the number of fires going at the same time, the usual snafus under similar situations loused up the meeting between flights and ground transport, requiring all sixty plus firefighters to remain at Burbank well past the lunch hour. And, as many of the men had not had the opportunity to have breakfast before joining up with me, I had a mean bunch of guys, who without much money, or no funds at all, were in no way able to obtain lunch at the airport cafe.

So, as flight chief, I took it upon myself to contact the hostess to explain the situation and offer her a government voucher in full payment for the meals that my troops would order. "No problem," she said, and my more than hungry men were soon satisfying one of the basic needs in life. At least, it would hold them until we reached our designated fire camp. But suddenly, a large and very nasty fly landed in the ointment of good cheer. The cafe manager, who not being privy to the agreement between me and his hostess, bluntly told me that a government voucher would not be accepted. He was not about to wait weeks before the voucher could be turned into real folding cash. So, what to do?

We could engage in a sit-in until a more reasonable mood prevailed on the part of the manager. We could also loudly proclaim the manager to be less than gracious to people who were giving up their weekend to assist the good people of Southern California, we could give the story to the media, or we could simply get up, walk out and thus be in the unenviable position of defrauding an innkeeper. All of these options would of course be entirely against Forest Service policy. Our one and only recourse was to determine just how much money all of us had in total and hope that it would cover the tab. Fortunately, or unfortunately, as the case may be, I had by far the most, nearly fifty dollars, which of course I had to declare under a gentlemen's agreement to abide by my own edict. And so, after some gentle arm twisting, we did manage to cover the entire bill, but did not have enough to provide a proper tip to the good waitresses and to the hostess who served us so well in our hour of need. With such a large group, individual tabs were not available, so there would have been no way that the total bill could have been properly apportioned to each individual, and with no declaration of just how much each man put into the pot, there was no way that we could ever reimburse those who paid more than their fair share. So, as is usual in such cases, the cheapskates got a free meal and the more responsible paid far more than their share. But, that is life, and even those of my forces who may have contributed more than their share, did not seek or expect to receive any more than I did in this matter. We avoided a sticky situation and though we might have made the manager very uncomfortable in the public eye, we did what had to be done and did it as gracefully as was possible.

BEIDEBACH'S GREAT HOSE-LAY

Robert W. Cermak

I worked with Don Beidebach only once, but that one time was unforgettable. The Stewart Fire on the Trabuco District of the Cleveland was started on December 14, 1958, by an off duty marine shooting tracer ammo. It became a huge fire, burning over 70,000 acres. It was also a miserable fire. The only time we were warm was when we were about to get burned up.

We managed to corral the central zone of the fire on the 18th after a spectacular backfire contained the last open line near the Stewart Ranch. After a day of on the job R&R, I was sent to the south zone camp at Tenaja Guard Station. This god forsaken hole was on the edge of a potrero, so dry that all water had to be hauled into the site. There were about 400 men in camp milling around in the usual state of mass confusion. I reported to plans which was located in the guard station office. There were so many people in that small room making so much noise that it looked like pictures I'd seen of the Chicago commodities exchange. It smelled bad too. Come to think of it, so did I. I hadn't had a bath in a week.

I managed to get the attention of the bleary eyed plans chief. He looked at his line up and said, "Go help Don Beidebach. He's in charge of the hose-lay." I went in search of Beidebach. Most of the Southern California firemen knew him and after awhile I found him. He didn't have much to say except, "Hello, let's get to work." He showed me the grand strategy that was to end the fire. The plan was to lay hose along a tractor line built on a ridge on the east side of Tenaja Canyon all the way down to the San Mateo River bottom. After the main hose line and branch lines were in and crews were in place the canyon would be fired from a helicopter.

My job was to do what Don told me which was to see that the hose was laid. Good thing because I knew from nothing about pressures, fittings and the techniques of hose-lays in general. We started not far from camp at the last point to which we could drive a water wagon. Here a pumper truck was stationed to pump from water wagons into the hose-lay. From this point we laid two, 100 feet of one and one half inch hose to a 4WD pumper stationed right at the break where the fireline dove off into San Mateo Canyon. The fireline dropped off into a deep saddle, then climbed a peak, dropped into another saddle, climbed another peak where we had a booster pump, then continued down without a break into the bottom of the San Mateo. The main line included about 5,000 feet of one and one half inch hose. At intervals we installed wyes ("siamese" valves) from which one inch lines led off at right angles to the main line. There was about 7,000 feet of branch line and a total hose-lay of about 14,000 feet.

The plan was great, now how were we going to get all that hose laid? Some genius in plans figured that this was the spot for a helicopter hose-lay. Sounded good to us but it couldn't do the whole job. We handed each crew member a section of hose to carry down the line with him with instructions to drop them every 50 feet. Meanwhile, the helicopter buzzed over and spewed out 2,000 feet of hose just as if it were on a training exercise. I was up and down the line all the while making sure we got the hose where we wanted it. Then I tried to hook up a wye to the helicopter hose-lay. No way, hose-lay! It was laid backwards! I trudged back up to the top, told Don who calmly sent me back to camp for adapters. Back down the hill, install adapters, check on the branch lines until all was ready. The line was tested. It worked! We were all surprised except Don who may have grunted once with satisfaction.

The real test came when the backfire began. I was stationed on the first peak down the line. The backfire was set too far down the hill and came ripping and roaring upslope. It swept across the 100 foot firebreak in the saddle below us and ignited the back side of the line. Fortunately a catskinner had wandered off into the brush at that point, so the crew stationed in the saddle was able to get to and put out the slopover with water from the branch line. Meanwhile, the fire roared up and over the peak occupied by me and a Zuni crew. It was hot enough to burn the main line in two and send all of us running for cover. I wasn't the only one to lose my eyebrows and a patch of hair that day. When the first blast had passed over I shut off the main line and directed the water into the branch line. The crew picked up the hose and soon had the slopover under control.

The backfiring stopped until more favorable conditions prevailed later in the day when the backfire was completed with no further incidents. The fire was contained after a week of effort and one death. I don't remember whether I had a chance to congratulate Don or not because I was off to Ontario that night and back home the next day. I suspect it was one of the larger backfires (2,000 acres as I remember) and longer hose-lays in Forest Service history.

Years ago I saw a notice of Don's retirement and wrote him a letter recalling our experiences on that humongous hose-lay. He replied with a note that it was indeed quite a feat. Harry Grace is right on target with his comment that Don Beidebach was one of the top firemen in Southern California, or anywhere as far as I am concerned.

AVOIDING SHANK'S MARE

Bob Bjornsen

The smokechaser of yore is virtually a vanished species. He (it was a male bastion then) traveled to fires afoot, or "shank's mare," as it was commonly referred to then. A hand compass, pulaski, shovel, rations and a bedroll strung atop his pack were the tools of his trade.

Plotting an azimuth from his departure point, he hiked across rugged country to the fire located by lookout cross shots, or sometimes only by landmark. It was a challenge for him to find a lightning hit snag that put out only an occasional puff of smoke. When this happened, he would use the gridiron method to systematically find the fire, often depending on his nose to sniff out smoke and follow the drift to its source.

When he had corralled ("contained" hadn't been invented then) the fire and felled the snag by hand, he set about putting it out—dead out. His instructions were "don't leave the fire until it's out." Sometimes this could be a couple days of bare hand feeling of suspected hotspots or of placing his head at ground level to see the tiny gnats which inexplicably swarm over a spot of buried hot coals.

Such was the legacy Leroy Merrick and I brought to the art of smokechasing on the Wallowa-Whitman NF one hot August day as we were hauling the "Speed" and "Smokey" horses up Sumpter Valley in a stakebody truck. That same piece of road today is under Phillip's Reservoir.

Suddenly radio silence broke with the excited voice of Huckleberry Mountain Lookout reporting a fire on the west slope of the Elkhorns. As he read off the azimuth, we were mentally calculating where it should be. I pointed out the window towards Bridge Creek. "It ought to be there," I shouted, and sure enough we could see the smoke.

Leroy spun the wheel, nearly putting the horses down and sped up Bridge Creek Road, dust aflying. He radioed we were headed for the fire to make initial attack and called for reinforcements. Reaching our departure point, we unloaded the horses, saddled up, grabbed our tools and struck out at a fast trot for the distant smoke.

Reaching the fire, we tethered our horses and went to line building/hot spotting. It wasn't long before we were gasping for breath, wondering where in hell that crew was. By the time they arrived, our posteriors were dragging our tracks out, to put it mildly.

It was a relief to mount our horses and leave the fire to the tough, young crew. Riding back to the truck, we reminisced about smokechasing in earlier years and how this was our first chase from horseback. We reckoned there wouldn't be many future fires where horses were used in initial attack.

Air attack and a better ground transportation network are enabling today's firefighter to hit 'em hard, hit 'em fast and keep 'em small. The objective remains the same, only the methods have changed. So here's to that smokechaser of yesteryear; his was a lonely, arduous lot that required a woodsmanship skill not readily found today.

A WEEKEND DOWN THE RIVER

Mac and Cleowa McEwen

Guess you could say we were a Forest Service family since L. D. Smith, my father-in-law was GDA for the Happy Camp District, Klamath NF, at the time I was courting my future wife. I was working on a survey crew on the Trinity Forest at the time. We were married in the "B" building on the Happy Camp Station. Smokey was not in attendance, however, we do have a dogs paw print in our guest book.

From about May 15 through Thanksgiving, we were on call 24 hours a day, keeping in touch with dispatch and all the rest of the good stuff. L. D. had passed away and Florence, my mother-in-law and her sister were living on the family mining claim at the mouth of China Creek about ten miles up river from Happy Camp. Neither Florence nor her sister drove. Cleo liked to check up on them from time to time. On one Friday I got the idea that we could slip off for a day or so and all would be well. Called Cleo and told her to get some stuff together and right after dinner we would head down river for the weekend. Thus the fun began.

Arrived at mother's house with the ladies off on a chit chat session and I kicked back with a book or something. Come morning, a good breakfast and plenty of coffee. All was well with the world. Clanks and clatter and a chug or two were heard coming down the lane and Florence said it was a FS rig. Sure enough in short order an old green pickup pulled into the yard and I was hailed by the station handyman. Seems the Gasquet District broke a fire and dispatch said I could/should/would go. I was to ride into Happy Camp and Doc Hall would fly me to Montague, pick up my gear, then fly me over to Gasquet. I left Cleo and my rig in place.

The plan was followed with few hitches and I made it to the fire. It was a great little place. Sixty percent or under was considered flat and then you got to the hills. In due time a couple of days of Oregon mist pretty well knocked down the fire and I went into Eureka to clean up and rehab the radio gear. About ten days after leaving, flew back to Happy Camp, got a ride up to the claim and right away could tell that this was not the happy homecoming I had figured on. Seems not all had been peaches and cream while I was away. Maybe it would be better if Cleo told you about that as I might leave out a fine point or two.

Yes, yes, there are several things that Mac has left out of this little story! A two year old and an eight month old baby with clothes, etc., for only a weekend. This included cloth diapers—no disposables. Then there was the matter of milk and baby food. One more thing—I didn't drive and never a stick shift, so there was no way to get from here to there. There was no electricity or phone, had to fire up the old wood stove for hot water. We had several kerosene lamps. It was sure fun when its time to check a fussy child at night.

There was no well. Water came via a ditch from the creek and would at odd times stop due to leaves plugging the intake or some critter stomping or digging a hole in the ditch. This invariably happened about sunset or later. An old gas washing machine for laundry, if it would start. For us it was kick, kick and kick some more, then let someone else try for awhile. A few days before Mac returned, the oldest child managed to slam his thumb in the latch of the pickup door—really mangled it. At that time there wasn't a doctor closer than Yreka, even if anyone could drive the old truck.

Looking back now, I guess we really didn't have it too bad, however, it sure was nice to return home with all the modern comforts.

OTHER WORK AS ASSIGNED

Roy Saarni

It was in the late '50s. Leigh Lint at the RO had phoned Ted Niehaus at Nevada City and stated that the photogrammetric grid for the new proposed Tahoe map was "shaky" in the Southeast corner of the forest. A photo location of a key township corner was needed to tie down the grid. The corner was somewhere east of the hellhole of the Rubican and north of the Desolation Valley Wilderness in rugged steep terrain.

Ted stopped by the Truckee Ranger Station and asked if I wanted in on his little project. It was, after all, "only one corner and shouldn't take very long." I'd had some experience in locating corners a decade earlier while on range surveys with Dell Fausett, Waldo Wood, Elmo Brough and Ralph Lessel.

We drove to the west shore of Lake Tahoe and up a rough Jeep trail to the Sierra Crest at Blackwood Pass. We then hiked about a mile north along the crest and unexpectedly found a corner yellow tagged two decades earlier by George Fischer, as I recall. (George really got around—found his tags on the Lassen and Modoc too—in the damnedest remote places!) We pinpricked this corner on our photos and then scaled off two miles west to spot a rough location for our elusive township corner. Murphy's Law was in full control, even in those years. If it was possible to be tough, it would be. The corner was, hopefully, somewhere on a steep south facing manzanita covered slope.

We descended several thousand feet, located the brushfield and started our search—no easy job in dense manzanita. After several hours and many scratches and rips, I heard a hootin' and a hollerin' away off in a thicket. Ted had crawled along an old deer trail and had actually stumbled across an iron pipe and a pile of rocks. The township corner was found!

Now we had to pinprick it on our aerial photos. With little variation in vegetation and terrain, this was a real challenge. We checked the original field notes again and found that three trees had been blazed and marked. But the slope had burned out decades earlier and no trees or standing snags remained. We measured out along two recorded bearings and found no trace even of stumps. At the third measurement, we lucked out and found a semidecayed fallen snag. Upon chopping into a fire scorched catface, we found faint but distinguishable bearing tree markings. This verified our position, since we were able to spot it under our stereoscope. By measuring in from other surface objects (rock piles, trails, emerging red firs), we were finally satisfied that we had correctly pinpricked the corner on our photos.

Now, it was nearing sunset and we had a long way to hike to the Sierra Crest. We were soon walking in the dark, uphill through brush and rocks and then timber and more rocks. It occurred to me about this time that I'd been "had" on the "only one corner—an easy job" story from Niehaus. Ted was always a superb storyteller. After hours of groping and aided only by a fledgling moon, we broke through to the summit about 11 p.m. Miraculously (or was it sheer instinct and uncanny skill), we ended up fewer than 100 yards from our Jeep!

We enjoyed a much needed meal at Tahoe City about midnight and reflected on the many odd and varied jobs foresters were called on to do. The job description category included it all under "—other jobs as assigned." All in a day's work. But we accomplished what we had set out to do, and that is what it was all about.

RETURN FROM THE DEAD

Bob Ziegler

Thursday, June 4, 1959, 2:15 p.m., Eagle Peak LO, Reserve Ranger District, Gila National Forest, reported a smoke on the adjacent Luna District, Apache National Forest. Luna District immediately dispatched a fire crew. The crew's first message to H.B. Birmingham, the Eagle Peak lookout, was the fire's location—near Bill Knight Gap; that it was small and had originated from a burning automobile.

The next few messages from the fire indicated a body had been found in the burning car. The body was so badly burned only the torso was left. The fire had been so hot that clothes, arms, legs and head had been burned completely off, There were not even any clothes buttons left. Then, the remains were positively identified as Manuel Armijo Jr., a Reserve, New Mexico school teacher. Identification was by relatives who lived at a ranch close to the area.

A request was made from the scene through Eagle Peak for Bob Ziegler, Reserve District Ranger, to notify Mrs. Armijo and her family in Reserve that Manuel Armijo had died in an automobile accident. Based on the reports received from the fire, Bob couldn't believe that such a positive identification was possible. Barry Flamm, Frisco District Ranger who shared offices with Bob was dubious that it was even a human body in the car. Much to the displeasure of some, Bob declined to make the notification. Later, Catron County Sheriff Barney Hickson returned from the scene to give the family the news.

The next morning, an investigation was made at the scene by New Mexico State Police and the district attorney from Truth or Consequences, New Mexico. The investigation determined that Armijo's car had left the road and smashed into a tree with enough force to render any occupant unconscious. It had burst into flame and the gasoline tank had exploded, setting a small forest fire. State police said it was impossible to make a positive identification since the body in the car was virtually cremated. However, the car was registered to Armijo and he had been in the vehicle earlier in the day.

The remains were released to the family and on June 9, a funeral service was held at the Immaculate Conception Church in Albuquerque, New Mexico. The remains were buried at Mount Calvary Cemetery in Albuquerque.

On June 11, one week after the "fiery crash," lo and behold, Manuel Armijo appeared to some loggers not far from the scene of the fire. It was said that some of the loggers who knew

Manuel were at first frightened because they thought they were seeing a ghost.

Manuel told the following bizarre story: When an ignition fire broke out under his dash board, Armijo realized he was carrying dynamite caps in the car. Rather than fight the fire, he jumped out of the window. He fell down a hill, getting up and running and hitting his head. That was the last thing he remembered until he came to a little over a day before he was found. Armijo said he had been carrying in the car an Indian skull he had found, a calf's head and a porcupine he had shot.

An investigation was opened, a court order to exhume the remains was obtained and the remains were turned over to the University of New Mexico anthropology department for study. The anthropologist's examination verified Armijo's explanation of what he was carrying in the car. Armijo was charged with arson and with carrying explosives in a motor vehicle. Both charges were later dropped.

To this day, Bob and Barry carry a vivid picture of a lucky porcupine that was placed in a bronze colored casket, had services said over it in a church, blocked traffic with a funeral procession through Albuquerque and was interred in a human's cemetery.

BLACK TREES AND SPOTTED HORSES

John Ed Tonnesen

During the summer of 1959, there was a project fire on the Sierraville District called the Cold Stream Fire. I was a PSO (project sales officer) on the Nevada City District and was detailed to Sierraville to mark and cruise the burned timber. Hal Turner was the forest Timber Management Officer and he had arranged for assistance from the regional office as well as help from other Tahoe ranger districts. There were seven districts at that time: Foresthill, Nevada City, Camptonville, Downieville, Sierraville, Truckee and Big Bend. It was quite a crew and we were bunking at the Sierraville barracks. A few of the names that I can still recall are Ted Nelson, Ted Gregg, Chuck Hardesty and Dick Lund.

Carl Schoelberg was the district ranger and Ted Gregg was the assistant district ranger/fire control officer. Chuck Hardesty was the district timber management officer. There was no such thing as "portal to portal" pay in those days. Employees were expected to be on the job at 0800 and your pay stopped when you left the job. Some rangers permitted the return trip on government time. Crews were expected to be loaded up, fueled up and out of the yard by 0730. If your crew was late getting out, Carl would have a pep talk with you that evening.

We were each assigned areas to mark those trees too badly burned to survive. We each carried two quart cans of Nelson tree marking paint and a hand operated squirt gun to mark the trees to be cut. Someone decided we needed a 20 percent cruise or measurement of the marked trees so we could estimate the kinds of trees and volumes of each to be sold. The system was simple; each man carried five marbles in his pocket of which four were white and one was black. Every time you stopped to mark a tree, you reached in your pocket and took out a marble. If the marble was black you stopped and measured the height to a 10 inch top, the diameter breast high, graded the logs in the tree and recorded all of this along with the species.

Marking fire-killed timber is a messy job. There are many smoking stumpholes, standing snags with fire still in them and ankle deep ashes. At the end of the day we were covered with black soot and our clothes had streaks of charcoal from rubbing against burned limbs and twigs. Our faces were black and streaked with sweat. Our legs were black inside our denims from the swirling ashes. The wind would come up in the afternoon and blow some of the paint we were trying to squirt on the trees back in our faces and on our clothes.

Many districts had saddle and pack stock for use in range, wildlife and trail administration. Frequently, when we were getting ready to leave the ranger station, Carl would saddle up his horse, load it in the horse trailer and drive up to the Cold Stream burn. That eastside country had been logged with donkey engines and railroads years before, so most of the large trees were gone. The fire had burned the second growth and thickets so clean that it was easy to ride a horse through the country. Upon arriving at the burn, Carl would load his saddlebags with lunch, water and extra cans of tree marking paint. He marked and cruised those burned trees from the back of a horse. When he drew the black marble, he would lean over with his Biltmore stick and estimate the diameter. Most of those trees were not over 60 feet tall so it was fairly easy for an experienced cruiser to estimate height.

I can still see Carl at the end of the day coming into the ranger station and unloading his brown horse, covered with spots of blue tree marking paint. In fact, the bridle, saddle, saddlebags and stirrups as well as Carl himself were all spotted! During the next thirty-one years of my career, I never saw another district ranger mark timber while on horseback.

HERE COMES THE JUDGE

Bob Gray

Judge Morgan was the justice of the peace at Junction City. A man very critical of the Forest Service, especially concerning "light burning." His wooded hillside above his home bore out his thinking because, indeed, it was a beautiful park-like area with no underbrush or litter. Each spring he let fires creep through the needles on his property, doing a thorough job of hazard reduction just as we of the Forest Service encourage home owners to do. His thoughts were good, but unrealistic from our standpoint, because of cost per acre and risks involved.

Though our philosophy was different, I liked the judge and determined to stay on friendly terms with him and also to continue my job in law enforcement within his area of jurisdiction. In fact, he was a challenge to me, so I made certain that I would get involved in all cases he handled regarding fire law violations.

The first case I brought before him was a licensed packer up Canyon Creek. Seems that one late afternoon his pack string stirred up a yellowjacket nest along the trail. In his frustration at having his pack string scattered and himself being stung several times, he set the yellowjacket nest afire. Well, I sometimes admire a man who takes charge and does something, even though it's wrong. He put the fire out, he thought, but a hiker coming down from the high country late at night discovered a small fire along the trail and at midnight he reached Weaverville, where he reported it to me. He had stopped its spread and said it was only about 10 feet in diameter, so I decided to go check it and finish putting it out by myself. On reaching the fire, about two miles up the trail and several yellowjacket stings later, I had the fire out cold. A little detective work like checking footprints of milling mules, a few horse apples, a chunk of lash rope and some gear scattered in the brush, correctly led me to believe a packer, probably from the trail head corral had something to do the fire. Sure enough, about daylight I stopped by the pack station in time to talk to the packer. "Yes," he answered to my question, " I burned out them damn yellowjackets. Every time I go by there they sting me and my mules."

I had everything for my case. An admission from the packer that he burned out the yellowjacket nest, a statement from the hiker that he discovered and worked on the fire, my own statement and four or five welts from yellowjacket stings on my anatomy attesting to my work on the fire. Armed with all this, I filed a complaint at Judge M's court.

On the given date to appear we all showed up, the packer, the judge and I. The packer plead "not guilty," and the judge confirmed his plea. I got a lecture on the beneficial effect of fire on the woods, including, "The Indians burned the forests and we didn't have all this underbrush that the Forest Service has allowed to grow." The admonishment to the packer was, "Now, next time be sure to put your fire out; we have to get along with the Forest Service."

I was flabbergasted, frustrated and mixed up. Who was on trial I thought, me, or the packer? Determined not to let this get to me, I asked the judge if I could talk to him after the trial. He then said the reason I lost was that the packer had done what any prudent man would have done in similar circumstances. Scratching my head I left his chambers. Next time things will be different, I mistakenly thought to myself. The next time wasn't long in coming. Mr. B., a small rancher up Connor Creek had a rather large burning job to do. He and I took about two hours going over the requirements on the burning permit. A day or two later, on checking his burning operation, I found a few details he was not adhering to. For instance, he didn't have adequate lines around his fires; they were not always attended during daylight hours and he paid little attention to the times prescribed to light new piles. In general, our two hour discussion, when preparing the permit was just so much smoke. I was pretty upset with his lack of compliance and poor attitude, telling him, "If you don't have all the permit requirements met in 24 hours, I'm lifting your permit and we'll visit the judge. Not only that, but we'll put the fire out and send you the bill." He seemed rather unconcerned as he grumbled some reply while I was departing.

Within 12 hours I had been dispatched to a fire on the Angeles Forest and had to tell Bernie Leas, the prevention man, of my threat to Mr. B. Bernie was to check on the conditions the next morning and carry out the issuance of a citation, if necessary.

When Bernie arrived the next morning, Mr. B. was nowhere to be seen; the fire had escaped and was on its way up the hill. About fifteen acres burned before the Forest Service brought it under control and Mr. B. had still not showed up. News via the grapevine travels fast, though not always accurately. Before the day was over, word got to me on the fireline of one of the many Angeles Forest fires I'd been on. Thankfully, the report I received was worse than the facts, as I was told several hundred acres had burned. They had just as well sent me home from the Angeles then, as I was not much good to them the rest of the shift. A telephone call to Weaverville put me more at ease when I finally got the facts from Bernie. "Get everything on paper, Bernie," I said, "and we'll at least 'hang' the guy in court." (Not a literal translation, you know, Mr. B.)

A week later, back home from the Angeles, we prepared our case. An airtight prosecution, with all the details on paper, plus the assistance of Don Kennedy, the Trinity County district attorney practically assured us of a conviction. Judge M. told me he would have to disqualify himself, as he was a friend of the defendant. This should really clinch our case.

At the appointed time, Bernie, the DA and I arrived at the judge's chamber. There sat Judge M. He had not disqualified himself. Neither had he told Mr. B. about the trial date, so he had to make a telephone call to his home in Redding. His conversation went something like this:

"Hello, Bill, this is Hank. It seems you're in some kind of trouble with the Forest Service. Something about a fire. When can you come over to get it straightened out? OK, I'll see you on ______."

Don shook his head in disbelief. My determination to be friendly with the judge was severely undermined. Well, on the new date, we met again, this time with Mr. B.

My diary, dated December 14, 1959, reads as follows: 12:30 p.m. to Junction City Court. Took Scott, Kennedy, Leas and McCaffrey. Presented our case. Mr. B, presented his side. Judge M. heard only one side of the story.

"Not guilty," sez the judge. Guess he didn't hear Mr. B say he lighted the fire that went up the hill. They should have this case on TV, it really sets a precedent.

Final comment, "I'll be damned!" With the final comment, Judge M. pointed his old crooked finger at me and said, "If you are interested in a contempt of court citation, one more word from you will do it."

I remained quiet as Don applied a little pressure to my shoulder.

My goal set in 1956, to develop a good relationship between the judge and the Forest Service had failed as I transferred to the Sacramento District a few weeks after this trial. This failure has bothered me ever since, as I feel part of my job was to bridge gaps of misunderstanding between our agency and the public.

DRY LAB CRUISING

Bob Bjornsen

It all started one fall at Thorn Creek Guard Station near the Imnaha River, Wallowa-Whitman Forest. Blen Holman, Chesnimnus Ranger, and I had been checking some winter C&H allotments and had decided to mix a little chukkar hunting with the work at hand, which was really to hunt elk up in the Red Hill country. Of course, you realize this was before purity codes were invented to spoil mixing work with pleasure.

For some unremembered reason we decided to saddle up the horses and head for Thompson Meadow Guard Station about dusk. We hadn't gone far up the trail when it turned darker than the inside of a cow and began to rain and sleet in buckets full. Our trusty horses had to pick their way up the trail without our guidance while we sat miserably in the saddle wondering who's idea this was in the first place. As lead rider my main worry was whether Blen's trail crew had limbed trees high enough to keep from sluicing me out of the saddle.

Eventually we arrived at Thompson Meadow with numb feet and hands so cold we could hardly unclaw them to take the rigging off the horses. Blen had been mumbling all the way about having killed the last of our bottle of spirits the night before at Thorn Creek. I let him grumble until the fire was lit in the old sheep herder's stove, before I brought out a reserve flask hidden in my bedroll. And so we whiled away a few hours listening to the wind whistle through the cracks in the old cabin.

During the night the wind picked up and really began to howl. We could hear branches breaking off the trees and an occasional tree going down. Our main concern was that one didn't topple onto the cabin.

Next morning the storm had passed and we decided to inspect the damage along the Thompson-Billy Meadows Road, driving a pickup truck that had been conveniently left at the guard station. We hadn't gone far before it became apparent there was enough blowdown of merchantable timber to warrant a salvage sale. A quick cruise was in order, so with Blen tallying and driving and me shooting the diameter/heights out the window, we sampled a swath along the ridge all the while marveling how the numerous hunter camps escaped being crushed by some of the falling giants.

Then it was off to Enterprise to call Owen Aydelott, W-W timber staff officer, that an estimated 10 million board feet of high quality pine had blown down and help was needed to put it up for sale by spring. Naturally, Owen wanted to know the particulars, especially how we arrived at the volume estimate. We danced around the latter when I said, "I'd seen some blowdown in my day, but this was the worst I'd ever seen and time was of the essence———."

Owen rose to the occasion and sent an SO crew out to help the understaffed district people get the sale ready. A year later Blen and I heaved a sigh of relief to learn the sale cut out within one percent of the cruise volume. After all, our reputation as pickup seat timber cruisers was at stake.

THE SIXTIES

THE ROUNDUP

Bruce Brown
(Submitted by James W. Caples)

April 14, 1992

Dear Jim,

You've asked for my recollections of the roundup and return of Frank Hussy's horses that occurred during the spring-summer of 1960.

I transferred to the Salmon Ranger District in March of 1960, from the division of timber management in the regional office. Other than summer work as a college student, the assignment to the Salmon District was my first assignment to a forest after graduating from college in 1956. Orville Engelby was the ranger at that time, and had been in the position for a whole week. To my knowledge, I was the first assistant ranger assigned to the Salmon District. Gene Powers was the supervisor.

Gene Powers was a man who believed that all people were honest and above board unless their actions proved otherwise and that proper range management and adherence to Federal Regulations was the inherent obligation of all citizens.

I knew little of Frank Hussy's past but stories relating to his acquisition of black angus livestock and the derivation of his "double meathook" brand would lead one to believe he was not an individual Mr. Powers would hold in high esteem.

While stories about Forest Supervisor's misfortunes are not readily circulated, it was generally understood that on some previous occasion Mr. Powers had single handedly rounded up a group of Mr. Hussy's trespass horses. While the details were fuzzy the circumstances of the roundup left Mr. Powers with no choice but to leave his personal saddle horse with the impounded stock while he traveled to town to prepare and deliver the paperwork necessary to implement the impoundment process. Upon completion of the legally required impoundment and billing processes Mr. Powers returned to the impounded livestock to retrieve his saddle horse. Mr. Powers found all of the livestock missing, including his saddle horse. A diligent search was made for Mr. Powers saddle horse over the next several days but the horse had apparently vanished from the face of the earth. The horse was later located some distance from the impoundment site and it had always struck the finders as odd that Mr. Powers horse had swam the Salmon River, traveled a considerable distance up a blind canyon and then pulled off and thrown away all it's own horseshoes.

At this point it may be helpful to summarize the setting. Here we have a crusty, self-made rancher who had enjoyed the privileged use of a significant portion of the national forest resources for a period of years facing the invasion of roads into most of his "domain" and the prospect of Forest Service personnel of all shapes and sizes swarming over his "private pastures" for the next three to five years. Entering the scene is the new steely eyed Ranger with instructions from the forest supervisor to get the range trespass situation under control and the new fuzzy cheeked assistant ranger. Sounds interesting.

I don't recall that I was aware in advance of the plan to round up Hussy's horses but in any event I was to appear at the Williams Creek Guard Station during the afternoon of the day of the roundup. The government cowboys were Range Staff Dick Hickman, Ranger Engelby, Ernest Marsing, and Jack Bunce. When I arrived at the guard station the die was cast.

The corral contained ten to twelve head of horses. Ranger Engelby would go to town to prepare and hand deliver the impoundment notice to Frank Hussy. Jack Bunce and I would remain with the impounded stock until they were reclaimed by Frank Hussy by payment of the trespass and impoundment

One of the first Southern California fire tankers. Fordson tractor with extended bed and six speed transmission. Supposedly could go up to 60 mph, but hard to steer. Towed trailer with 500 gallon tank - 1928.

Pony Peak Lookout, Klamath NF, California - 1928.

fees, or moved to another site for public auction. The Rangers instructions were clear—"Don't leave the horses."

The Guard Station had a small barn and corral in addition to the standard small dwelling. On this day the dwelling was occupied by Paul Fritz, the forest landscape architect, and his nervous Weimaraner Linda.

Evening approaches. Jack and I eat a meal of fire rations and prepare to bed down for the night. We bed down on opposite sides of the corral, each with his own thoughts of what the night might bring. The horses don't like us, the corral, or each other and it's long after dark before it's quiet enough to think about sleeping. Somewhere along about 3:00 or 3:30 a.m. the horses become a little more than restless—quite a little more. Jack and I talk to the horses, yell at the horses, and I think throw a rock or two at them. Jack believes a bear has passed across the hillside above the corral. No eyes are reflected in the light of a flashlight so we pitch a few rocks into the brush but no sounds of a critter leaving occurs. Things finally settle down and we doze for awhile but we're up and moving before first light. Except there's not much place to move to. Circling the corral is the best you can do and that gets old after awhile. A breakfast of cold rations does little to change the pattern. Full light arrives and no untoward action has occurred.

About 6:30 or so Paul Fritz walks down from the dwelling and offers an invitation in the form of hot coffee on the stove. Jack and I discuss the opportunity and here is where the young assistant ranger makes his first error. I believed that if anything was going to happen it would have occurred during the night or shortly after first light. With that erroneous decision in mind I elect to have both Jack and I grab a quick cup of coffee from the pot on the stove rather than going up one at a time, or asking Paul to bring us down some when it's ready.

The first cup is barely poured before Linda begins to growl and bark vigorously, all the while remaining very close to the back door. I stepped outside and walked a short distance beyond Linda towards the corral to see what Linda was seeing. Nothing was moving that I could see but the corner of the barn prevented me from seeing the entire corral. By this time Jack and Paul had exited the backdoor and Paul was finally able to quiet Linda. Error two. I conclude that Linda being the nervous dog that she is has been barking at some small creature and nothing is amiss. Jack and I finish the poured coffee and return to the corral. As soon as the corner of the barn no longer hides a portion of the corral from view I know I've had a very expensive cup or coffee.

The horses are gone. No dust in the air, no movement on the hillside or down the drainage, no sound of a horseshoe hitting a rock. Nothing. A quick swing across the hillside above the corral shows a bunch of horses have moved through the area in a hurry. A quick circle of the corral reveals the upper and middle rails have been kicked off between two posts and the tracks lead straight to the wire gate in the small fenced in pasture adjacent to the corral. It's at the wire gate that I begin to get a measure of my adversary. The wire gate is still upright and wired snugly to the post. There are no human footprints near the gatepost or the gate. Whoever opened the gate did so without leaving his horse and rewired the gate to the post after the impounded stock passed through. Through my anger and chagrin I'm impressed.

Jack's not saying much and I silently thank him for that. Now it's time to call the ranger.

Back to the guard station and crank the phone.

"Orv, I lost the horses."

I'm not sure what to expect but the silence is deafening. I no longer remember his response other than it was restrained and appropriate.

Aftermath:

The following spring Ernest Marsing was doing some work on the hill above the corral and found a slightly weathered Model 94 Winchester leaning against a down log overlooking the corral.

I've often wondered how things might have turned out if only one of us had gone for coffee. Did Frank have a pocket full of dollar bills or just a handful of 30-30 cartridges?

I got a "minus" in range management on my annual review that year.

/S/ Bruce Brown

FIREARMS AND MULTIPLE USE MANAGEMENT

H. M. Lilligren

During my stint as ranger from 1954 to 1966, firearms were tools used in the operation of what is now the Tiller District of the Umpqua National Forest. We used firearms at Tiller for erosion control, conifer seed collection, rodent control, campground safety, and in one instance, for fire suppression.

In the early '60s, a 300 foot section of the South Umpqua Road up river from Boulder Creek Campground began to move downhill towards the river. It was sinking just a matter of a few inches a day. The timber purchasers were hauling heavily, so we allowed them to grade the edges of the breaks and continue to haul.

One afternoon, Del Godbey, an engineer on the district, told me the earth movement was accelerating and recommended we close the road until repairs could be made. I asked the office force to inform the timber sale operators by phone of the road closure. I went upriver with Del and we placed warning barriers. Then, we started getting hell from the operators. I refused to budge. By daylight the next day, the entire section had dropped six feet and the criticism stopped.

It was an interesting earth movement to watch. Some trees on the slide area below the road had their roots anchored in solid rock, and some were in soil. By marking trees and making simple transit readings, it was found that some trees were "walking" gradually downhill between other trees that were standing fast with all trees still standing in a vertical position.

Del and I were examining the top of the cut bank above the slipping road section several days later, searching for a possible source of surface water that may have provided the lubrication for the slide. Machinery was working both ends of the slide area when the whole mass slowly moved at the rate of about 10 feet per minute down into the South Umpqua River canyon. The trees slowly wrapped themselves around each other and when the strain became great enough, broke like wooden match sticks.

The earth mass pushed across the river and up the other bank for a short distance, temporarily creating a dam and a small lake. The raw slide was an eyesore and I thought there must be some way of getting some kind of vegetation back on it without bringing in a helicopter. The slide area was too sloppy to walk across. It was almost like quicksand.

So, I used a shotgun—my own! I got some "roadside revegetation" grass seed mix from our supply; got some of my own clay pigeons and filled them with seed. A paper disc was glued over the rim to hold the seed in the saucer. I then took my son, my shotgun, my automobile, my hand trap (target thrower) and a box of shotgun shells purchased at government expense and headed upriver. When we reached the slide, my son threw the targets and I tried to break them. The paper disc destroyed the aerodynamic properties of the clay targets and they flew like wet pancakes. When I hit one, it spread seed over an area about 40 feet in diameter. When I missed, the target hit the mud and broke, later resulting in a very dense patch of grass about 18 inches in diameter. The grass cover didn't readily solve any erosion problems, but it made them a little less noticeable. It cost Uncle Sam a few pounds of grass seed and a box of shotgun shells. I don't recommend this as a general practice, but there are more uses for firearms than just committing crimes.

All Forest Service people know that when an injury or fatality happens on FS land, all rangers eventually get a directive to correct the cause, whether or not there are any funds allocated to do the job. One time the Umpqua rangers got directives to remove all the dead and dangerous limbs from the trees in the campgrounds. Tragically, somewhere some member of the public had either been injured or killed by a "widow maker," a dead limb falling out of a tree.

I didn't have a major problem "tree limb wise," but the danger was there. I didn't have any money to hire a climber, so I thought, how about using a rifle? I made sure to use it in the off season so some alert citizen camper couldn't write his congressman and report that an irresponsible forest ranger was shooting hell out of the trees in the campground.

I worked up a hand load for a 30-06 Springfield (World War I vintage) with a 110 grain soft nosed bullet. Arriving at the campground with the most numerous dead tree limbs, I started shooting. I soon discovered that by shooting at the base of the limb from a safe distance, the wood exploded and down came the hazardous limb.

So, firearms can be an economical and useful forest management tool in special cases.

Then, there was the case of the illegal use of a firearm in the South Umpqua Falls Camp area, by a person unknown.

I transferred to the Rogue River Forest in the spring of 1966, where I became the liaison officer representing the forest in mutual planning with the Corps of Engineers on the three proposed dams involving the Rogue Basin Project.

Several months after my transfer, my friend Homer Oft, the venerable fire control staffman in Roseburg called me. He wanted to know if I had ever stored dynamite at the South Umpqua Falls Camp. I said there might be one or two cases in a sheet metal shed back against the base of the hill. Our nearest explosives storage building was in Roseburg and it wasn't very practical to drive in for a few sticks of powder, when needed, and return the unused explosives to Roseburg every night—a total of 260 miles a day.

All my previous requests for a powder magazine on the district had been denied.

On a Monday morning, a Tiller employee driving past the South Umpqua Falls Camp noticed a change. The metal storage building was gone. Further examination showed the surrounding trees had bits of metal imbedded in their bark. There were also many small pieces of hide and flesh spread over the area and a dead deer badly mangled in the vicinity.

It appeared that someone had gone down the road on Sunday night, had spotted deer over by the metal shed and fired at one of them. The poacher may have hit a deer, but he also hit the dynamite. From the evidence, it appeared that one deer was completely fragmented and the second one only partially so.

No one ever admitted the shooting, but it must have been a big shock to the poacher to see his targets disappear in a bright flash and loud explosion. I think someone was trying to give him a message!

ALIENS

Chuck Smay

It was one of those typical Sunday evening responses sometime back in '61-'63. A call through the dispatch channels from the Shasta-Trinity dispatcher gets you going down canyon from Junction City, to the Lower Trinity Ranger District office at Willow Creek. After the normal standing around waiting for directions, we are loaded into a covered Jeep for the ride into the back country. Naturally, since 4x4 vehicles are in short supply, we compensate by putting twice the designed vehicle capacity of firefighters in the vehicle for the trip to the fire. Finally, we reach the "real" 4x4 road, which is actually little more than a couple of ruts that someone in the distant past considered "passable." Fortunately, it is pitch black by this time so those of us who are serving as "filling" for the inside of the Jeep cannot see the roadway hazards that the cramped driver may be encountering. The extra human mass inside the vehicle does little to quiet all the squeaks and rattles of the old Jeep as it claws up the set of tracks called a Jeep road. We finally learn that our destination is a place called Red Cap Prairie. After an hour or so, we start to see a red glow off in the distance and have at least some assurance we are on the right road. Just as we reach a small clearing, which is to be the fire camp, we hear a loud snap from the underside of the vehicle. After a bit of "fiddling around," the driver determines that we have just lost the drive line and this is the end of the road for us and the vehicle. We try not to think about the consequences of a similar drive line failure back on some of the steep stretches of the trail.

Even a casual reading of Forest Service history in Northern California soon makes it evident that the Red Cap area has provided several generations of firefighters with some memories for books like the one you are now reading. Little did I realize that this Red Cap Fire was going to provide me with one of the strangest sights I have ever seen in a fire camp. The fire was rather uneventful, just a few days of hard work on the side of the mountain without any public fanfare, since it was in the outback even for the remote Six Rivers National Forest. It was late fall and after about a week we got a weather change. One morning I woke up just as daylight was making its contribution to the day. Something did not seem right. Perhaps it was the fact that the inside of my paper sleeping bag had a definite damp feeling. Normally there is some hesitation before leaving the sleeping bag on a cold morning, but not when the inside of the sleeping bag is wet. About then my eyes and brain registered there was a very dense fog and light drizzle engulfing the area where the base camp was set up.

The next thought was to get some hot coffee to help get warmed up! It is amazing how fast one can get dressed, even into cold damp clothes and boots when motivated by the need for hot coffee. Hasty steps took me toward the kitchen area, where I stopped dead in my tracks. There before me was a large group of alien beings! What my cold damp mind finally registered was a scene of dozens of cold damp firefighters with their damp brown paper sleeping bags inverted over their heads. There were ragged holes torn in the improvised cover to permit at least one arm to extend to the outside. In it was clutched a cup of coffee that the occupant had attempted to direct to the torn mouth hole in the bag. Artistic brown stain patterns ran down the front of the bag indicating the occupants were having some coordination problems getting the warm liquid to the intended location. These brown rain suits also contained eye holes to enable the occupant to negotiate around the camp, primarily to once again locate the coffee pot. Some of the bags were so wet they were like formfitting uniforms, but it was impossible to recognize even close friends. The intensity of the beady eyes in each rain suit made it clear this was not at the moment a funny situation or the time to make smart remarks. So, I dutifully improvised a brown sleeping bag raincoat for myself, went and got my coffee and became one of the aliens at Red Cap Prairie.

After some thirty intervening years, this is still one of the most humorous sights I have seen in fire camp.

BIRTH OF THE WINEMA NATIONAL FOREST

Dan Abraham

The first I ever heard of something called the Winema National Forest came in a phone call to me at Missoula, Montana, in January 1961. About 30 of us, mostly Forest Service types, mostly rangers from all over the country, were there for a month long attempt to lay a veneer of management skills and couth over the rough grain of our general natures as bushy tailed young foresters more comfortable talking to trees than to people. At any rate, someone in R-6 personnel called and asked if I would like to move from Gold Beach, ASAP, to Klamath Falls and be fire, recreation and lands staff on the Winema NF. I said I had never heard of such a forest and was told it did not yet exist, but was about to, and they wanted to get people in place and organized before fire season started and so forth. I said I would rather be timber staff and was told take it or leave it, so I took it. Rode the train to Portland (Northwest Airlines was on strike), flew to Medford, met my wife there, drove to Klamath Falls and bought a house. That was scary, being our first venture away from Forest Service housing. The cost was $14,400, and the monthly payment was $65. We were worried about being able to make it. Apparently the bank thought we could, so we moved in around mid-March.

The first SO was in a small storefront place on a side street half a block south of the main drag in the old business district. The office was just south of the old Winema Hotel, which was near the east end of the old business district along the main street (which may have been Main Street, but you can tell it has been a long time since I have been to Klamath Falls). The one room former store was chockablock with desks and such, and the new SO was in business. Alex Smith, the new supervisor was there, and Bill North, the new AO. I guess the rest of us showed up at about the same time: Norm Gould, timber staff; Kjell Bakke, forest engineer; Chuck Waldron, range staff; John Grove, assistant fire staff; Bob Crittenden, personnel officer; and the many others whose faces I can picture, but not the names. The office soon expanded into another storefront next door. Not too long after that it was moved further downtown, west along that main drag near the Klamath County Courthouse. This was a second floor place, the former meeting hall of the Knights of Something. Most of the space was a large meeting room and dance floor, with a stage at one end and some small glassed in offices at the other. It was crowded. I have a vivid memory of a visit by Chief Ed Cliff. He took the time to go to every desk on that dance floor and shake hands and say something to every person. He charmed us all.

While we were making do in storefronts and dance halls, an addition was being built onto the post office on the south edge of the business district. The SO moved there probably sometime in 1963 or 1964. We thought it quite palatial, but it was very modest compared to the present edifice on the north edge of town.

Back to the birth: The Winema was declared a new national forest by the President in July 1961. Prior to that time we may have been operating as an adjunct to the Rogue River NF, or something like that. Most of the land came from the Klamath Indian Reservation. Added to that was the Klamath District of the Rogue, essentially all the land east of the Cascade Divide. Also added was a slice off the south edge of the Deschutes and a piece off the west edge of the Fremont. In all, it was about a million acres.

After WWII, national policy shifted toward allowing Indian tribes to become more autonomous and less dependent on the BIA. In the late '50s, Oregon Senator Richard Neuberger shepherded legislation through Congress which allowed the Klamath tribe to sell their reservation if they so wished. The law required that the land could be sold only in large blocs, and that the purchasers agree to continue management under the principles of sustained yield and multiple use. Land not bought by private interests would be purchased by the federal government and become national forest land. About three-fourths of the tribe voted to sell and one-fourth to keep the reservation intact. So about a fourth of the land, mostly lying just north and east of Chiloquin was set up as a management unit with the U.S. National Bank of Portland designated as trustee to manage that land for the benefit of the remaining members. The rest of the reservation was put on the market. Crown-Zellerbach bought 90,000 acres lying along Highway 97 south of Chemult. All the other became national forest land and the heart of the Winema. Years later the remaining members sold out and that U.S. Bank management unit became part of the Winema.

If I remember correctly, the Forest Service recommendation in 1960, was that the former reservation land be added to the Deschutes and the Fremont and business would go along as usual. The people in the Klamath Basin did not want that. They wanted "their own" national forest, with headquarters in Klamath Falls. From that grass roots political energy, the Winema was created.

The name Winema honors an Indian woman who served as an interpreter between the Modocs and the U.S. Army during the Modoc War of 1872. She was related to Captain Jack. The name "wocus" was also considered, which is the Indian name of a type of water lily used as a food source by the natives. We joked at the time that whatever name chosen had to come after Willamette alphabetically, so as not to upset the numeric listing of R-6 forests for the punch card system, that archaic forerunner of the present electronic magic.

It was a big advantage to have the local people involved in the creation of the forest. It was "their baby," and we had good support by the local paper, and various civic groups and the county commissioners. There was some grumbling by cattle permittees and other users about our higher standards of land management, after years of dealing more casually, apparently, with the BIA.

Three ranger districts were created. The Klamath District came intact and functioning from the Rogue. Earl Karlinger was the experienced ranger. The Chiloquin District covered the central part of the new forest. Homer Faulkner was the ranger, with headquarters in the well-worn buildings at Klamath Agency near the north end of Upper Klamath Lake, former headquarters for the BIA. A new ranger station was built later on Highway 97 on the west edge of Chiloquin. The north portion of the forest became the Chemult District. Doug Shaw was the first ranger. The ranger station was in the abandoned one room little red school house on the north edge of Chemult until a new station was built next door.

That first summer was a hectic time. Somehow we got enough people and resources in place to get the various programs into operation. Timber sales were made, cows were grazed and there were no disastrous fires. We did spend a lot of time being lost on the former reservation land. The tribe did not want visitors on the land, so there were no road signs. But there were roads galore and all unsurfaced. Kermit Lindstedt, in an inspection report, described it succinctly as "a bewildering multiplicity of roads." Alex came up with the idea of equipping every vehicle with a supply of box shook (thin pine boards used to make shipping cases for fruit), shingle nails, lumber marking crayon and a hatchet. Whenever you identified for sure a specific point or intersection you were supposed to put up a sign, hopefully with arrows and distances to help wanderers find the way back to the highway. The first map we had was barely adequate made by pasting together parts of maps of the Rogue, Deschutes, Fremont and the reservation. And of course, the BIA map showed very few of the roads that existed on the ground. By the following year we had a new map, roads numbered and standard signs replacing the box shook.

Looking back with nostalgia, those of us lucky enough to be pioneers on a new forest can pat ourselves on the back and tell these tales of how great it was. Not so at the time. Then we were just charging ahead, working hard, doing what had to be done in the normal can do spirit of the Forest Service. It never entered our minds that there was any task too tough for us to do and do well.

THE NEAR MISS

Gary Munsey

In 1962, Ranger Joe Schmeller and I were cutting firewood above the Westfall Ranger Station and above state Highway 41.

We had just completed a cut on a downed black oak approximately 40 inches in diameter that was lying on a steep slope parallel to Highway 41, some 30 yards below.

Allow me to digress from the story momentarily to ask if you have ever entertained the question, "why do objects roll?" The answer is, "because they can." And that's what happened—the freshly cut oak round began its inexorable roll down the steep slope, gaining speed and momentum, while we watched with the sinking feeling that only comes with the realization of certain disaster. As the rolling round neared the cut bank above the highway, a car passed by at 60 miles per hour. The oak round then hit the highway behind the passing car and oblivious driver and continued its trip across the highway—two seconds before another innocent driver passed by.

I have never crunched the numbers in whatever physics formula is applicable, but I would guess that a 300 pound oak round at full speed striking a vehicle traveling at 60 miles per hour could ruin your whole day.

DEALS, MEALS AND WHEELS

Don Bauer

Section 33 at the head of Fish Creek was owned by Dale Gentry. Fish Creek Meadow is within it and two thirds of it, the southwesterly portion, is all within the San Gorgonio Wilderness. That was the key. We wanted to consolidate the land ownership within the wilderness because we could see the wilderness bill coming down the road. We wanted that section very badly and I got authorization for some limited funds to buy it if we could negotiate its purchase. Dale Gentry was a multimillionaire and was agreeable to considering an offer from us.

Warner Hodgdon, a local land developer, wanted Fish Creek Meadows to put a subdivision and mobile home park on. The road to Bean Flat had been built when Southern Pacific had a timber sale above Heart Bar. In the dickering that followed it was like a high-low poker game with Gentry in the middle having us bid against each other. I had a lid; the maximum we could pay for it was the appraised value of $500 an acre for 640 acres. I made that as our top and final offer. Gentry was inclined to accept it as a fair price.

We arranged to have dinner at the California Hotel to continue negotiations. Following dinner, Gentry opened the discussion by addressing Hodgdon with, "The ranger says he'll give me five hundred bucks an acre. Now, it's up to you, Warner. What's it worth to you?" Warner stated he could beat that figure, but his offer involved a house at Lake Arrowhead, along with cash. Gentry was having fun; he wouldn't cut this thing off with just one meeting. No, he'd drag it out because he enjoyed the bargaining process, especially since he'd paid $5 an acre for it from Southern Pacific many years before.

Gentry had been president of the state fish and game commission under Governor Warren and for several governors before him and down in his heart he was a true multiple use man. When it came down to weighing the bids that Warner was able to offer, compared to our offer, topped out as about equal. He'd say, "Is there anything else you can throw in Warner?" Well, Warner would throw in a boat dock at lake Arrowhead. Then, Gentry would turn to me and say, "It's up to you, Don, what have you got to throw in?" I said, "I'm out of string, but I can assure you that if you convey that to the Forest Service, it will be managed under the laws governing the national forest, pretty much like the land that surrounds it. That means multiple use, water, timber, recreation, grazing and wildlife." He felt very strongly about that, so he said he would consider the two offers.

I came back to him a couple of days later. The RO was bugging me about whether I was going to close the deal or not and I took Gentry on a trip. His favorite country was Sansevaine, where he grew up as a small boy with his family, as tenant farmers under the Sansevaine people who owned a lot of land. Gentry had come from Missouri where his father came down with tuberculosis and was given six months to live.

He had three younger sisters and an invalid father, who just spent his time sitting in a rocking chair on the porch waiting to die. The doctor said if he moved to a warm, dry climate he might live a little longer. The old guy lived to be about 80, but never did a lick of work. Gentry thought a great deal of his mother. He would have been about 13 or 14 at this time, in the early 1890s. He told the story that he had an old, gray horse and his mother would give him two bullets to get meat for the family. They were very poor. He would go up Sansevaine Ridge, from the mouth of Sansevaine Canyon, and he would hunt for deer and bighorn to get meat for the table. Once while he was up there, he had shot a deer, but he crippled it—didn't kill it, so he left his horse and tracked the deer. While he did this, a storm had moved in and there was a clap of thunder and the horse got frightened and ran off and he found himself up there on foot when it started to snow and hail. He spent the night curled up between two sugarpine trees, covered himself with pine needles, and he survived. He told me, "Don, those trees saved my life when I was a youngster and I ask you as a favor, if you could do whatever you could to ensure that they are never destroyed." I said, "You have my word that those trees will never be cut as long as I am with the Forest Service, and I will pass that word on to my successors." They are historic shake trees. Incidentally, they're twins; they're right together. So, I built on that. I said, "Dale, you have my word on that. How about Fish Creek Meadows for $500 an acre?" He said, "Now, that's one of the things you can do that Warner could not do. One other thing," he said, "I'd like to have a trailer site where I could come up here with my little trailer." So, I said, "Is that a deal then?" He said, "Yeah, you give me a trailer permit, where I can bring my little trailer and come up here and spend an evening and enjoy the country here and think of my days as a boy, and maybe do a little band-tailed pigeon shooting." So, I swallowed hard and said, "Let me go to work on that."

We had very rigorous rules about issuing permits for trailers, which are viewed as recreation residences and should only be permitted in approved campgrounds or areas approved by the RO. This would have been considered as a recreation type use. There's no way you could have got that through without three copies of an approved plan and answered a hundred reasons why you can't do it.

I was struggling and Everett Jensen, who had helped me on land acquisition when I was a district ranger and was one of my friends in court in the RO, gave me this advice. He said, "Well, Don, you as supervisor don't absolutely have to consider that a recreational use." I said, "What do you mean?" He said, "Well, you tell me that Gentry is a great student of wildlife who is a past state commissioner of fish and game. Maybe you could consider a special use permit for a wildlife study area or observation post, which would not involve the division of recreation." I said, "Enough said." My friends in the divisions of lands and wildlife management had helped me out with a good idea.

I issued a heavily conditioned life tenure permit to Dale Gentry to expire with his death, authorizing him to have a trailer for wildlife observation purposes at Sansevaine Well. When I delivered and reviewed the conditions of the permit with him, he signed the agreement, we shook hands and Fish Creek Meadows became national forest land. That's how we got Section 33. This consolidated the San Gorgonio Wilderness in public ownership.

Dale Gentry never did put a trailer on the pad, but I would take him up there every spring when the tiger lilies were in bloom and he'd reminisce. He just loved Teddy Roosevelt, the Forest Service and what we stood for. He was the cattleman and accused as a "raper of the land" and all of that, but when the chips were down he said that the public deserves Section 33 to be managed for multiple use forestry rather than just another private resort.

STORY OF A WATER TROUGH

Benny G. Parmele

A guy by the name of Gus Zoglman lived in the Silver Creek Valley, 25 miles west of Burns, Oregon. He had two and one half sections of land which joined the Ochoco National Forest, Snow Mountain District. It had a small house and a well.

In the late '30s and '40s, Gus made trees into water troughs for cattle and wildlife. The trees he used were 30 inches to 36 inches in diameter and the troughs were 25 to 30 feet. He used the old crosscut saw. He would find a good spring, clean it out and make a spring box. At first he had to use a team of horses to pull the log up to the spring. Then he would drill holes in the log, about six inches apart and about half way into the log. He would then place rags that were soaked in oil and kerosene into the holes. They would burn slowly and some would burn together. He next cut away the center wood and pipe in the water. He worked with a galvanized pipe. As long as water was kept in the troughs they would last for years. Some are still working today.

Around 1963, I asked District Ranger Cal Weissenflint about making some water troughs. Frank Barnhart, the timber officer marked the trees. At Hollenbeck Spring, Elmer Hurd and I worked on the troughs. We had it a lot easier than Gus. A tractor was used to pull the logs to the spring. Power saws were used to cut the logs. We cut halfway into the logs and walked backward to cut both sides about 12 to 18 inches wide. Then we chipped it out.

I left the district in 1966 but came back in 1970 and the water was still running. It was good to see. Came back in 1981 and someone had thought a metal trough was better and had left the log troughs to rot away. Was back in 1996 and no water was coming out of the pipe. Looked as if no one had cleaned out the spring box for years. So now even the metal trough has no water. Of course, the spring was still running. It's so sad to see things that have gone to hell.

THE FLOOD OF '64

Bob Irwin

(Helped by memory joggers from
Bill Dean, Gary Munsey, Ed Tonnesen and Bruce Van Zee)

I write this in January 1997. Big time flooding has occurred in central California since the first day of the year. Cities and towns have been evacuated because of levee breaks and associated inundations. TV reports equate this experience with the floods of 1995 and 1986. Some historical references are made to 1955. Very few recalled the trauma of 1964 that struck the northern parts of California that year. Here are recollections about one small part of that flood.

I took the district ranger job at Gasquet, Six Rivers NF, in July of 1964. Having served the two previous years as assistant ranger on the Mad River District (same forest) I knew that Gasquet was pretty wet. Some past years, 120 inches of rain had been recorded at their weather station.

When I asked the old-timers on the district about the rainfall they said, in effect, "Those figures are correct. You won't believe how hard it can rain here. It will get bad and then it will get worse."

The house at Gasquet was the only two story ranger's house in Region 5. It had been built when Gasquet was a part of Region 6, before the Six Rivers was formed in 1947. The house had been built from a *Better Homes and Gardens* magazine plan during a ranger-staff meeting. Merle Lowden was on the staff at that time. He came by Gasquet in 1966, and told me the story of that house raising. Supervisor, staff and rangers all pitched in and the house was finished in two weeks. (Can you believe that people actually worked at meetings?)

I tell about the house because of the night of December 21. I was on annual leave because the 22nd was our eighth wedding anniversary and the 23rd was my birthday. We had big plans for those days but they were not to be. Jean and I were in our downstairs living room watching the only TV channel available. It was raining hard (as it had been for days) and then the intensity increased. We turned up the TV volume so we could hear the program over the din on the second story roof. Then it rained harder and we turned up the volume again. We repeated this exercise one more time and then gave up.

The local folks had told us that we wouldn't believe how hard it could rain and now we thought we knew what they meant. We didn't know we were into the worst storm of the century.

Early in December snow came to the district. It left about six inches at the ranger station, a good deal more over higher portions of the district, and even covered Crescent City (19 miles west at sea level) with an inch and one half .

Then it froze. And it snowed some more. The Forest Service radio relay on Red Mountain quit working because of icing. That cut radio communications to the supervisor's office and some points on the district.

Then it warmed up. Snow melted, rain kept coming. The rain came in deluges and denuded most of the snow cover. On December 20, things slowed down. There was a lull in the storm and all of us relaxed a little bit. Assistant Ranger Ed Tonnesen and I went down to the bridge over the South Fork of the Smith River. We had had about 17 inches of rain on top of the melting snow by that time. We saw whole trees (crowns, roots and all) moving down river and the water level rising. We knew that we were facing problems upstream but there was not much we could do at the moment. We had already seen moderate bank and fill failures on several roads, including Highway 199. Other district personnel had reported similar failures on timber sale roads.

I didn't know how much more snowmelt was upstream, and I didn't know how much more rain was ahead. I started pulling all of our people out of the woods. When I did, I felt like Chicken Little (there is a vulgar expression about chickens and their life processes that fit my feelings better at the time). I said to myself, "Here you are, a new ranger, and you are afraid to get the work done."

Then came the night of the 21st. According to the *Humboldt Times* (Eureka, CA, 12/23/64) and my notes, another 17 inches of rain fell in 24 hours.

At six o'clock in the morning on the 22nd, the district clerk woke me up with a phone message. She said I had better get over to the office because there was "big time trouble." She was right. The bridge over the Smith River on State Highway 199 east of the ranger station was gone and the highway west of the station was under water. Gasquet was isolated. Several of our employees were voluntarily trying to save some mobile homes in lower areas along the Smith River adjacent to the highway.

It continued to rain through the 22nd. Everywhere I went to check conditions (which wasn't very far in any direction) the rain beat down on my old Bullard wide-brim hard hat so hard that the person next to me had to shout to be heard. Sometimes we had to get back in a vehicle to talk. That wasn't much better.

The 22nd was a day of confusion, consternation and impotence. We knew we had limited communications at best. There was no travel over most roads, power and telephone outages were happening on a regular basis and close to 300 people were essentially "trapped" in this situation. We did have a California Highway Patrolman, a Del Norte County sheriff's deputy and a state game warden locked in with us in our small area. The California Division of Highways (now Caltrans) was located just to the east of the ranger station, but they were as immobile as the rest of us. A couple of loggers were also in this mix. So what? What do we do now and how do we do it?

I called a meeting of my staff and all the others for the morning of the 23rd. We decided that given the situation, we could develop an effective group by utilizing the Forest Service "Large Fire Organization" model (now replaced by the incident command system), adapted to our situation. There wasn't any choice. We were constrained by events. We all needed to do something, even if it was wrong, and you couldn't do very much wrong in a three mile square area. The fire organization and guidelines gave us a chance to do something right.

I don't remember the exact organizational structure, but we set up law enforcement, recon and intelligence, planning and a line boss to get the work done. Law enforcement went to the state and county folks, recon went to the Gasquet timber organization, planning went to Ed Tonnesen and I, and the work went to Hank Mostovoy the district fire control officer, assisted by Bill Dean and help from district personnel, loggers and other volunteers. District Clerk Joanne Dean became our "message center" and all agreed to get any news, no matter how insignificant to her.

Nothing went well the first days. Confusion and feelings of impotence in some people still prevailed. And then good things began to happen. The river went down a little bit west of the station. We got emergency vehicle access to Crescent City. Mostovoy and Dean forged their way into Big Flat over steep, muddy, and partly washed-out French Hill Road. They brought supplies and medicines to the 25 or 30 inhabitants of this small community on the south part of the district and extended our limited access.

Forest Supervisor Wes Spinney took advantage of the first safe day to fly over the district. On the direct radio channel (Red Mountain relay was still out) he said to me, "Do what you have to do, don't worry about constraints. I will back any decision you make. Your people will get paid for all hours expended on flood work."

Mr. Spinney was true to his word. We did all kinds of things within our local organization that, under different circumstances, would have caused heart failure amongst forest and regional staff who were trained to detect the slightest deviation from policy on the districts. There were no gripes and no requests to explain actions. After Wes gave those directions, the next three weeks were the best I'd spent in the Forest Service up until that time.

Spinney also told me that three US Army helicopters would be assigned to the Gasquet District. Three Hiller 12-Es arrived the next day. They were two-person aircraft, the pilot and one passenger. They came from an Army contract compliance testing base in Texas. The pilots took the copters right off the assembly line and put them through minimal performance tests and then came to Gasquet. The process had put about 100 hours on each ship. Time for a maintenance check, but no way to get that done now. I was not thrilled by this news.

The flight commander was Captain Billy Mitchell, a name out of history. I sat down with him and reviewed Forest Service flight safety rules. He listened patiently and then said, "Don't worry about me, I am a 'DFC'." I thought he meant, "Distinguished Flying Cross" recipient, but he said, "No, that means 'Devout Flying Coward'." After that, I knew we would be OK.

Mitchell said he had only one problem. His "turn and bank indicator" had ceased to function during the trip to Gasquet. That was not safe in cloud and fog conditions, especially in the canyon conditions we were to fly in. He asked if we could get him a large bolt and about two feet of string. We did that, and he made a plumb bob that hung from the center overhead strut of the canopy. Mitchell kept his eye on that device, always kept it in a vertical position and we never had a problem.

Two of the three assigned copters were diverted to other missions in short order. We did get one flight into Big Flat from copter #2, and then it went to Crescent City or thereabouts.

I flew with Mitchell two days. Our first trip scanned Highway 199 for stranded motorists and highway damage. There were no stranded motorists but there was a tremendous amount of damage.

As we went upriver we also examined the existing habitations along the river and Patrick's Creek. The first cabin east of the station (also east of the westernmost bridge washout) belonged to an old miner whose wife was in the hospital in Crescent City. Mitchell told me he could not get down there because of a large cedar tree intruding on the only landing site. When we got back from the recon (no radio compatibility available to communicate from the copter), Bruce Van Zee, Phil Dean (Bill's brother) and another fellow started toward the miner's cabin to fell the cedar. They went up the old Patrick's Creek Road which was rapidly disappearing and then hiked down a steep and crumbling ridge to the cabin. They felled the cedar and made a helispot. (More on this later.)

We landed at three other spots where I knew there were people. At each one we dropped off a case of military "C" rations and asked what other help was needed. None of these landings was routine. Mitchell wound his way down through tangled power lines, downed trees, flying snow coming off steep banks and light weight flood debris that got thrown about by the rotors.

At Patrick's Creek Lodge the caretaker said nothing was needed although the lodge was awash in about two inches of water. At our second stop Mitchell made a one skid touchdown on a small ledge. I jumped out and ran to the cabin of another old miner. He said, "If you can't bring back a case of Scotch, then don't come back." We didn't.

Ed Carlson lived on the east side of Patrick's Creek about a mile upstream from the lodge. Mitchell started to set down on a sandbar near Ed's cabin and I started to jump out. Mitchell grabbed me by the strap on my hard hat and pulled me back. The copter was sinking in quicksand! He powered out of the goo and tried two other places before he found solid ground. Carlson had come out of his cabin to watch the air show. He, like many others, had never seen a helicopter jumping up and down in his front yard. When I asked Ed what he needed in addition to the rations he said, "A case of dog food and some snoose." We delivered the dog food and four cans of Copenhagen the next day. (I paid for the snoose.)

Our recon on that first day was cut short by bad weather but we were up again on the second day. Confirming the "quick and dirty" observations of the day before, we found that Highway 199 just wasn't there anymore. Along the 22 miles between Gasquet and the Siskiyou Summit, four major bridges had washed away, slides were present almost every mile and roughly twenty percent of the pavement had gone into the river. Forest Service and county roads had faired no better. (Later ground recon found more than 50 slides on 70 miles of FS roads and every culvert plugged or washed out.) The Gasquet District lost all or major parts of four campgrounds.

After our second trip, Mitchell voluntarily went back to the helispot made by Van Zee, picked up the miner and took him to Crescent City to be with his wife.

No local federal, state or county government was equipped to deal with the challenges that existed. Still, progress began once the rains diminished.

By mid-January, 1965, the California Division of Highways repaired Highway 199 west to Crescent City. They began working with the Army Corps of Engineers and installed temporary bridges and "pioneer" road grades that provided access to the east. The Corps also established a "ferry service" to replace the Klamath River bridge that had washed away. That opened access from Crescent City to the supervisor's office in Eureka and ended my short period of independence. From then on, ranger meetings were held every two weeks in Eureka. Standard regulations and bureaucracy crashed back into our procedures.

On the ground, forest and district personnel worked to re-establish access to current timber sales, refurbish damaged campgrounds and repair the miles and miles of damaged watersheds. It was slow going for a long time. When I left the district four years later, we were still doing some type of flood repair work on roads, trails, timber sales, special uses and campgrounds.

Wes Spinney had planned to retire in early 1965. When the whole Six Rivers suffered so badly from the flood he elected to stay with us until we were back to some semblance of normal, which took close to two years. He had served his time, his wife was ill, but he did not want to hand off a seriously injured forest to a new supervisor. I remember him for that.

I also remember Billy Mitchell. I rode with many copter pilots in the following twenty years. If I had to grade them all, Billy would come out on top. I lost contact with him when the Vietnam war heated up. I hope he is still alive and well.

LAUNDERING MONEY

Joe Church

A few days after the December 1964, storm and flood, the Orleans Deputy Sheriff Slim Seward, drove into the Orleans Ranger Station to our house. He got out and grunting, pulled out a great big bucket full of muck and dirty water. He said the owner of the Orleans Hotel had finally been able to dig his way into the ground floor of the hotel and, among other things, got the safe open. He had emptied the contents, muck, water and all into the big bucket and asked him, Slim, to take it someplace safe until he could clean up the hotel. So, Slim brought it all to the ranger station. "But Slim," I said, "We don't have anyplace on the station to keep anything valuable." He said it was the only place he could think of for it. With that, he got back into his vehicle and drove off after saying that he'd be back for it when the hotel was ready.

Ginny was out there by that time, so I suggested that we get a bucket of water and start cleaning up whatever was in the bucket so we could get rid of the muck and its smell. So we did, then went fishing in the mud. In no time at all we were fishing out money, lots of money. Little bills and big bills. We finally were able to move to the kitchen where the kids could help. There was no school for them because the school had burned down at the height of the storm and the flood. So I fished the money out of the mud and washed it off in the bucket of clean water. The kids dried it with towels and Ginny hung it wherever-she could find a place. When we were done, bills were draped on curtain rods, chair backs, lamp shades and so forth. Our house was decorated with money, including a long string of $100 bills. I hadn't seen that much real live money since the summer of 1944, when the torpedo squadron was being paid off by a lieutenant paymaster escorted by two great big SP guards and he had those tall stacks of fresh bills in front of him, ready to deal. The last thing that came out of the bucket was an old-fashioned silver watch fob chain, about ten or so inches long and from which hung, evenly spaced, five placer gold nuggets about the size of a small fingernail. A really beautiful thing.

The money dried pretty quickly so we took it down, stacked it by denomination, then counted it. The total came to about $5,000, which included seventeen $100 bills. What a pile! All that plus the watch fob. We then bundled it altogether in a sack and set it aside in the house until Slim came back. A couple of days later he did come back, but instead of picking up money, he brought more; this time only about $500 in a coffee can full of mud and money. We quickly cleaned it up—we were experts by now—and added it to the big bundle. Later Slim brought a huge pitcher full of change from the automatic laundry from the hotel. It was mostly quarters, but did have some halves and surprisingly, a few silver dollars. That catch too, joined the hoard. We had become a bank.

Within a week Slim was back. He said the hotel was now cleaned up enough to put the money back and he had come for it. So, we gave it to him—bundles of cash, pitchers of coins and a watch fob. He put it all in the car and, after thanking us for our help, drove off with it and that was that.

When talking about that time even now, the kids who are no longer kids, always mention the money draped all over the house. I remember that too, but it's really the watch fob that sticks in my memory, a real beauty. And wasn't that a time!!

SANTA CLAUS (?) COMES TO COFFEE CREEK?

Donna D. Heilman

In the 1960s, my husband Ed Heilman was the fire staff officer on the forest supervisor's staff of the Shasta-Trinity National Forests at Redding, California. We had a pleasant house at Redding, our kids were about eight and ten years old, Ed enjoyed the job, and it was a good time for us.

One December in the mid 1960s, there were heavy snows in the mountains of Northern California, but just before Christmas there was a sudden warm spell, heavy rains, and therefore serious flooding, particularly in the upper Trinity River area northwest of Redding. Many roads, bridges and some buildings were washed out. Power and telephone lines were down, and many people had been isolated without power for several days at a time. Without power, freezers defrosted, water system and gasoline pumps wouldn't work, and there was no relief in sight—not the situation that would make for a Merry Christmas for these flood victims.

Included among those isolated were the Coffee Creek Ranger Station people—the ranger, assistants, their families; and also some ranchers, loggers and other area residents north of the Coffee Creek bridge, which had washed out. Coffee Creek is a fairly large drainage, so a lot of water was flowing. Since Trinity County agencies couldn't help at that time and place, any help would probably have to come from the Forest Service. Forest Supervisor Paul Stathem authorized a mission to take several electrical generators and other emergency equipment from the Redding fire cache to the Coffee Creek area, to refreeze freezers and otherwise provide some relief.

On Christmas day Joe Noble from the supervisor's staff drove a truck to be left, with several generators at the accessible end of the Coffee Creek crossing, and Ed drove another vehicle to bring Joe back to Redding, even though both men were using up year-end annual leave. Ed invited me to ride along (this was acceptable back in those days). I arranged for our kids to stay with neighbors and we left on what was expected to be about a four or five hour round trip.

As it turned out, there were still some snowy patches and numerous trees down across the roads, so it took more than four hours just to get to where we were to leave the generators. It was almost dark by then and it was raining steadily. There was a small grocery store/bar at the accessible end of the washed out bridge, and a small group of locals was there, some of whom had apparently been at the bar for some time. Outside were several trucks, some front-end loaders, and a big, beat up old logging bulldozer (I learned later it was a Caterpillar D-8).

Ed and Joe knew several of the locals. After a few minutes of explanation of why we were there, the owner of the D-8, a really rough hewn logger who couldn't speak a sentence without several swear words, volunteered to take the generators, tied down on the hood of his D-8, across flooding Coffee Creek to the people on the other side.

By this time it was dark. With a few swear words thrown in for good measure, the logger asked me if I'd like to ride across the creek on the D-8 —just for the fun of it. He'd been across the XX!#**# creek several times earlier in the YY!X## day, and it wasn't a MM#!#** problem. I was somewhat more adventuresome in those days, so I said yes, I'd like to ride over and back on the D-8. Ed had known this logger for several years and considered him more or less reliable, so Ed didn't try to stop me. So, with generators secured on the hood, I sat on top of the big winch at the rear of the operator's seat (there was only one seat for the operator), held as best I could to the safety canopy, and away we went down into the rushing water.

Some of the trucks had their headlights aimed across Coffee Creek and we could see whole trees, logs and other debris rushing past in the really swift current. The rushing water made a lot of noise, in addition to the diesel engine of the D-8. Before we were very far into the creek, the water was well above the top of the tracks, and I was beginning to wonder just what I'd gotten into. Light from the truck lights behind us faded away. It was really dark. We couldn't see across the creek to the far bank.

Somewhere out in midstream the D-8 got hung up on a big boulder in the streambed. It seemed to me the logger's movements with the controls showed that he was concerned that we might really be stuck. At this point I was trying to figure out, if I were to die in this situation, who Ed could marry who would be a good mother to our children. Finally, with a great lurch, we got free and kept going. After what seemed like hours we reached the other side. The creek bank had washed away and there was a nearly vertical wall to climb. The logger had some trouble getting up out of the creek and I had to hold on to my awkward perch for dear life to keep from falling into the rushing water but eventually we made it.

The people on the other side were expecting (hoping?) to see us and they were delighted that we and the generators had made it. Under the circumstances, we were probably more welcome than Santa Claus. We unloaded the generators and other gear, bade the ranger and others farewell, and headed back down into Coffee Creek, which seemed like it was deeper and swifter than before.

We could barely see the truck lights on the far side, and after what again seemed like hours, we made it safely back to the store area where Ed and Joe and the locals were gathered. Ed had begun to worry about what was taking us so long and he was much relieved to see the big old Cat come up out of the water. I thanked the logger for getting us back safely and he said it was no X!X!### big deal, but I was welcome.

By now it was long after we had expected to be back home but the phone lines were out. Ed's car radio wouldn't reach the FS mountaintop radio repeaters, so Ed couldn't talk back to forest dispatcher Rip Ripley in Redding. There was nothing to do but head back to Redding. We got back well after 10 p.m., retrieved our worried kids from our neighbors and really counted our many blessings.

It had not been the Christmas Day we expected, and certainly the rough logger with the colorful vocabulary did not at all look like Santa Claus, but together, in our own ways, we had brought to those beleaguered people what must have been to them some very welcome cheer.

And it had really been an adventure, after all.

THE OLD CAJON STATION

Don Bauer

The original Cajon Fire Station site was right at the junction of what is now Hwy. 138 and I-15. That location was right next to the railroad and ideally suited for fire.

When the new highway was being built, which is now I-15, the state determined they needed the station site for an interchange. We were approached; we had no objection and we were committed to high safety standards with any new highway that was built. But, we had to insist that if the station was being relocated or rebuilt we would have to have the same or equivalent facilities at another location. State law limited reimbursement for facilities to the actual cash value, which was the depreciated value, of the improvements. This created a problem. Since the Cajon Station had been built as a CCC station, there was very little capital in it. Most of it had been depreciated. The state felt that this was very good, but I told them I couldn't accept that. It took an act of the state legislature to make an exception here. They had to change their own rules. But they did build us the station that we needed. I used to try to sleep at the old station on fires and the diesel trucks would sound like they were going right through the barracks.

The little hill right to the north of the old Cajon Station, where the interchange is now, was named Beacon Hill. When I inquired how it got that name, Dave Tucker who had been the ranger there, told me that they used that hill originally on the early airmail flights across the country. When it was foggy the planes coming from Chicago and the east side would have to come through the Cajon Pass. They would go up to the top of that little peak and build bonfires to help guide the planes through. The planes navigated by following the railroad tracks. Then they replaced the bonfires with an electric beacon. Today we have air traffic, gas lines, railroads, highways, probably every form of transportation known to man, short of space travel, going through Cajon Pass. That's how Beacon Peak got its name.

TO RESCUE A HORSE, OR OTHER DUTIES AS ASSIGNED

Dwain Nebeker

It was early winter in the middle '60s. Bill Lunsford, Calaveras District Ranger, Stanislaus National Forest, was asked by the county sheriff to help in the rescue of a horse. The horse was supposed to be somewhere around Stanislaus Meadow. The sheriff had tried to catch the horse earlier but was unable to do so. Now the snow was getting deep and the horse would not be able to survive the winter if left in the meadow area. Bill asked me to assist because I was familiar with the Tucker Snow Cat which I used to make snow surveys.

We left early the next morning. The party consisted of Bill Lunsford, Dwain Nebeker, a few county mounties and a few tag alongs with small snowmobiles. We made real good time and soon found the horse right where it was supposed to be. The horse didn't want to be caught so we chased it with the Snow Cat. One of the county mounties rode on the hood of the Tucker with his rope at the ready. It didn't take long because the horse was soon in deep snow. We caught him. Trouble was we couldn't lead the horse out because the snow was too deep. He was belly deep in snow and gave up very soon. It was suggested that I pack a trail with the Snow Cat. It didn't work. The Snow Cat did more floating than packing. By this time everyone was obsessed with making it happen.

We could use a sled of some kind to put the horse on. We went looking. I saw a forked tree that turned up near the base and I thought it might work as the runners and frame for a sled. The tree was cut down to be used for our sled. We needed a platform to put on the runners. We found a bed spring. With some rope we lashed everything together. Now we had a sled to put the horse on if we could get the horse on. The horse was finally upended, hog tied and placed on the sled. With the Tucker I tried to pull the horse and sled uphill to the more gentle grade of the road. Either the horse would slide off the sled or the Snow

Cat would spin out, dig down or both. I couldn't keep it going forward.

It was now getting late. We were running out of time. The sheriff said that if we couldn't get the horse out he would have to shoot it because it would only starve and be coyote food. We would try one more time and if not successful, the horse would have to be shot. This time I really put the pedal to the metal. The Snow Cat moved forward. The horse was part on the sled and part on the snow but I kept it moving forward. We reached the road grade and could now proceed down the snow covered highway. We returned by moonlight. One of the snowmobiles broke down and the operator wanted to have it also towed behind but Lunsford felt that we would be pressing our luck too far to do this. It could be retrieved later. Finally we arrived at the point of winter closure just above Cottage Springs. The horse was untied and was able to get up after a long, cold ride down Highway 4 on a crooked tree and a bed spring.

SHEEP ON THE ANGELES

Jack Horton

In 1964, the Saugus District of the Angeles National Forest experienced a serious brush fire which burned entirely through the district from south to north. The burned area included Drinkwater Flat. This vicinity included a valley of perhaps two and one half or three miles long, oriented in a north-south direction. The ridges on either side of the valley were not nearly as steep as topography typical of the Angeles Forest. Vegetation was primarily chamise, a brush species common in Southern California. This species sprouts prolifically following a fire.

Following the fire a suggestion was made to graze sheep on the burned area to retard regrowth of chamise. This suggestion was studied and it was decided to give the proposal a try. We hoped to establish a fuelbreak to aid in the control of future fires as well as to create a more attractive and open area.

The burned area was sprayed with 2-4D (at a time "herbicide" was not generally considered to be a bad word). This treatment greatly suppressed regrowth in the months following the fire. The following year a permit was issued to a local rancher to run sheep on the area, about 3,000 animals I believe. The experiment was still in progress when I transferred from the district in 1969. Without retreatment with chemicals or prescribed burning, chamise will come back eventually but grazing will delay the process. The project convinced me that grazing could be a useful tool in the management of vegetation in Southern California under some circumstances.

The project offered some interesting and sometimes amusing (and revolting) experiences. The rancher hired Basque sheepherders. None spoke English. I used to stop at one herder's trailer and have lunch with him. He was a big guy—maybe 250 pounds and good natured. The menu was usually the same; lamb stew, cheese, sheepherder bread and red wine. Since we could not speak each other's language, there wasn't much conversation. We sat there and grinned at one another, occasionally attempting some sign language, like pass the stew. The food was always good.

Each year we counted the sheep when they were turned onto the allotment. I took the district clerk along on one of these occasions. She stood next to the chute as the sheep were unloaded and tried to pet sheep as they came off the trucks. She wound up with sheep stuff splattered all down her front. A bad clothes day for her, no doubt.

Male lambs were castrated before being turned onto the range. Basque herders harvest the "oysters" with their teeth, a method Americans don't have the stomach for. The sack is slit with a knife, the herder takes the "oysters" in his teeth, extracts them and drops them into a bucket. Since several other herders bring him the lambs, one herder can harvest an impressive number of "oysters" in a day. These so-called "mountain oysters" were later cooked and eaten at a lamb barbecue. They were yummy.

EARLY JOB CORPS

Dave Filius

The Cedar Flats Civilian Job Corps Conservation Center was activated in 1965, as part of President Lyndon Johnson's Great Society. It was located along the Selway River in central Idaho, on the Nezperce National Forest. It was staffed by career Forest Service and newly hired private sector personnel and populated with 16-21 year old corpsmen mainly from down south and back east.

The unusual feature of Cedar Flats was its isolation. It was located in what started as a cedar swamp several miles up a dead end dirt road that provided trail access into the Selway-Bitterroot Wilderness. The terrain was mountainous and heavily forested with conifers. The center was located on a narrow sloping "flat" with steep dense forest behind the center and a similar high, steep, timbered mountain slope across the road and river.

If you are still wondering about our isolation, we were told that there had been a prisoner of war camp for German soldiers even closer to town than Cedar Flats twenty-one years earlier. We were not only far out but winter in the Selway Canyon was mostly dark and gloomy and wet.

Cedar Flats got its water from the river and we pumped it uphill to the center. Our heat was from a really big bottle of propane and our electricity came to us on poles up the road along with the telephone. There were many times that first winter when lots of things could and did go wrong. Usually all at once.

It was one of those times that I remember most. It had rained for several days and the road to camp was a quagmire. Then it started to snow. The food truck, also bringing the mail, was our essential lifeline, and happened that day to slide off the road and over the bank into the river shallows, knocking out the power and the telephone in the process. The mail contained the corpsmen's pay checks, essential to their anticipated liberty in Lewiston, Idaho, about 100 miles down river.

Without power the water pump didn't work, so our camp was without both power and water. Without power and water and food, the kitchen was pretty much out of business, and without lights the school was really dark as well. Without paychecks, the corpsmen were grumpy.

Well, there was nothing much to do but try to make the best of it, so the corpsmen came to the river with their towels and tooth brushes to wash up until we could get a portable pump in operation to provide handier water. Many of the boys thought that was fun and pretty novel. The setting was also a scary wonderment to many of them.

With snow coming down in big downey flakes and darkness looming, our cooks were desperate to feed and soothe a hungry and sort of milling bunch of corpsmen and staff. Someone suggested we have a hot dog roast, so we built a big bonfire in the road and the whole camp gathered to roast hot dogs, sing songs and tell stories.

Before long, one by one, the problems got solved, but that hot dog roast in the road by the river did more to blend all of us at Cedar Flats into one big family than all the new policies and training from OEO and the FS put together.

There wasn't much that first winter we didn't have to improvise for, because moving a fifty person staff and a two hundred person enrollee contingent into a new, wet, muddy, isolated camp with untested utilities, new buildings and a weird assemblage of personalities, surely tested everyone's skills, temperament and often sanity.

But the day the food truck went in the drink and everything else went on the fritz was a day everyone pulled together.

MAD RIVER MEMORIES

John Ed Tonnesen

The year after the 1964 flood, I left Gasquet and became the district ranger at Mad River. Mad River was called the "banana belt" because it was the southernmost district on the Six Rivers

English Peak Lookout Station, Salmon River District, Klamath NF, California - 1931.

Mules hauling telephone poles for construction of line from Oak Grove RS to High Point Lookout on Palomar Mountain, Cleveland NF, California - 1929. Each mule carried tow poles up narrow mountain trail which is still in existence.

National Forest and it didn't get as much rain as the other five districts. There were six districts on the Six Rivers in 1965. The Tish Tang District with headquarters in Willow Creek, was combined with Lower Trinity and the Redwood District became trading stock in the creation of the Redwood National Park.

The Mad River District was located 51 tortuous miles from Fortuna. The road was State Route 36. It was mostly paved, though barely wide enough for two vehicles to pass. The district contained the headwaters of three of the six rivers; the Eel, Van Duzen and Mad rivers. (The other rivers are the Klamath, Smith and Trinity.) It was beautiful country and one of my favorite districts. The community was mostly ranchers, loggers and millworkers and I had many friends among them.

During my stay I had two FCOs; Don Spence followed by John Johnson. I also had two Resource Assistants; Martin Warne followed by Bill German and two TMOs; Dick Bundy followed by Rich Claybaugh.

At least two of our logging roads forded rivers because the floodplains were so wide it was not economical to build a bridge. They were both less than a hundred feet wide in the summer but two or three times that in the winter. One was the road to Buck Mountain across the Van Duzen River near Dinsmore and the other was by the Ruth Guard Station on the Mad River. In the spring, you could often ford the rivers in the morning but by late afternoon, they would sometimes rise a foot. This would be due to melting snow in the watershed. Many a Forest Service employee guessed wrong and got stuck in the loose gravel in the middle of the river. Your choice was to either radio for help or wade to shore and find someone to winch you or pull you out.

There was a guard station above Ruth Lake that was manned by a fire suppression crew. The foreman's name was Roger VanBuskirk. Roger and his wife, Deanna, lived in a government trailer on the station. Tourists would frequently stop by for directions and campfire permits. Roger told me about the time a deer hunter from somewhere south of Fortuna stopped by and asked where he could find a good place to go hunting. Roger was busy so he told the guy to go up the first Forest Service road that came to mind. He then told him to take a right turn on such and such a road until he reached the end.

There would be a fine place to find a buck, said Roger. Roger and his crew then got back to the task at hand. About an hour later, the deer hunter came back with a nice three point buck in the back of his truck. He thanked Roger for giving such precise directions to the buck, which was apparently standing at the end of the road when the lucky hunter arrived. After validating the deer, Roger mumbled to himself as the happy hunter sped down the road with his trophy.

A CAMPFIRE IN A CATFACE

Les Joslin

I coaxed the pale green U.S. Forest Service patrol truck onto U.S. Highway 395 and drove southward. Adjusting the "squelch" knob on the radio, I pulled the mike from the dash, completed a radio check with the ranger station four miles northwest of Bridgeport, California, replaced the mike and began to scan the eastern Sierra landscape.

I had just left the ranger station for another day on patrol. It was a clear, crisp mid-July morning in 1966, my fifth and final year as a Toiyabe National Forest fire control aide. The early morning sun warmed the cab and the air carried the exhilarating fragrance of sagebrush and pine. I rested my arm out the cab's open window and continued my scan.

To the southwest, the granite summits of the Sierra Nevada soared to meet a brilliant blue sky. Many of these peaks were old friends now. Streams rising there rushed through a geologic jumble of ancient volcanic ridges and more recent glacial moraines to join near Bridgeport as the East Walker River. North of town, at the foot of the Sweetwater Mountains, the Bridgeport Reservoir briefly detained the East Walker's run to Mason Valley and the Nevada desert. Ahead, east of town, the lower Bodie Hills, wooded with gnarled and brittle pinyon and juniper, huddled beneath Potato Peak. Diamonds of dew, covering the meadows of Bridgeport Valley, sparkled in the sunlight.

At Bridgeport, the diminutive seat of Mono County, I turned toward Twin Lakes, a popular summer recreation area nestled in a glacial canyon dominated by the steep Sawtooth Ridge escarpment. Here, where mountains and semi-desert meet, wildfire is a real threat. A total fire prevention effort included frequent patrols and intensive public contact, a big part of my job.

Following a freshman season on the fire crew, I had been promoted and appointed district fire prevention guard. To the public, I was "the ranger"—the U.S. Forest Service officer in the field. In fact, I worked under one of the district forest ranger's primary assistants, the fire control officer or FCO. Although it has been more than three decades since I last wore the pine tree shield for the Toiyabe National Forest, I recall those patrols as if the last were only yesterday.

A brief trip through irrigated summer pastures brought me to rising ground and the national forest boundary. Ahead, several hundred acres of blackened brush and timber marred the eastern slope of Sawmill Ridge and reinforced my belief in my job. Two summers before, while returning from a patrol, I had spotted the smoke, reported it, and raced to the scene—only to experience defeat by an inferno raging far beyond control by the district's meager assets. One day, two hundred men, and tens of thousands of dollars later, the Robinson Creek Fire of June, 1964—in reality a relatively small project fire—was under control. But its scar would take decades to heal. If only, I mused, I had been able to prevent that one.

The sudden sight of white smoke snapped me back to the present. I hit the radio.

"KMB-652, D-4-5, over."

The ranger station responded.

"I have a smoke up here toward Robinson Creek Campground, but I can't see what's burning yet, over." Although the base of the thin white column was obscured from view by a low moraine, I knew it wasn't coming from the camping area.

The ranger station asked my location.

"Honeymoon Flat, just crossing the bridge, over."

"Keep us posted."

"Ten-four, D-4-5, 10-8."

The ranger station signed off.

A few minutes later I turned the patrol truck off the Twin Lakes highway and followed a little used dirt road. It wound through blue-green sagebrush and granite erratics toward a small grove of aspen and pine. An ancient Jeffrey pine, still living though hollow from bottom to top, dominated the grove. At first I was confused. The smoke puffing from the top of the old pine didn't make sense. As I drew closer, however, the situation became clear. I decided to relieve the anxiety at the ranger station.

"What's up, D-4-5?" The FCO, attracted by my previous transmission, had come to the radio.

"You wouldn't believe me if I told you. But it's no sweat."

"Try me."

"Really, it'd take too much time. I'll fill you in tonight."

"Suit yourself. Let us know if you need a hand. This is 652, 10-8."

"Thanks. I'll be off the air for a while. I'll let you know when I'm back in service. This is D-4-5, 10-7."

I halted the patrol truck behind a large, late model station wagon bearing New York license plates. I blinked at what I saw. A short distance beyond was a half pitched umbrella tent. Shiny new camping gear lay strewn about. Before the tent stood a woman and a boy of nine or ten, admiring a large crackling fire burning in a catface—an old open fire scar—at the base of the old pine. Rocks piled in a ring around the catface completed their "fireplace." The hollow trunk of the tree served as a chimney.

Apparently oblivious to my arrival, a stocky balding man in flowered Bermuda shorts and tennis shoes was hustling another armful of dry branches and cones toward the hungry blaze. I got out of the truck.

"Good morning, sir," I called pleasantly, smiling as I ambled toward the man.

"Huh? Oh... hi-yer," he mumbled in reply, dropping his burden. He appeared both startled and annoyed by my arrival. After gazing at me for a moment, taking in the bronze badge and green-gold shoulder insignia, he asked the question. "You a cop?"

"I'm not a policeman, sir. I work for the Forest Service. How are you today?"

"Uh ... okay, I guess ...." He continued to eye me warily.

"What's going on here?" I was still smiling, but getting down to business.

"Uh... campin', yer know .... Dat against th' lawr er sumthin'?" He was at once defensive and aggressive.

"Campfires are here, sir. According to that sign you passed when you left the highway ...."

"Sign?" he cut me off. "Wha' sign? I din' see no sign."

"That sign, sir." I pointed back toward the highway at a small brown signboard with a peaked roof on which a standard yellow and black Forest Service notice proclaimed NO CAMPFIRES PERMITTED HERE. "Campfires are allowed only in designated campgrounds."

"Look," he flared, informing me that he was from the Borough of Queens in New York City, "I know all about fawrests an' th' fawrestwry! I got heah foist an' I'm campin' heah!"

"I'm sorry, sir," I countered pleasantly, "You can't with that fire." He was going to be hard to win over. "That fire's dangerous. It's in violation of national forest regulations and California law." I bit my tongue to keep from remarking that I couldn't believe he'd built his fire in the old pine. "It could get away from you, spread quickly through this dry brush and timber, and do a lot of damage. A fire starting here could burn into the campgrounds and endanger a lot of people." Hands on my hips, I glanced again at his Jeffrey pine fireplace and barely controlled the urge to laugh. Now, I thought, I'd seen everything. But I was young, and I hadn't. Turning back to the man, I made my best offer. "Let me help you put this fire out and find a place in the campground ...."

His further protests, including comments about "bureaucwats" and the United States still being a "fwee countwy," evoked my patient restatement of the wildfire danger. The Robinson Creek burn, clearly visible less than a mile to the north, provided useful graphic support. I avoided suggesting that, if necessary, the judge almost certainly would see things my way.

"Well, awright," he finally relented. "You cops always win." I ignored the "cops" remark. "But I ain' got no water!" That didn't surprise me.

"I have," I said, swinging onto the rear of the patrol truck to start the pumper. For a moment, I considered handing my friend from Queens a fire tool and soliciting his assistance, but then thought better of it. A shovel or pulaski in the hands of an inexperienced person could be dangerous.

While this particular situation was unique, several seasons of patrol experience had taught me not to be surprised by what some national forest visitors may do. Many of them are, after all, in an unfamiliar environment. This time, however, I was in for a big surprise. It came as I began extinguishing the fire.

Throughout our parley, the man's wife and son had stood by as quiet observers. But suddenly, apparently not comprehending what was happening or why, the little boy bolted toward me, fists flailing as he shrilled. "Father! Father! He's ruining our beautiful campfire!" Never before had I been attacked by a child. I shot a desperate glance at the man. He, too, seemed genuinely surprised. Finally, he restrained his son and hustled him off to help pack the station wagon.

I was still mopping up when, without another word, the family roared off in a cloud of dust. Issuing the man a citation—a rarely used last resort—would have served no useful purpose. He was out of his element and needed education, not embitterment. I hoped my explanation and suggestion of a camping place had helped.

A BEAVER'S TALE

Don Bauer

I was supervisor and Carl Hickerson was ranger at Mill Creek on the San Gorgonio Ranger District. He phoned up and wanted to alert me to a problem. I said, "What's that?" He said, "Well, Horace Hinckley is going to be in your office in about 15 minutes raising hell." Saying that Hickerson gave him a bad deal and he ought to be fired. I said, "What's it all about?" He said, "Well, he came in and he wanted a blasting permit." This was in fire season. State law required that you get a permit for blasting. So Hickerson asked him what he wanted to blast and Hinckley said he was going to blast those goddamned beaver dams out of the Santa Ana River. It was a low water year and they were at the peak of the irrigation season and the Forest Service ought to do this anyway because the dams were on national forest land, but they were too lazy to do it and he was going to do it on his own with his crew, but he needed a blasting permit. So Hickerson denied him the permit, telling him that it was not a proper use for a blasting permit. Hinckley wanted to know who would have authority to overrule Hickerson, so Hickerson told him, "The forest supervisor."

Hickerson was helping me out by letting me know that trouble was on its way. Hinckley came steaming on in, red faced and upset, just as Hickerson had predicted. Saying that Hickerson is being unreasonable and don't we know there's a water crisis on? Water was being rationed in San Bernardino. The city had passed a law that you couldn't water your lawn or wash your car and all that. Redlands was going through the same thing. So, I sustained Hickerson's decision and said, "The Multiple Use Act specifically includes wildlife as one of the resources of the forest, along with water."

Under the 1897 Organic Act, only water, stream preservation or water quality and timber were spelled out. All the things like grazing and recreation never really had any statutory foundation, except in budget bills, until the Multiple Use Act of 1960. This was a good law in which Congress clearly directed that the forest shall be managed for combinations of these uses, each in harmony with the other. So, I told him that wildlife, including beavers, was an important value here. The law also required, very specifically, that we cooperate with the states in the matter of wildlife management. The state owned the "pink slip" on each of those beaver and they were a protected species in Southern California, and the ranger acted properly in denying him the permit. He said, "I'll get you overruled on this Bauer. I'm going to San Francisco." I said, "You go right ahead."

As a relatively new supervisor I needed to find out how the local public would view my decision. I knew it was going to be a controversial decision. The Bear Valley Water Company was an important cooperator. We needed their cooperation in fire, land use, recreation and other areas. I wanted to try to resolve this amicably, but there was no easy way to resolve the confrontation that Hinckley had started when he threw down the gauntlet.

I went over and talked to my old friend, Jim Cram. I went out to seek the counsel of an old-timer. Jim Cram was a highly respected citrus farmer who had served in the state legislature. I described the problem to old Jim, who at that time was in his late eighties. He was as lucid as could be, but physically he wasn't able to get around very much. I asked him for his counsel on what I was doing. I described the problem of the beavers and the water. It turned out he was one of the major stockholders in Bear Valley Water Company. He said, "Well, I don't know Don. I think the beaver have got some rights, too. The Bear Valley people, they're trying to do their thing, but they don't own that national forest. Everybody owns it. I think you're doing the right thing in protecting a reasonable number of beaver. That's my advice. If you'd like to, why don't I get together several of my friends, old-timers in Redlands, and we'll have a little steak fry. Maybe you could arrange a place up in the woods where we could meet, sit down and talk; maybe look at the beaver dams and barbecue a steak or two, have a drink and we'll discuss the matter." I said, "Great." So, the stage was set for what was to be our first Forest Wildlife Advisory Committee meeting.

Things weren't as cumbersome in those days. All I had to do was call Carl Hickerson, the ranger, and say, "When can we get together for a meeting at Horse Meadows?" I'd like to bring up some key people and have you sit in with us as our host to discuss the beaver and water issue and maybe look at the timber program there on your district and get their advice." Carl enthusiastically supported the idea. So, all it meant was a barbecue grill with makings and a table or two.

That first advisory committee was mainly the Redlands bunch. There was Frank Adams of the county fish and game commission and the Redlands Sportsman Club; Frank Moore, editor of the *Redlands Facts*; Dale Gentry, past chairman of the state fish and game commission, Jim Cram and Lloyd Yount, who represented the Fruitgrowers and Sunkist Orange Company; Wes Break, chairman of the county board of supervisors and Gordon Cram, who was Jim's son and director of the soil conservation district.

The meeting turned out to be most helpful. We talked about several issues; the high risk sanitation timber program, the beaver, deer, bighorn and predators. It wasn't a heavy meeting, just very pleasant. A discussion among friends in an attractive setting next to the horse corral. The ranger put on a demonstration about how he went about and identified high risk trees. They were very receptive and gave us the green light on the two key issue areas; the beaver issue, which involved wildlife, water and recreation, and to push forward on our timber management program but do it right and keep the public informed. That was all it was. We had a pleasant steak lunch; those who were inclined had soft drinks, those who wanted bourbon or scotch had it, too. We just split the bill amongst us all equally. It lasted about three or four hours. (Horace Hinckley was not at this meeting. I knew I would be accused of sandbagging him. But I wanted to find out how other people felt; I knew how he felt about the beaver.)

So, armed with the information developed at the meeting, I went down to Jim Guthrie, who was the chairman of the state highway commission and owner of the *Sun-Telegram*, and asked him how he felt about the beavers and gave him the story. I guess news was pretty slack at that time. He said, "Well, goddamn it! The Bear Valley Mutual doesn't own that mountain. The national forest is owned by the people and you're hired by the people to manage it for the public. I think you're doing the right thing in protecting those little beaver. It gives people a lot of fun seeing those beaver dams. What about all these city kids in the youth camps that don't know what a beaver is?" He said, "Let's teach Bear Valley a lesson. I'll send a reporter over to talk to your ranger and, if you'd like, he'll talk to you and we might even run an editorial."

The next day a reporter came out to talk to Hickerson. He took him out and gave him all the facts. The news article was run. Then, the following Sunday, the editorial sledgehammer dropped. Pointing at Bear Valley Mutual Water Company as a greedy villain trying to kill all the beaver, and by blasting, of all means! They just covered them with bad press. It only took 24 hours and Hinckley was back in my office expressing the desire of his board of directors to try to work out something on the beavers. I set the stage for an agreeable resolution of the beaver and water users conflict, based on mutual respect and recognition of the need to manage the national forests in accordance with the multiple use principle.

PHIL, WHAT DID YOU SAY?

Chuck Smay

We live in an age with such rapid technological change that we are no longer surprised when the tools of last month are replaced by some newer gadget (undoubtedly controlled by a computer chip) before we have fully mastered the fine print in the instruction manual. Dealing with the newest gadget has always been a challenge to every generation of workers. Sometimes the involved workers who didn't have a friendly encounter with the latest invention would just as soon forget the humbling experience. However, in the close family that existed in the Forest Service in the late 1960s on the Stonyford Ranger District, Mendocino NF, there existed another group of equally dedicated coworkers who were not about to let the learning experience go unrecorded!

Before today's ever present word processors, there existed papers, pencils and typewriters. Often, those with the paper and pencil were blessed with marginal handwriting skills and could not type. A group of dedicated employees, known as the clerical staff, were often asked to translate what was written on the paper into readable text. By some magic they were expected to fill in what was intended, and to convert all this into a logical and professional sounding product which would impress the writer's boss. One day, the latest gadget salesperson was able to convince the forest and district management people that a dictation system would make this process much more enjoyable for everyone. Usually, even those who could not write very well could do a better job of talking, since they had a great deal more experience in this area. Soon, several very convenient hand held dictation units and a master transcribing unit arrived on the district. The timing was right since the forest had just experienced another of the 100 year storms that seemed to come about every five years.

Much inventory work on the forest road and trail system needed to be done in short order. Certainly the new hand held recorders would be great for gathering the field data. So, Vivian Flurry, District Resource Clerk, gave Phil Shimmel, District Engineering Technician, one of the new recorders. Phil had several full days of inventory work to identify the damage to roads on the district. A couple of days later, Vivian found a note on her desk with one of the tapes from Phil indicating the tape was ready to be transcribed. Vivian sat down at the new transcribing machine, with assurance this job was going to be easier than trying to decipher the longhand field notes of the past. Vivian hit the play button with her hands poised to start typing. Coworkers noticed a strange look on her face, then a roar of laughter. "What's so funny Vivian?" they demanded to know. She rewound the first part of the tape and switched to an external speaker. When the tape started playing the listeners heard a truck door slam, the motor start, the engine engage and the truck drive down the road. Then slow down, stop, the engine shut off and the door shut. Then silence. You get the picture? Seems Phil had gotten a bit confused on the position of the on-off switch position on the microphone. He had spent the day talking into a mike that was turned off, turned it on when he got back in to the truck, and wound up with a day's worth of driving sounds for all his hard work! Knowing Phil, I bet he had the day's site visits memorized since he did get his reports in on time. However, he did endure a bit of ribbing about his first encounter with the new technology. You can bet the rest of us checked and rechecked the mike button position and watched for tape movement every time we used our field recorders after that!

EARLY OUTFITTER ADMINISTRATION

F. Carl Pence

In Idaho, in the late 1960s, both the state and Forest Service were heavily involved with trying to get a rapidly expanding outfitter industry under proper state licensing and federal special use permits. I was assistant ranger on the Cobalt Ranger District of the Salmon National Forest. This district was involved because of the district's outstanding big game hunting opportunity and especially because of river boating experiences on the Middle Fork of the Salmon River.

During this era, the Cobalt District administered the floating use of the river. At that time, administration of the more then one hundred miles of river was shared between the Boise, Payette, Challis and Salmon National Forests. There were a lot of differences in management attitude and the normal competition/jealousy between Forests. This eventually led to unified management and formation of the Middlefork District a few years later.

Since white water boat floating was just beginning to be "discovered," we had an especially difficult situation on the middle fork. Everyone with a boat was becoming an outfitter. We were just realizing this use should be under permit. Due to the size of the problem, we decided to have a "light" compliance program the first year (there were about seventy folks trying to be outfitters on the river). We had sent all known outfitters a letter explaining the need for permits and state licensing and were personally meeting with those who were using the river as a business but didn't have permits, allowing them to pay for a permit and a late fee when we found them.

Since the Salmon Forest administered the end of the river, we had an agreement with the other forests that when a suspected unpermitted outfitter put in at the upper input sites administered by the Boise or Challis National Forests, they would phone us on the old ground return phone line that was still operating and an important communication system in remote central Idaho. We figured how many days it would take for them to get to the end of the trip. Then we would meet the outfitter and discuss the permit need and collect appropriate fees, which amounted to less than $100 for most outfitters.

On this particular occasion I got a call that one of the most active unpermitted outfitters was on the river and I was there to meet him. I really didn't know him but knew one of his boat guides, so I had the guide introduce me and I carefully called him off to the side and explained the situation.

He seemed properly contrite and wanted to pay promptly. Normally I would have had him stop at the SO in Salmon because it was on his way and I could avoid the accountability of handling the money but it was so late in the afternoon that the office would be closed by the time he got there. So I told him that while he loaded up I would drive the few miles up river to the Indianola Ranger Station, get on the phone and find out what the exact costs would be and drive back, meeting him at the take out or somewhere along the road and he could pay the fee.

I did just that and found out the fee would be $65 plus $10 for a late fee. On the way back I met his school bus full of guides and clients along the road a few miles from the take out site. We both pulled off and got out of our rigs. But somewhere along the way his disposition had obviously changed. I sometimes wonder if he had a lawyer as one of his clients. He jumped out of the bus like a mad bull followed by his bus driver, who was equally ugly, and came charging over to me saying they were not going to pay any fee and were going to throw me into the river.

Both he and the bus driver were much bigger than I was and I wouldn't have had a chance in a fight with the two of them. Also, the river was very high and I am not a good swimmer. I didn't know what to do other than try to reason with him that throwing me in the river would just make his situation worse. To this day, I can't remember what I said, but I remember talking fast. However, he wasn't listening and was calling me every name in the book.

I finally thought a retreat would be the best strategy and said, "Eldon, I can tell we aren't getting anyplace with this discussion. I think it would be best for us to deal with this another time when there is less emotion and I am leaving." I went to get into my rig but he grabbed me and threw me against the truck yelling I wasn't going any place. He was set to hit me but I made no effort to protect myself to keep from giving any impression of aggressiveness on my part. I just said, "Guess you are right," leaned against the pickup and let him go into another tirade of name calling.

Finally, he wound down and said he would pay the use fee but no late fee. I responded that I didn't care if he paid anything, that I just wanted to go home. He made out the check, threw it at me and stomped off. I was relieved.

On getting home I shared the story with Ranger Andy Finn and we called the Forest Supervisor, Gene Powers. We decided to call the region's only law enforcement officer in Ogden. We thought that if any situation deserved a "threat and intimidation" charge, this one did. But the response we got was less than favorable. The RO person asked me if he had hurt me physically, which he hadn't, and stated that in court it would be his word against mine. I didn't think that would be the case since about twenty folks witnessed it as it all occurred in full view of his guides and clients who were in the bus. But, guess we were expected to take that kind of abuse in those days.

We did formulate a plan that called for me to go up to Stanley the next day, where the outfitter's headquarters was, tie in with the Stanley Ranger Tom Kovaliky, and get the late fee and an apology.

When we got there, the outfitter had left the area and we only got a general answer from his wife about his whereabouts. Instead, she accused me of accosting and embarrassing him in front of his clients. That got my dander up and I asked her if that was really what he said, that I knew a couple of his guides and wanted to get them to describe what really happened. On that, she admitted that her husband had a bad temper and it probably occurred as I described. She apologized for her husband and wrote a cheek out for the late fee.

I got transferred to the Humboldt National Forest at the end of that summer and never had the opportunity to see him again. However, he only operated one more year. That next July he suffered a heart attack while floating through Haystack Rapids and died. I guess his temperament got the best of him. Considering that, maybe it was okay that we didn't prosecute.

THE SEVENTIES, EIGHTIES AND NINETIES

LAW ENFORCEMENT BEFORE AGENTS AND LAW ENFORCEMENT OFFICERS

F. Carl Pence

My career with the "outfit" has been one of the best. I started out in the sixties when things were "like it used to be." Now, I'm winding down in a much changed new Forest Service. I would be the first to say that some of the changes have been good. For example, I can't see us dealing with today's law enforcement challenges in an outfit that didn't have highly trained law enforcement officers prepared to deal with the pot gardens, meth labs, gangs, etc. But before, we also had our enforcement challenges and, since we weren't armed and trained to deal with the potential of serious resistance, we learned to improvise and bluff when appropriate.

Following is a story that happened to me when I was on the Humboldt NF, Ely Ranger District in 1970, living at Lehman Caves National Monument with Park Service employees and administering the Wheeler Peak Scenic Area. This area is now the Great Basin National Park. The area had been the old Baker Ranger District but had been combined with the Ely District about ten years previously.

When I got there in early fall I kept hearing about Christmas tree thieves and how they were ripping us off. I had visions of folks coming in, cutting two or three trees for family and friends and not getting the proper tags. So I didn't take it very seriously.

However, I got a rude awakening when right after Thanksgiving I visited an area up Snake Creek where we were planning a commercial Christmas tree sale next year. It had been creamed! The Humboldt had an active commercial Christmas tree program. We tried to design sales that included a combination of piñon pine and white fir. Piñon was a mediocre tree on the market but the area's white fir was a premium tree and could be sold in the Las Vegas or Salt Lake area for good money. We could sell a lot of piñon if there was a mix of fir in the sale. The thieves pulled a truck and trailer in and cleaned out all the good fir and some of the best piñon and got what looked like several hundred trees. They not only stole a lot of trees, they wrecked the potential sale. After that, I was ready to try very hard to catch them the next year.

I found that this was a way of life in this area that was somewhat similar to the bootlegger areas of the South. Instead of bootlegging white lightning, folks stole juniper posts from the BLM in the summer and Christmas trees from the Forest Service in the fall and collected unemployment in the winter and took whatever temporary mining and agricultural jobs provided.

To make enforcement more difficult, there apparently was a Utah sheriff who was involved in a loosely organized crime network. This made it particularly difficult because we certainly didn't have authority to stop vehicles with trees on state highways and depended upon state police and sheriff department folks to cooperate. All this required communications, and when the crooked sheriff heard something was happening, he could frustrate our efforts.

There also was some violence associated with these actions. A few years previous, a BLM employee caught some of the folks stealing juniper posts and he got beat up and his vehicle was shot a few times. A FS employee stopped a truck with trees headed toward the Utah line. A firearm was pulled and he was told to get out of the way. Which he did.

In my particular case we were making weekend patrols around the Snake Range on weekends when they expected we wouldn't be out and the historic thefts had occurred. The range tech was on patrol on one weekend and came upon a truck loaded with trees. They were driving a new Chevrolet truck and he was driving one of our old Dodge 4x4s. They took out through the sage brush, drove around him and easily out raced him.

We got hold of the state police and sheriff department in Utah, where they headed. We eventually found what we thought were the trees but the butts were cut off, so we couldn't make any butt matches with the stumps on the ground. There was a strong likelihood the crooked sheriff helped them.

It was snowing as I patrolled the same area the next weekend, Thanksgiving weekend. As I hit the south end of the Snake Range, I ran into some fresh truck tracks and got a sinking feeling. Why would someone be out in this Godforsaken area on this date unless they were stealing trees? I pulled in on the tracks and followed them up Big Wash, came around a corner and right in front of me were a couple of guys loading up about 200 white firs.

They ran up the hill into the woods and hid. I looked around and found I was parked in a good place. There was a big gully on one side and a steep hillside on the other. There was no way they could get around me with their rig like they had the range tech the previous weekend. So I decided to just stay where I was. It was snowing hard and I wanted to see how long they would stay in the woods. The only problem was that I was in a radio blind spot. There were a lot of them in those days.

I worried some about what I would do if they got violent. There were two of them and only one of me. They stayed in the woods for about 30 minutes and then cautiously came down to their rig and were talking together. During this time I made a big show of talking on my vehicle radio. I was calling in their description, vehicle description and license number in spite of the fact no one could hear.

After talking for several minutes, they came over to my rig, looking angry. I got out. One of them said they weren't going to let me do this to them. I told them that there wasn't much they could do, that they had seen me calling in their description, and that it would go much better for them if they just cooperated. In doing so, they may find a sympathetic judge. This changed their tone significantly.

I wrote them a Notice of Violation and told them they needed to follow me to the guard station I worked out of in Baker, Nevada, with their load of trees where I planned on impounding the trees. They complied well. You can bet that as soon as I got out of the radio shadow, I got hold of the dispatcher and informed them what was going on.

As soon as I got to the main highway, just outside of Garrison, Utah, the local sheriff, who was a good cooperator, pulled in behind them. This really gave me a sense of power. Up until then, I was so nervous that I had a hickey sucked on the seat cover with my backside.

They followed me to the station where the sheriff and I had them unload the trees. Then, we let them go home to Minersville, Utah. Afterwards, they did get appropriately charged and fined. I got transferred about six months later but heard later that my catch led to additional action which started the breakup of the ring that had plagued the area. And, eventually the crooked sheriff got out of office.

THE LITTLE YELLOW SCHOOL BUS

Chuck Smay

If you have been to a Forest Service party given for someone who is transferring, you know the routine. At some point after the pot luck or dinner, the individual providing some direction to the evenings activities attempts to gain control of the groups attention. Gaining some degree of attention, they then provide the opportunity for those attending to tell some embarrassing or other tale that will properly humble the one departing and let everyone know the facts about them. One of the basic rules of such oral history is that they don't need to worry about any of the details they may have forgotten or that may need to be added to help the story. In May 1, 1970, I found myself the focus of one such story on my departure from the Saugus Ranger

District, Angeles NF, to the Stonyford Ranger District on the Mendocino NF.

It seems that Chuck and one of his coworkers (who wishes to remain anonymous), were returning from the north side of the district cruising down Interstate 5, about five miles above Castaic, when they came upon a large school bus stalled in the uphill fast lane of the freeway. A quick check with the driver of the school bus indicated it had vapor locked and he could not get it off the roadway. He was stalled just above the through cuts in the freeway where the "racetrack" drivers were required to do some fancy driving to miss the stranded bus. About this time the local CHP officer arrived on scene and asked our two aspiring CHP "trainees" to assist him in putting out a flare pattern around the turn to alert the uphill traffic of the hazard.

Not desiring to argue with an officer with a gun, our two forest workers dutifully grabbed an ample supply of road flares and started working with the CHP officer to lay out a series of warning flares downhill in very fast traffic. About halfway around the sweeping freeway turn, the CHP Officer received a call on his radio directing him to some higher priority assignment. With a, "sorry I have to go, will you guys finish this for me," he departed to bigger challenges. Our willing civil servants continued with their assigned task. Several minutes later, the job done and feeling relief at still being alive and having helped prevent an accident, they proceeded back up the roadway to the point of origin. About halfway back, they both stopped and stared up the roadway in disbelief. There before them was a perfect flare pattern ending at absolutely nothing! Seems the bus driver at some point was able to get the bus started and drove on!

I hope the drivers on the freeway did not get a good look at our industrious worker's uniforms or the shield on the Forest Service truck. We have an agency reputation to protect! Chuck, as a symbol of your work on the Saugus, here is a little yellow school bus to help you remember your accomplishments!

P.S. They did not have to make too much up on this one, that is pretty much what actually happen.

THE EMBATTLED MONONGAHELA

Jay H. Cravens

In 1970, I was promoted to regional forester in Region 9, with headquarters in Milwaukee.

The Monongahela was only one of those national forests, but it demanded a great deal of my time. I had to learn what was going on and what had gone wrong in West Virginia. I had to meet the public, the users and get to know our Forest Service people, including the outstanding Forest Service research scientists and the state and private forestry experts. I spent a great deal of time looking at our cutting practices and engineering work. I had to agree with some of our critics. The clear cutting did not look good.

The Monongahela was a brush patch when the Forest Service acquired it for a few dollars an acre almost half a century earlier. It was given fire protection and then management. Timber management started slowly and consisted of individual tree selection, essentially taking out the trees passed over during early cuts. It was really economic high grading. The more valuable trees from a timber and wildlife standpoint were the intolerant species, trees requiring full sunlight to reproduce and grow. These species were suppressed and disappearing. Small research cuttings made earlier demonstrated that small clear cuts achieved this objective and stimulated intolerant species.

But we created problems on our way to federal court. The small clearcuts responded so well in obtaining regeneration of desirable new trees that our timber people made the cutting areas larger. Five acres looked good, so 50 acres, or 500 or even 1,000 would be that much better. Release of intolerant species required cutting large trees and small poles and pulp size trees of less desirable species. But there was no market for the small stems and 15 to 20 cords per acre of felled material was dropped to slowly decompose in some very scenic, sensitive areas for all to see. Foresters were so proud of what they were doing they had attractive signs erected around the forest explaining that a timber harvest was really no different than cutting a field of corn. It could be cut, replanted and would grow back. Try telling that to a 75 year old state legislator who as a small child sat under large oak trees with his daddy hunting turkeys. One fall he returned and his favorite tree was gone and in its place was a field of stumps and logging debris. He could see one of those oaks growing in his lifetime and 60 to 80 crops of corn. There was no comparison in his mind. The people of West Virginia did not like clear cutting. They particularly did not like to see lovely hardwood stands being clear cut and replaced by plantations of disease free, fast growing white pine.

We supported the besieged Monongahela Forest people and the Chief backed us. Generally, what we were doing was biologically and silviculturally sound. We were the professionals. We knew what was best in the long run. The only problem was the local people did not like it and we were poor listeners. We changed our cutting practices. Clear cuts were reduced in size, some were eliminated, others placed in less sensitive areas. Utilization was improved and we did not waste any merchantable wood. We used more shelterwood cuts, which created only small openings in the forest, but these were enough to obtain increased growth on the residual trees and regeneration of new desirable species. But the local people would not acknowledge we had changed. They had us on the run and on the defensive. So we entered a period of injunctions and court battles. Federal District Court Judge Maxwell ruled against us. It did not matter that cutting on private lands within and adjacent to the forest was disastrous in all aspects. The Society of American Foresters and state wildlife biologists supported us, but the timber industry sat on its hands while we were being crucified. Judge Maxwell's decision to stop clear cutting was overruled by the Court of Appeals. While we won on an appeal and the Supreme Court refused to hear the case, that was not the end. The opponents were successful in changing the federal law. This resulted in the National Forest Management Act. The important outcome of the Monongahela case was that we changed our timber management and utilization practices and improved our environmental sensitivity. We listened. We got the message, but it was too late, the environmental movement gained headway and the National Environmental Policy Act (NEPA) also became law.

BEAR HUNTING SECRETS REVEALED!

Patrick Harrison

The Marine's have a saying, "There's nothing as dangerous as a second lieutenant with a compass." Much the same can be said of a forester with a rifle. See'n as how it's hunting season and all, I thought I'd squander the reader's time with useful instruction concerning bear hunting. Most of this deep insight derives from my first, and last, hunt for the aforementioned quadruped in the wilds of northern Idaho.

The Red Ives Ranger Station is a remote ninety miles up the St. Joe River from not all that unremote St. Maries, Idaho. It's the sort of place they threaten to transfer you to if you mess up in some noteworthy way. By volunteering to live there, prior to being caught in mid-mess-up, I figgered I'd pulled a fast one on the ol' Forest Service. After my bear hunt I was convinced that someone had indeed pulled a fast one. The only certainty was that I was the pullee.

The Red Ives Ranger Station, as with most outposts of humanity, was characterized by unique inmates.

My initial contacts with the residents showed them to be predominantly big, mean, tough characters .... and that was just the Forest Service wives! The husbands appeared to live, eat and breath hunting. Since this was early May, the spring bear hunt was the chief topic of conversation. As the newest initiate to this bunch of bullet fixated foresters, I felt it necessary to portray a similar interest in bear procurement. Before I knew it, I'd been invited to accompany a bunch of the guys on a bear hunt the very next day, but not before I'd been carefully questioned about my physical stamina, woods craft and ability to endure physical pain and the vagaries of Northern Idaho weather. "We don't stop for nothing," declared one bearded guy who looked sorta like a bear himself. "If you're put off by a little rain or wet brush, ya'd better not bother getting outta bed."

Next morning dawned gray and drippy. I was later to find that this qualifies as a glorious morning in northern Idaho. I waited outside the ranger station, amusing myself watching the water pour out of the roof gutters and occasionally looking at my watch. The guys had talked about getting started at "O-dark-hundred." On a hunch I went into the station and up the stairs to the offices. Sure enough, there was the bunch of 'em, eyeballing timber sale maps, working up timber cruise data on little calculators and generally looking diligent and office bound. "Hell, ya didn't expect us to go bear hunt'n on a day like this?" said the bear look'n guy.

At this point I made one of those pivotal decisions that determines the shape of your self image for the rest of your life. Mine's been lousy ever since. I'd show this bunch of nimrods that the new guy wasn't afraid of a little moisture.

Six hours later I was hunkered in some wet ceanothus brush on a knife edged ridge top, totally fogged in, listening to the roar of the St. Joe River a gawd awfully long way below. The rainwater trickled down my collar, over the watershed of my shivering back and hence through a drainageway who's mention is inappropriate in mixed company.

I was in totally unfamiliar country, half hypothermic, and following gravity down to the river would only serve to put me at the bottom of a damnably steep canyon about nightfall. Naturally, I started downhill immediately. I will spare the reader the "Field and Stream" version of the actual bear dispatching. I blundered onto a real monster (sorta medium sized, actually), and coolly dispatched him as he charged. (He was charging away from me and I was definitely cool. Remember the cold rain?)

I stashed the quarters and the hide in a snowbank. Night was falling, rain was falling, the St. Joe River was near at hand and in spring flood, and I was a savy, woods wise forester. I would prudently construct a warm fire under the boughs of an alpine fir, and morning would find me hiking out of the canyon to the eventual respect and admiration of my fellow timber beasts and loving wife .... or maybe not.

I immediately held my rifle high over my head and waded the St. Joe, getting across on the luck reserved for fools, little children and foresters.

My wife is both perceptive and very inquisitive. When I came through the door at midnight, soaked, redolent of bear guts and mud she was ready with an insightful question, "Are you outta yur flip'n mind?" Lacking a ready answer I went directly to bed.

I sure hope you've benefited from the bear hunting lessons and observations herein presented. Summary ... 1) Avoid the company of hunting companions who look like the intended prey. 2) Be suspicious when the hunting party prefers working up timber cruises on little calculators to going outside. 3) Don't bother trying to think up answers to impertinent post hunt questions by wives who just don't understand bear hunting.

PLANNED FOR IN ADVANCE

Gordon Walker

One year, about Christmas, Supervisor Harv Seeley and I got together in his office and picked a date to inspect the Pacific Crest Trail. The days would be the 22, 23 and 24 of August 1973. I felt mighty important that day as we sat there making plans for a trip so far in advance. But one thing for sure, every activity between then and later would fit around these three days, not in them.

On Wednesday the 22nd of August, we headed east from Red Blanket trail head toward the Sky Lakes country. Harv rode Spook; I rode Patches; and, of course, Big Red tagged along behind with his big ears flopping, carrying about 200 pounds of things we needed.

We covered the first two miles rather quickly. The stock clattered across two or three narrow bridges and their shoes caused clanging sounds as we passed over rocky areas. Pumice dust rose from our horses' feet and caused us to squint our eyes and breathe with our mouths closed.

Now and then, the trail passed through a rock cliff, a perfect place for yellowjackets to build their nests and I told Harv to watch out.

About three miles up trail we started across one of these sites. Spook stopped by a ledge of rock for just a moment. He had bumped against a loose slab of rock and riled up a nest of the little yellow bees. I saw them boil up and shouted at Harv to take off and move out. The nest was right under Spook's belly. I couldn't get past Harv and Spook and it was nearly impossible to turn around and take off down the trail. The bee's nest literally exploded. The last I saw was a yellow mass about two feet across and Spook started moving out with Harv riding like a veteran.

I pushed Big Red out of my way. Patches was beginning to get the message. The big horse plunged off the trail down the mountainside with long, high jumps. Yellowjackets were beginning to make their point. Big Red caught about 20 or 30 somewhere around his britches and lower belly. Patches was being stung some place because his mouth was open and his eyes as big as saucers. He carried me down the steep hill for about 70 feet and then jumped a down tree about four feet high. He was kicking and bucking by now and, to say the least, it was nip and tuck to stay in the saddle. Tuck danged near got it.

Anyhow, when Patches jumped the log, two yellowjackets stung me on the back of my neck. Now, I am allergic to bee stings. The last time I got stung, my body swelled up about one half inch deep all over. Immediately, I took a pill to counteract the poison.

In all the hundreds of miles I've ridden and the dozens of yellowjacket nests my pack string has stirred up, I've never been stung.

I told Harv about what might happen. He really got concerned and kept close watch on me. He was perfectly willing to stop the trip, call in a helicopter and take me out and I assured him everything was OK.

The worst thing that happened was that I fell asleep in the saddle somewhere on the Oregon desert and never came to for several miles. Harv's voice finally woke me.

That night we made camp at the west edge of Cliff Lake in Seven Lakes basin. It was about two hours before dark. I unpacked Big Red while Harv unsaddled Spook. We were the only ones at that lake that night so far. It took me about a half hour to pile the saddles and rigging at the spot where my bed would be and take the horses and Big Red up the mountainside to a nice little hidden meadow. There was a 30 foot stake rope for both horses and a drag rope for the big red mule.

The smell of the campfire smoke reached my nostrils. The aroma of fresh boiled coffee filled the evening atmosphere. Soon I was at Harv's side watching as he moved from one side of the fire to the other preparing our evening meal. Harv is the best camp cook I've ever camped with.

Our beds were a good 100 feet apart. My sleeping bag felt so good that night. There was one big star that seemed to shine right at me as I closed my eyes.

The next thing I knew, Harv was starting a morning fire. (When I first wake up on a cold morning back there by one of the mountain lakes, many mixed feelings go through my mind.) My bed felt so good and I was in no mood to get up. However, here was the forest supervisor already up and ready to go. I was supposed to be tough but here was a proving ground. So, nothing else to do but roll out, put on my hat, haul on my pants and meet the day with a big fake grin on my face.

Harv greeted me with a hearty "Good Morning" and we set about to start pulling loose ends together. I went up to look over our horses and mule. It was sort of dark and I had to watch my step in a place or two. Patches nickered and Big Red snorted when I walked up. Each animal had spent a good night. There was evidence of a lot of grass being grazed. They stood still as I put their halters on them and coiled up the stake ropes.

At the edge of the meadow I looked down toward the lake and saw the reflection of our campfire. A few more steps put me in sight of our camp. Harv was working busily around the flames preparing breakfast. I stood there for a moment to watch. The smell of boiling coffee and frying bacon dominated all other odors. My stomach felt empty and hunger really began to set in. I sorted out the different smells coming to me from Harv's breakfast fire and one began to stand out above all the others.

It was potatoes, eggs and a little onion mixed in for flavor. Even Big Red inhaled as though it was the sweetest thing he had ever sniffed.

A few moments later, the horses and mule were eating oats nearby and I walked over to the campfire. Harv had put plates out and poured me a cup of coffee. He was stirring a frying pan filled with the most wonderful breakfast I have ever seen! The potatoes and eggs were approaching a golden brown. Boy, did they smell good! Harv kept asking if he should call them done and didn't seem to pay any attention each time I told him yes.

For the next couple of moments the only audible sounds were the crackle of our fire and the scrape of the pancake turner across the bottom of the frying pan to keep the food from sticking. Harv moved the pan toward the edge of our rock ring fire and instinctively gripped the handle a little tighter. Suddenly, the most awful thing happened right before my eyes. The collapsible handle released. The pan tipped forward and spilled its contents into the fire. We both stared in disbelief and it took a moment for the sad truth to register. Harv was the first to break the silence with, "Oh, God, I think I'll cry." Quickly, he handed me the nearly empty pan and said, "You take what stuck to the bottom." He grabbed a fork and started stabbing some of the chunks that hadn't been consumed by the fire. After the impact of the initial shock wore off we had a good laugh, ate what few other things we had prepared and broke camp.

THE COUP DE GRACE

Avon Denham

Some years ago, while we were hunting elk in the Wanaha area, we were returning to camp at almost dark. Wright Mallery and Jimmy Wilkins signaled to Al Oard and me, who had just climbed out of a very deep canyon, of some elk grazing on a hill just ahead of us. I had the only scope on my gun so I tried to look through my scope to see if they were bulls. It took some time to get breath enough to hold a gun fairly steady. I finally saw that they were both bulls, one a spike and the bigger bull at least a five pointer.

The shooting then started, and after several shots, the big bull went down and the other acted like it was wounded. We gathered ourselves and took off to where the elk were last seen. When we arrived some 15 to 20 minutes later, the big bull was dead as a mackerel and was on his back under a windfallen tree. All you could see were feet and horns.

Mallery rushed in to him and shot the dead elk between the eyes, practically blowing the horns off his head. I asked Mallery why he shot the dead elk and ruined my trophy, to which he said, "I gave him the Coup de Grace." From then on, that area was known as the Coup de Grace Ridge.

WILDERNESS IN THE NATIONAL GRASSLANDS OF NORTH DAKOTA—YES OR NO?

Dave Filius

In the mid-80s the Custer National Forest was really wrestling with whether or not to recommend a wilderness on the Little Missouri National Grasslands in western North Dakota.

These national grasslands were managed by the Custer, and because they consisted of intermingled acquired lands, dating back to dust bowl days, the rules of management were much different than traditionally proclaimed national forest land.

The prevailing local mood on the ranches and in the small communities was that federally defined big "W" wilderness didn't fit. In the bigger cities of the state, however, the grasslands were viewed as unique, and there was growing pressure to have some wilderness classified.

As the Custer's supervisor, I felt that our job was to try to sort that out so the FS could make a reasoned recommendation one way or the other in our forest plan.

On the "for" side, there were thousands of acres of remote, little traveled picturesque badland and rolling prairie landscapes. Compared to the rest of the state to the east, these western grasslands were still wild and woolly. They not only contained some of the most scenic land in the state, but they featured some of the most unique wildlife habitat, some of the best sites for paleo-Indian and pre-ice age animal artifacts, and a wealth of pre and post settlement history.

On the other side of the coin was the reality that the grasslands were established to promote grassland agriculture. The intermingled private lands were dependent on the associated federal lands as pasture, essential to the economic livelihood of ranchers and their communities. Several large grazing associations in fact, leased the federal lands for grazing by their many members and had since the thirties.

The other main factor was that almost all the grasslands had underlying oil and gas structures of high geologic potential, most of which were already leased and very likely to be developed. Some of the public lands were surface only ownership, the subsurface was private. Oil and gas development had started in the fifties, was booming in the eighties and could continue to produce for decades.

In that environment could these lands provide quality, untrammeled wilderness without permanent improvements, with primeval character, where man was a visitor with outstanding opportunities for solitude?

These were lands where cattle had the right to graze freely amidst fences, windmills and well pads, and where ranchers and mineral leasees could have motorized access while a wilderness visiting public could only travel on foot. It was a heavy sack full of contradictions.

The other looming factor at the time was the threat of leafy spurge, a noxious weed that was spreading voraciously across even the most remote grassland areas and could only be contained by aggressive chemical treatment.

With all of those legal and practical considerations, we recommended no National Grassland Wilderness classification in the Custer Forest Plan. We did, however, identify critical habitat areas, and those areas that were most scenic and sensitive for either no surface occupancy or other protective sitting lease stipulations.

Today, many of those lands still remain as the most remote and scenic as any in the state, well managed in the true sense of sustainable multiple use by a dedicated cadre of Forest Service employees.

BILLEE, RAIDER AND MULE

Tom Davis

It was a wet and misty morning in August 1985, when the Mt. Baker Ranger District trail crew met Bill, the packer, at the Tomyhoi Lake Trailhead. The bossman, Scott Paul, a chain smoking passionate defender of wilderness and trails, had hired Bill to pack the crew's food and group gear to Coyote Camp, three miles in on the trail.

This was something new to the ten person crew. The district pack string had been disbanded decades ago, and we had been using our backs to haul our gear around all summer. But it had reached the point where we had to pack in for a seven day camp out to finish the force account reconstruction job.

The crew took pride in cooking the best crew dinner, and since we didn't have to carry our food, we got a bit extravagant. All of us packed big boxes of grub for the long camp out. The foodstuffs included three watermelons, gorp, 120 cans of beer, and a hodgepodge of canned and dry goods. Add the group equipment of Coleman stoves, lanterns, pots and pans, and tents and the total poundage soon reached an exorbitant figure.

Bill arrived early to pack the saddle bags and load the stock. Probably an all too common occurrence, the number of stock ordered was not enough for all the gear without overloading the animals. But Bill wanted to prove his worth on this FS purchase order, so even though the loads were probably too heavy for the two pack horses and one mule—Billee, Raider and Mule—Bill loaded everything up. The driving rain had turned the trail into a slick mudbowl.

While Bill was packing our equine friends, the crew started hiking up the trail. As I slogged up the steadily climbing trail with my backpack full of extra clothes and personal gear, I remembered some of the events of this summer with the trail crew on the Mt. Baker-Snoqualmie National Forest, located on the west (wet) side of the North Cascades in Washington.

I had worked four summers for YCC and YACC and was a forestry student at the University of Washington. Scott Paul hired me. Scott was a great boss—fair, knowledgeable, and instilled a sense of quality and pride in workmanship among his workers. He was a man of letters and words and loved to converse. He would stay up till the wee hours of the morning talking of arriving in Washington from his home in New Jersey after numerous wanderings including a stint at Naropa Institute in Colorado. He started with the FS on a trail crew in 1976, on the old Baker River RD, and said he showed up with hair halfway down his back. Scott was quite a poet, and had a number of poems published about his experiences and passion for trails and wildlands. One of my favorite writings of Scott's is "...this ground, this work, burns like a fire in the backcountry of my heart."

The crew assembled in June. All told there were ten of us, but four of the crew also did wilderness ranger work, so the core crew was me, Maria "Cookie" Cook, the crew leader, a five foot tall dynamo; Robert, a SCA (Student Conservation Assoc.) volunteer from Virginia who was legally blind even though he wore thick spectacles; Dan, a big football player; Layne, a SCA volunteer from Texas, and Rowena. Tom "Chumpos" Langley was Scott's right-hand man who figured out how to take care of the details that Scott dreamed and planned in his head; Barb, Kathy "Cat," and Carol were wilderness rangers half the time and worked with us the rest of the time. Scott was still a seasonal, so he worked with us quite often when he wasn't doing paperwork.

Our first major job was reconstructing and relocating the Sauk Mt. Trail, which overlooked the Skagit River Valley and switchbacked up a steep alpine meadow full of wildflowers to the site of the Sauk Mt. Lookout (burned down in late 1985 or 1986 by the FS). Over the years the steep meadows had slid and sheared in many places leaving the trail tread nearly nonexistent. We eliminated a few of the switchbacks by building some steps, and also relocated some trail segments at a steeper grade. This was rocky country, so during the course of building trail we unearthed many refrigerator sized boulders. Scott hadn't got his blasting card yet, so we dug and dug the rock out and then got two or three rock bars under the rocks and five or so of us pushed them down the hill. The large rocks would bound and bounce, flying over the switchbacks before finally stopping in the trees below the meadow. We closed the trail so as to not steamroll some hiker while rolling rocks. One day some hikers walked past the trail closed sign at the trailhead and started hiking up the trail. We yelled that the trail was closed but they kept hiking. So we waited until they were out of danger in the timber and let a big rock go. They got the message and hurriedly returned to the trailhead.

A lot of the excavated dirt and rock material tended to collect in one chute, and the dirt soon turned to dust. We returned to camp each day covered in grime from the dust clouds generated by the rolling rocks.

Bayard Tucker, then nearing the end of a 30 year career in FS trails, would visit and inspect our work. Bayard was a stickler for quality, and his words of wisdom and stories of trail crews of yesteryear, instilled a sense of pride in our work. Once we were nearing the completion of one leg of a switchback, when Bayard came up with his clinometer and said the grade had too many wows in it and needed to be five feet up the hill to maintain a steady grade. So, knowing he was right, we bebuilt the trail.

We then moved on to reconstructing the four mile long Tomyhoi Lake Trail, which entered the newly established Mt. Baker Wilderness (designated 1984). On day trips we had completed the first two miles to spec on this often hot and fly infested trail. We removed a lot of roots. Robert had a hard time with this because with his poor eyesight he had difficulties aiming his chops, but at least he could see his feet well enough to always miss them. We also did a lot of brushing, log out, waterbars and ditch work. The black flies were terrible, especially on hot days—swarms of hundreds buzzed and crawled all over you.

Now, back to the horse pack trip. As all ten of us hiked up the trail, Bill, the packer, followed behind. I had hiked about two miles and was a couple of switchbacks below Gold Run Pass, when the horses finally caught up with me and passed me. Tom Langley was ahead of me. As I approached the pass, Bill yelled, "Grab that horse!" to Langley, who was already holding one horse in a level area about 25 feet below the pass. I dropped my pack and scrambled to see the three other horses tangled in the small subalpine trees. Their reins and harnesses left them helplessly tied. Bill calmed the horses down, told everyone to move out of the way and cut the horses loose. One horse slid and tumbled about 15 feet to the level area, hitting some rocks on the way.

Finally, the animals were standing in the level area. Some of their packs were loose, one had a horseshoe dangling, and all had cuts on their legs. The horses were badly shaken, but Bill wanted to fulfill his contract and reloaded them.

By now we had all reached the pass. The rain continued. We waited, getting chilled at the pass while Bill reloaded the horses. Bill explained what had happened, "I was stopped at the pass and the end horse stepped back pulling the rest down." Bill's horse was tied to the first horse, and when the horses started going down he quickly cut the line, and luckily came out unscathed.

After a short ten minute rest for the horses, Bill, this time luckily was not tied to the first packhorse but held the lead rope in his hands. He began leading the pack train down the steep backside of Gold Run Pass. This was a steep section of trail, constructed in a nearly vertical cliff, and a 49 percent grade in one section. I was the first to follow. As Bill neared the first switch back I noticed the last horse's pack was loose and leaning to the right, and yelled, "Bill, you're losing the pack on the end horse!"

"What?" he said, as he slowed for the switchback. As he slowed, the last horse stepped back, lost its footing and fell pulling the other two pack horses down too. They slid down the steep embankment without making a sound. We all watched in

District Ranger Smith Taylor at Horsepasture Saddle Camp, McKenzie RD, Willamette NF, Oregon – 1934.

Lumber for Lake Mountain Lookout after a 10 mile pack by mule back. Klamath NF, California – 1933.

horror as Billee, Raider and Mule kept tumbling and rolling for almost a quarter mile, nearly to the bottom of the valley. Once they stopped rolling there was no movement and Bill said, "Does anyone have a pistol?" as he started riding his horse down the trail. We all knew the horses were dead, and just stood there for a while in a daze, mourning the loss of these big animals. As we hiked down to see what could be done, we saw our food spread out in a shotgun pattern across the mountainside from the point of impact, a 20 foot drop off a rock ledge. They probably died then, breaking their vulnerable necks. The hillside smelled of beer and there was little remaining of the watermelons. Most of the canned goods were dented, but some of the dried foods in plastic bags and cheese, as well as 12 cans of beer survived. We picked up the trash, stashed the salvaged goods and hiked back out.

The next day some of us returned to further organize the salvaged goods. The air smelled of putrid horse, and one horse's legs were sticking straight out like in the cartoons, about 25 feet below the trail. The other two were further down from the trail in a gully. Bill also returned to salvage his packs and bags.

Bill didn't receive his contract payment. We lost all told about $500 worth of food, and Bill had to spend one more day disposing of the horses. He used explosives to blow them apart so the coyotes and maggots would dispose of them.

We eventually returned to finish the trail work, camping at Coyote Camp—so named by Scott because we would hear coyotes at night. By the time we returned it was September. One half the crew was gone (went back to college), and the weather had turned wet and cold. There were skiffs of snow a couple of mornings, but at least in one of the few clear moments the snow dusted mountains—American Border Peak and Tomyhoi Peak were splendid.

We had ample firewood. Though this was wilderness where motorized equipment is prohibited, Scott was more into efficiency than primitive tools at this point, so we cheated and had a chainsaw to cut up firewood. We rationed the 12 salvaged beers, drinking a couple around the campfire each night. Scott talked and we listened.

Finally, Scott called it good enough, even though we weren't quite done and we packed out. So the five of us packed everything up on our backs—gear for 10 people plus a bunch of tools and trash. We all looked at each other when it came time to load the chainsaw on someone's pack, which were already way too heavy. Scott took the saw.

We hiked out, passing the final resting place of Billee, Raider and Mule on the way.

Postscript—Scott was hired permanently in 1987, and became a master performer in trails for the Pacific Northwest Region (R6). He became manager of the wilderness and trails program for the Mt. Baker RD, which grew to almost 30 workers by 1988. I left the district in 1989 to become district trails coordinator at the Skykomish RD. Scott died on September 25, 1993, while working on a trail bridge on the Hebo RD, Siuslaw NF. At least he was doing what he loved when he died.

THE BWCAW—A CONTINUAL CONTROVERSY

Dave Filius

When I became the supervisor on the Superior National Forest in 1988, the Boundary Waters Canoe Area Wilderness (BWCAW) had already been a source of local, state and national controversy for the better part of the century. The Superior, in NE Minnesota, was established in 1909. It is a huge area of forests and lakes and wolves up against the Canadian boundary and adjacent to the huge Quetico Provincial Park in Ontario.

The early battles on the Superior were about protecting the unique recreational lakeland values from dams, industry, excessive logging and mining. Later, recreational float plane access and interior resort development caused another few decades of controversy.

Finally, after an air ban, wilderness designation, the cessation of logging, purchase of private inholdings, and the almost total ban of motors, the controversies, by 1988, had simmered down and were mostly local, mainly about quotas, illegal motor use, fire management, party size, trail development, blowdown and campsite impacts.

But the one that got the pot reboiling had to do with three remaining motorized portages. These were portages that utilized motor vehicles to haul heavy motor boats from one motorized lake to another. There were only ten motorized lakes out of nearly two thousand, so it didn't seem like too big a deal. But typical of the BWCAW, it stirred up a big fight generated by the crowd who wanted no motors at all, despite the wilderness legislation of 1978, that had allowed a few to remain.

To resolve the issue the FS decided in 1989, to do some test hauling over these portages of typical combinations of motorized equipment, using as haulers typical combinations of recreationists. We had men and boys, old men, and women and men, loading their assortments of motor boats and equipment onto portage wheels and then pushing and tugging the quarter mile to four mile distances over the portages. Some exercise physiologists from the University of Minnesota at Duluth, measured the extreme exertion this took.

The question in this case was feasibility, not possibility. Based on the exertion required, the FS maintained that it was possible but certainly not prudent, safe or feasible. We were taken to court, and in the initial court case the judge ruled in our favor. On appeal, however, the plaintiffs won and that marked the end of motorized portages in the BWCAW, but not the end of controversy.

Today, the battle goes on with a local effort that appears to go off the deep end in the other direction with legislation that proposes to restore motors to even more lakes.

Thus the BWCAW is assured a continued, high level of controversy probably well into the next century.

Will these issues ever be resolved? If almost a century of it is any indication, I wouldn't bet on it.

THE WHITE PICKUP

Jackie Johnson Maughan

I work a fire lookout on the Nez Perce National Forest in Idaho. My lookout is called Coolwater and is located on the big east-west ridge between the Lochsa and Selway River drainages. I'm two miles from the Selway-Bitterroot Wilderness to the east and due south is the Frank Church—all told, 4.2 million acres, the largest expanse of wilderness in the lower 48 states.

To my north is the Clearwater National. Forest. I call it the "clearcut" because that's basically what I'm looking at—clear-cuts. The closest towns are Kooskia and Kamiah, population circa 1,000.

My first week or so I was glad for visitors because I was lonely and disoriented. It was nice to have that reassurance that if I wanted down, I could get down, pronto. I'd had to leave my car down at the ranger station because it couldn't navigate the road.

At any rate, there's a certain pattern that emerges when you study something long enough. For example, the blue birds and finches appear early mornings and evenings, the hummingbirds at midmorning. A pair of kestrels (sparrow hawks) make their rounds in the early morning, at noon and again about 6 p.m. A pair of rough legged hawks usually visit at noon and late afternoon. I also see an occasional eagle and I've twice seen a pair of goshawks. Actually, the eagles like stiff winds and I've only seen the goshawks right in front of storms.

Like the birds, the human beings have their patterns. First off, they come in either four-wheel drives or on dirt bikes. I see them long before they reach me as they snake along the ridge road after it emerges from the trees. Usually I see their dust or, in the case of the bikes, hear the high pitched whine of their engines.

Once they're within a half mile of the lookout, they disappear under a cutbank, a place carved by the road which is constantly eroding and full of rocks and boulders. This is where they really have to slow down and I know it will take them at least 15 minutes to get to the lookout. This lag gives me time to put on a shirt, straighten up the maps and run a comb through my hair. Usually they don't come early in the morning, for which I'm grateful. If they do show up before 8 a.m., I simply don't let them up on the lookout. I keep the trapdoor latched from above. I mean, I may be up at six, but that two hours is mine. I'm not officially at work until I check in at eight.

But usually they start showing up about noon. They park down near the outhouse and there's always a pause while everyone hits the head and I can hear the door swinging open and closed. (There's a really fine view of Fire Lake as you sit there; that is if you leave the door open.)

Then I hear their voices and their breathing if they're out of shape as they climb the last 100 feet. The first and biggest wave usually peaks about 2 p.m., then there's another about four, and then another about seven. This last group has it timed to make it back down the mountain before dark.

Now in addition to time of day, there are days of the week. Thursday through Saturday are the biggest with the density reaching the top of the curve on Saturday. It drops off a bit on Sunday and Monday is the day I can just about count on no visitors, especially in the morning.

Monday morning is the day I do my laundry and get cleaned up. There's a spring 200 feet down the mountain. Here I've placed a wash basin filled with water and put this inside a black plastic trash bag and set it so it's in the sun. That way I have warm water. At first I packed the water up to the lookout. This is not easy—hauling 50 pounds of water in a QVC container up a mountain while it sloshes back and forth on this kind of Trapper John packframe they give you. No, I found out it was easier to take my underwear and socks and whatnot down to the water rather than vice versa. I don't even try to wash my jeans; they just have to stay dirty till I stuff what I can into a daypack and hike the six miles and 5,400 feet down the mountain to the ranger station where I can pick up my car and drive into town to the laundromat.

After everything's been through the wash cycle, I quick soap up my hair with the gray water, then dump the water so it courses down the road and not into the path of the spring. Then

I refill the basin. And let me tell you, that water's cold. Anyway, I rinse out my hair, then swab off my body in sections and finish up with the clothes. It takes four or five basins full to get all the soap out.

This is an important ritual for me. By the time Monday rolls around, I'm dirty and my clothes are dirty. The hardest thing about being a lookout is the scarcity of water and how hard it is to keep clean. At least that's my view of it. This whole process takes about 45 minutes. I'd enjoy it more if I had a clear view of the road. As it is I can't see more than 50 feet in either direction and the spring is pretty noisy and I'm a little nervous about someone driving up on me while I've got my hair full of soap and my shirt off.

I was really irritated one Monday morning when I took my binoculars and glassed the road before going down to the spring and saw some jerk in a white pickup coming up the road. I watched as he crept along, cursing to myself because he was ruining my day.

I was expecting my husband who was driving 600 miles from the other end of the state. I hadn't seen him all summer and this was August.

My husband drives a blue truck. When I saw the dust down there, my heart took a leap, but I could see even without the binoculars that this truck was white. I'd been watching for days for that blue truck. A green truck would have been okay too because that would be Forest Service and they might be bringing my mail and news from down below and we could talk about important matters such as a merits of White versus Red Wings, or why propane tanks thread backwards or how to refold a forest map without making lumps in it.

Well, hell, there he was, coming at me as relentless as a swarm of horseflies and I supposed I'd just have to go down and do my laundry after he left. After that I'd have to settle for a spit bath up in the lookout because I didn't want to risk someone coming up on me later in the day. I was not happy and was frankly skeptical that I could be very nice or accommodating, naming mountains and such for whoever it was in that truck.

As the truck came closer, my irritation grew. I wanted too much to see someone of my own, not another stranger. When the truck went under the cutbank, I decided I'd just put the chain on the trapdoor and make it look like I wasn't there. I'd have to keep low so they couldn't see me inside. Then I'd wait it out while they rattled around down there taking pictures or whatever.

Well, that's what I did. I could see before he went out of sight that it was a single driver, probably a man. He couldn't see me at this point unless he had binoculars. Even if he did, he'd be too worried about the road to think of looking up. I got some satisfaction thinking about that.

I had the place all chained up and was crouched there on the floor. I heard him pull up down below. The quiet after he shut off his engine. Then I heard the kind of noise hiking boots make on rocks. Whoever it was was moving fast and not panting at all. Then I heard someone call out, but I couldn't quite make out the words because the wind came up and the flag started flapping. Then he called out again.

Then he came closer, right below me on those big, flat rocks. Then I heard my name. That was a shocker. I rarely hear my name because on the radio I'm referred to as Coolwater. And, naturally, the tourists don't know me from Adam or Eve.

I'm sure you've guessed it by now. The man in the pickup was not a jerk; he was my husband. He was driving a new truck, a white one. Actually, he'd bought it because I'd written how horrible the road was and he was worried that the old truck wouldn't make it. And that's how things go sometimes, but sometimes they go better than others, especially when you're riding in a new pickup.

TALAPUS BUTTE FIRE

Les Joslin

After two summers patrolling the Three Sisters Wilderness as a volunteer wilderness ranger, Les Joslin, a retired U.S. Navy commander and former U.S. Forest Service firefighter, pioneered the Green Lakes Trailhead Information Station as a contract wilderness information specialist in the summer of 1992.

The morning of Saturday, August 29, 1992, began about the same as 41 previous mornings at the Green Lakes Trailhead that summer. It was sunny, clear and warming. There were 29 cars in the parking lot at nine o'clock, and more hikers and backpackers were arriving at the popular Three Sisters Wilderness entrance. Sixty-one visitors entered the wilderness there between nine and noon. I had greeted them all, answered the usual questions, sold one Three Sisters Wilderness map and one copy of William L. Sullivan's *100 Hikes in the Central Oregon Cascades*, and helped one visitor make an emergency radiophone call home.

At noon I helped a group of seven from a Portland suburb who had shown up at the trailhead without the required overnight wilderness permit and then, just a couple minutes late, strode over to the parking lot to make my noon car count. Another routine day I thought. Back in the trailhead cabin, I was entering the 53 cars I'd just counted on my daily tally sheet when a radio transmission ended the routine day.

"Redmond dispatch, Round Mountain, fire report," crackled veteran lookout Roger Miller's steady voice.

I automatically copied the legal description—township, range and section—of the smoke on a piece of scratch paper. As I reached for my map, it occurred to me that the location I'd just copied was mighty close to my own. A glance at the map showed it just south of Devils Lake, not more than a mile as the crow flies—two by road—from my trailhead.

I ran out the door and peeked around the cabin wall at Talapus Butte. There it was, rising behind the butte, just the faintest wisp of white smoke against a clear, blue sky! I ran back into the cabin, closed the door, changed from trail shorts to green jeans, grabbed my hard hat, shovel, pulaski, emergency pack and radio. As I hastened through these preparations, a flurry of radio transmissions confirmed the fire's general location and revealed that the nearest Bend Ranger District engine crew was at least half an hour south. So was the Elk Lake recreation guard. I was the closest person to this smoke. Maybe, I thought, just maybe, I can pull off a save!

I scribbled a note—"Gone to fire"—for the door, locked the cabin, jumped in my Jeep and turned west onto the Cascade Lakes Highway toward Devils Lake, Was I excited? Sure! I hadn't fought a fire for the Forest Service since 1966.

A few minutes later, after rounding the bend at Devils Lake, I could see the smoke clearly. Several cars were stopped along the road, their occupants watching the smoke puffing out of the trees near the base of the butte. I parked where I had a clear view and picked up my radio to make my first transmission of the incident. I wasn't sure I could reach Redmond, but knew Round Mountain could hear me. And Roger knew me.

"Round Mountain, Joslin."

"Go ahead, Les," the lookout responded.

"Roger, I'm on the Cascade Lakes Highway south of Devils Lake. I estimate the fire at about one half mile south of the lake and about a quarter mile east of the highway. I can make initial attack if you like."

"Go ahead. Scout it, anyway, and protect the point of origin."

"Copy, Joslin clear." I stuffed the radio into my pack. As I put on my hard hat and shrugged into my pack, Redmond dispatch called Round Mountain.

"We don't recognize the person you've been talking with." Redmond meant me. I'd planned to go in and get acquainted but never had. And I rarely talked on the radio.

"That's Les Joslin," Roger explained. "He's the guard at the Green Lakes Trailhead."

"COPY."

Pulaski in one hand, shovel in the other, I crossed the highway, climbed the bank and plunged into the woods toward the smoke. There was no trail, but the lava flows, pumice flats and pine thickets weren't too rough going. Navigating toward the smoke was easy. That was the good news. The bad news was there was a lot more of it, and it was darker. That told me the fire was spreading into heavier fuels.

As I walked toward the fire, I began to think. Am I doing the right thing? After all, I'm just a contract wilderness information specialist, not part of the fire organization. I hadn't been to the

Deschutes National Forest's fire school. I didn't have any fireproof clothing. I didn't have a fire shelter. Except for a canteen, I didn't have any water. And I'd turned in my last red card—the fire qualification card that rated me as a crew boss—in 1966, when I'd left the Toiyabe National Forest to go on active duty in the Navy. That was 26 years ago! Should I be doing this?

Sure, I answered my own question. I'd started out in "the old Forest Service," when everybody did everything. I'd done this many times before, in the days before fire shirts and nomex britches and fire shelters. I'd had five seasons of fire experience on the Toiyabe, been on a couple big fires. More recently, I'd fought several wildland fires—including the 3,300 acre Awbrey Hall Fire on the west side of Bend two summers before, as a Sunriver Fire Department volunteer. I knew this business. And, most important, I knew how to keep out of trouble.

About the time I'd resolved that little issue, I was at the fire. Two things were immediately obvious; the point of origin and the fact there would be no one man save here today. I found a safe spot on a lava flow to stash my tools and pack, then got on the radio to tell Redmond Dispatch the score.

"Redmond Dispatch, Joslin."

"Redmond Dispatch."

"I'm at the fire. I estimate it at between three and five acres. Trees are torching and it's spreading up the butte fast. It's no one-man job. I'm going to scout the flanks."

Redmond acknowledged my transmission and advised that the first two engine crews, still en route on the Cascade Lakes Highway, were 15 to 20 minutes out.

I began to scout the northern flank of the fire I hadn't a prayer of stopping. It was running up Talapus Butte faster than I could walk up the butte. I advised Redmond dispatch of that fact and that, given the terrain and distance from the road, this would be an air and hand tool show.

Back at the lava flow, I got my shovel and pulaski and—careful not to destroy any evidence—started building fireline around the lower end of the blaze where it was creeping from the point of origin through lodgepole pine thickets. After a while, I could tell from radio chatter that a couple of engine crews had started walking in. Also, air tankers and smokejumpers from the Redmond Air Center were on the way. Redmond dispatch had told them to "tie in with Les Joslin at the fire."

"Hi. We're engine four-nine," the first crew of two introduced themselves.

"Welcome aboard," I responded.

Within a few minutes another crew of two, led by a young woman named Jackie, arrived. About the same time, the plane that had been circling the fire searching for a drop zone, dropped smokejumpers some distance north of the fire.

Jackie scouted the fire and the other firefighters began building line up the flanks. When she returned, I suggested I might as well get back to my trailhead job. She'd soon have all the firefighters, equipped with all the right stuff she'd need, and she agreed. And so, less than a couple of hours after leaving the trailhead for the fire, I left the fire for the trailhead.

Air tankers were plastering the head of the fire, a couple of hours old and at least 50 acres in size, as I trudged back through the forest toward the highway. I hadn't made the "save" I'd hoped for, and a similar incident 28 years before came to mind.

On June 29, 1964, when I was fire prevention guard on a Toiyabe National Forest ranger district, I had spotted a smoke, reported it and raced to the scene. That time too, my efforts were to no avail. After a futile, and probably foolhardy, attempt to head off a fast moving fire, I had to pull back to keep from being burned over. That fire too, spread rapidly uphill and eventually burned 350 acres. Twenty-eight years later, I'd been defeated again.

I hit the Cascade Lakes Highway about 50 yards south of my Jeep. About the same distance north of it, the just arrived incident commander—the guy we used to call the fire boss—was putting together his show. I quickly checked out with him and drove back toward the trailhead. As I did, I met a convoy of green rigs bringing firefighters to the fire.

Back at the trailhead, the fire that had turned the brilliant blue afternoon sky a murky orange was the main topic of conversation. My job there changed from providing wilderness information to narrating the fire from which I had just returned. Through the smoke, air tanker drop after air tanker drop was observed as retardant was applied to the head of the main fire and numerous spots. Soon, more than narration was needed. When the Cascade Lakes Highway was closed to westbound traffic and campgrounds were evacuated as a precautionary measure, I advised visitors on alternate routes to their destinations. Wilderness hikers and backpackers, cut off from their cars or pickups, also needed help.

The fire's southeasterly advance toward Sparks Lake was halted at the Talapus Butte ridgeline by 22 retardant drops over seven hours. About 340 firefighters eventually controlled the fire at 88 acres, and mopup continued for days. Some spotted owl habitat and parts of a research natural area were burned. It was, of course, a human caused fire.

And it was another fire I wish I could have stopped.

MULTIPLE DECADES

Editor's Note: In each of the following accounts a time period of several decades is covered. Each story represents only portions of a much longer narrative.

EARLY STANISLAUS FOREST RECOLLECTIONS

Joe Elliott, Jr.

My father, Joe Elliott, Sr., joined the Forest Service as a forest guard at Sonora, California, on September 1, 1908, just five days before his twenty-third birthday. Not having a formal education (about three years of grammar school), he was considered a "poison oaker" in contrast to the limited number of professional foresters at the forest level in that era. One of those professionals was Dave Rogers, who stayed on the Stanislaus about a year before moving to the Plumas for the balance of his career.

From all accounts, Dave was not bashful about letting the poison oakers know that he was a professional forester. The poison oakers had their ways of striking back. It was a long time before they stopped reminding him that, as an Easterner, he had gotten off the train at Sonora on July 4th carrying an umbrella. On another occasion, Dave and two or three poison oakers were in the field cruising timber. The party was camped in an abandoned miner's cabin on which most of the roof was missing. One night a heavy lightning storm developed. Very soon everyone was wet and miserable. About three in the morning one of the poison oakers began to laugh. When Dave asked what was so funny, the poison oaker replied that he was just thinking what a hell of a place this was for a professional forester to be.

Sometime during the winter of 1915-16, Dad was transferred from the Riverside Ranger Station to the Center Camp Ranger Station. On the day of the move we traveled with two saddle horses. Most of the time I rode with my father while my mother rode a gentle mare that Dad had acquired for her use. We followed trails that likely do not exist today from Riverside up past Mt. Provo to Confidence and thence down the other side of the main ridge to Center Camp. There had been a recent storm that had left a layer of snow up to two feet deep at the higher elevations. At these locations my father rode ahead to select the route and to break a trail for mother and I to follow on her horse. Center Camp had originally been built as a forest supervisor's headquarters, but it became a ranger station when it was decided that Sonora was a more logical place for the forest supervisor to be.

Our personal belongings consisted of a very limited amount of clothing, a few cooking utensils, some everyday silverware, an alarm clock, possibly a couple of thunder mugs, a limited amount of food, a few blankets and quilts, possibly a few sheets and pillow covers, regular towels, some bleached flour sacks for dish towels and some personal trinkets. Since there was no electric power in the mountains, there were no appliances; for that matter, there were very few appliances in those days even where electric power was available.

The reason we did not have very many personal belongings was that many of our needs were met by the government. Some of the items generally found at all stations consisted of a cooking stove containing coils connected to a water tank, a heating stove, kerosene lamps and lanterns, straight chairs, tables, beds, folding cots, large wash tubs and mattresses. The mattresses were not necessarily of the Beauty Rest variety, but in those days no one knew about Beauty Rest mattresses. A gravity water supply was available at Center Camp by tapping an irrigation ditch that contoured the slope a few hundred feet to the south. This ditch, like many others in

Tuolumne County, was originally constructed to provide water for hydraulic mining purposes, principally in the Columbia area.

There was no bathroom. The Chick Sales was located about thirty feet from the kitchen door. A large washtub was stored in a cabinet under the sink. When it was time for a bath, the tub was pulled out into the kitchen and filled with water at the temperature desired by the bather. Another tub and a scrub board on a small porch just outside of the kitchen served as the laundry. Depending on the season the wash was hung outside or in the attic to dry.

At the time of the armistice that ended World War I, the operators of the sawmills at Tuolumne and Standard and the box factory at Sonora blew their respective whistles steadily for about an hour. It produced a very strange effect, as the continuous sound traversed the adjacent canyons punctuated by sounds from aroused dogs.

Another event may explain why my father spent the balance of his career in administration as opposed to research. Somewhere it is recorded that he was engaged in developing a special plow pulled by horses to be used for building fireline. The Center Camp Station had a fine crop of bear clover which served as landscape. It was also a fire hazard. Dad used his plow to construct firelines preparatory to doing what we would later call control burning. Everything worked out reasonably well except that his fire somehow reached the tool shed and destroyed it and consumed or damaged the contents stored therein. The building was about six feet wide, twelve feet long and eight feet high. The contents included hand tools such as axes, mattocks, rakes, logging saws, sledge hammers, carpenter tools, telephone line construction gear and other small hand tools. Until shortly before World War II, in connection with annual inventories of property, it was necessary to account not only for all major property used by the Forest Service but small items such as hammers, pliers, axes, etc. If the responsible party could not account for all the items he was charged with, he found it necessary to fill out a "lying statement" otherwise known as form 858. There was space to explain how each item were lost, stolen or destroyed. If the story turned out to be good enough, the responsible party was relieved of liability. If not relieved he might well have to pay for the missing items.

During the time we were at Center Camp, all telephone lines on the forest came to that location and there was a line from there to Sonora. Thus the district rangers could have a telephone meeting without the benefit of the supervisor or his staff's participation. In this case, and as a result of such a meeting, the other rangers backdated transfers of missing property to Dad. This, of course only worked once. As the story was retold over the years, it would have taken a major warehouse to have stored all the items.

From time to time, Dad would bring home animals such as pigs, turkeys, chickens and the like to be fattened for use in our diet. Often times the animals escaped and there would be quite a flurry before they could be located and recovered. One time he out did himself. Since he felt they would be good company and protection for mother and me while he was away on field trips, he acquired and brought home a couple of bloodhounds. Compared to today's conditions, there were relatively few people living in the area and there was a lot of sheep grazing activity. These two conditions seemed to have been favorable for the development of a large coyote population. At times it seemed that after dark there was at least one on every hill. Coyote calls were not received favorably by the bloodhounds. The first time they heard them they promptly crawled under the house from where they answered call for call in their deep baying wails. The yelps were so bad the building would seem to vibrate. After a few nights of that, mother was fit to be tied. Upon Dad's return he was informed the bloodhounds were leaving promptly, or she was. Since we continued to live there, it can be safely assumed the dogs left.

In what was the September 9, 1920, issue of Collier's, the National Weekly, appeared an article, complete with pictures, based on an actual fire. This man caused fire started in the South Fork of the Stanislaus River. The article traces my father's travels from Center Camp to the fire in a Model T Ford in the company of an assistant. There is also an account of how the fire was controlled and an explanation of "gum shoeing." Dad collected the remains of a newspaper from Stockton, made plaster casts of tire marks and foot marks at the camp and found other evidence. He then started a journey up the Mono Road that led to Sonora Pass asking questions as he went. Somewhere on the grade between Kennedy Meadows and the top, he overtook the suspects, arrested them, returned them to Sonora and was successful in obtaining a conviction.

Somewhere in the story, there is passing reference to the fact that my mother received and forwarded messages, ran the switchboard and did other chores that were clearly of a dispatching variety. What she was doing was not unique for the Forest Service. It was the norm for wives all over the Western United States. In later years we were to learn about "perks" granted to employees. In the teens, the government was the beneficiary of perks that were just considered a part of the life of a forest ranger's wife. Paid telephone operators would come on the scene later. It can also be stated that most of the wives were well qualified to be dispatchers. Part of it was self defense in order to get their husbands home as soon as possible to cut wood, perform other chores and to get rid of bloodhounds.

During our stay at Center Camp my father had to spend much of his time on horseback trips. Since I was too young for this type of outing, my mother and I remained at the station. However, from time to time, my father's duties would take him to Long Barn, what would become Pinecrest, Cow Creek, Brightman Flat and other places along the Mono Road. Some of these trips would provide an opportunity for the family to go on what would amount to combined business and camping. We would travel in a Model T Ford and might be at a field location for several days or weeks. Travel was slow and difficult. Normally more time was spent in trying to travel rather than actually moving. Boiling radiators and tire troubles resulted in frequent delays. In the "teens" it might take four hours from Sonora to Strawberry and probably another four hours to reach Brightman Flat. When radiators and tires were not the problem, burned out bands or fouled spark plugs might cause delays. When anyone went a considerable distance without such delays, he was considered to have been very lucky. Most of the travel was at speeds of less than twenty miles per hour.

Except for stretches from Kennedy Meadows to the summit of the Sonora Pass, one of the longest and most difficult grades on the Mono Road was known as the Patterson Grade. It extended from near the confluence of the Middle Fork of the Stanislaus and the Clark Fork to just beyond Niagara Creek. Going up, the driver experienced a burning sensation on the ball of his left foot from holding down the low gear of a Model T. He had the same experience on the way down to make use of compression and thus save the brake band and possibly the reverse band as well for a real emergency (short of banking the vehicle). In any event, as he reached the bottom, he could look forward to a stop to clean the carbon out of the spark plugs that would have become fowled after a long downhill run in low gear. Since the gasoline tank was in the cowling just in front of the windshield, it paid to have the tank full before you started up such a grade. Model T Fords depended on gravity feed to the carburetor; if the tank got too low the lizzie might sputter and stop. While Fords could be driven in reverse under such conditions, this was real tough on the neck.

Borrowing from the experience of the Army, the Forest Service very early developed its own telephone system. While such telephone lines were initially built to serve government purposes, additional demands soon developed. Individuals, permittees and companies who would be valuable as cooperators, especially in the field of fire control, were often admitted into the system. In addition, there were demands from the general public to use the lines for personal reasons since in most parts of the western mountains during this era, there were no other telephone systems available. Thus not only was the Forest service the "Ma Bell" of the Stanislaus, it played a similar role on many, if not all, the rest of the national forests in the Western United States. This situation continued into the thirties or even later in some areas. A telephone had been installed on the porch of the Forest Service cabin at Pinecrest where members of the public could take or initiate calls. It amounted to an open air telephone booth where it was possible for one and all to listen to one side of the conversation.

In 1916, the predecessor of the Pacific Gas and Electric Company built the dam that created Strawberry Lake on the

South Fork of the Stanislaus River to store water for use with the Spring Gap powerhouse. Several years later the name would be changed from Strawberry Lake to Pinecrest Lake. In an area now occupied by the Pinecrest Resort, a two story crew house was built to provide living quarters for the dam construction crew. The building later became the initial Pinecrest Resort. My father's father was a member of the construction crew and became the first dam tender following construction of the dam. My grandfather was an expert in building log cabins and in making shakes. Since family relationships were no problem in that era of the Forest Service, my grandfather was hired to build a log cabin and to fence in a meadow where the Forest Service employees could keep horses when on official business. During this period my grandfather built several other log cabins for the Forest Service.

It was at both Pinecrest and Brightman Flat that a future lands officer, who would consider summer homes improper use of public lands, sinned at an early age. In those days and even later, summer home tracts were surveyed on the front and sides with a staff compass and a chain, but no survey was made of the back side. This took care of any error of closure. (In a few years, after the summer home permittees moved the corners to their liking, the original error of closure became moot.) The sin occurred in this manner. My father would establish a front corner of a proposed summer home lot by driving in a wooden stake. He then stood me (age of about six) over the stake with instructions to hold the zero end of the chain. He would then extend the chain to obtain the distance to the other corners he had established while using me as a surveyor's rod. His name appears on the original plat, but not the name of his helper.

THIRTY-TWO YEARS WITH THE USFS

T. J. "Tom" Jones

During the winter of 1909 and 1910, with the thermometer hovering around twenty below zero, Henry Atcheson, who had by this time passed the ranger's examination, Charlie Shillingford, who was ranger on the Bridgeport District and I, started a boundary survey on the east side of the Sweetwater District. We used a team and spring wagon and saddle horses. This boundary, some 200 miles, ran from the southern end of Nine Mile Valley north to the crossing of the Walker River and around the rim of Mason and Smith Valleys. We spent the entire winter with temperatures as low as twenty-five below zero, and I doubt if for two months the temperature was higher than zero at night. The boundary was marked largely by monuments and, there being plenty of rocks, this was easily accomplished, provided you could get the rocks loose from the ground. We had to use a sledge hammer or other implement to pry them loose from the frost. We camped in a tent and slept on straw placed on frozen ground. However, we didn't suffer and were just as happy as the present day ranger in his modern quarters. There was absolutely no complaint.

I eventually became ranger-at-large on the Mono Forest until early May 1918, when I was transferred to the Plumas Forest as deputy supervisor. One of the highlights of the transfer was the fact that, although I had met Supervisor Rogers once before, he not only did not recognize me but didn't even know which one of the Joneses I was or where I had come from. There was a Ranger Jones on the Eldorado at that time known as "Deacon Jones" and Dave Rogers' first greeting to me was, "Oh yeah, you're Deacon Jones." With a somewhat suppressed laugh, I replied, "No, not Deacon Jones. Just merely God-fearing Tom Jones from the Mono." I was assigned to take charge of grazing as I had expected. Just how fully Dave wanted me to take charge of grazing I never knew, but I always surmised that I went further than he expected. The season of 1918, was spent with a pack outfit and in many places on foot or with a burro where a regular camp outfit could not get into the country. At any rate, when the season had ended, I think I knew every grazing permittee and range and how the stock were being handled, if at all.

To begin with, the valleys which contained practically all of the meadowlands were owned by sheepmen and so used. Many hundreds of cattle were allotted to the ridges between the valleys. This was particularly true of the east side of the forest. Difficulties experienced were that the sheep, especially the lambs, were not doing well down in the meadows, and there weren't enough cowpunchers to keep the cattle on the ridges. I, therefore, originated a plan of land waivers by which the hill range was occupied by the sheep and the valleys by the cattle. The plan, being to the advantage of all concerned, appeared quite satisfactory.

On the southern part of the forest, there were a few prosperous cattlemen, practically all of whom were enjoying better than average summer range, mostly in the high country. Along the foothills were many small ranches, what Lou Barrett used to term "poison oak" ranches, with anywhere from 25 to 100 head of cattle, inbred and looking as though they had never had enough to eat. The boundaries between these ranges were not observed. The cattle ran where they ran because of having been there for generations and not because of any applied scheme of management. Their range boundaries and capacities as applied to units meant nothing. It was in conjunction with this problem that I got Supervisor Rogers into considerable hot water, at least for the time being. During his absence to San Francisco, I was acting in his stead, and I broke the news to the people using the Merimac and LaPorte Districts that some method of management was going to be put into practice which would be quite different from what was then in vogue. This resulted in the permittees getting together immediately to form a protective association—to protect themselves against the Forest Service. I well recall Supervisor Rogers' reactions upon his return. Convinced that some method of cooperative management was going to be necessary, I started to prepare the ground by writing to permittees on the Mono for whom I had organized a plan of community management, whereby they ran the stock collectively as if of a single brand. They elected one of their number to hire and fire the necessary labor and actually administer the stock, regardless of ownership. I received a large number of letters from these people acknowledging the benefits they had derived from the plan and their unqualified approval of the method. These letters, I believe, were finally printed and bound into little pamphlets which were distributed to the Plumas permittees who were on the warpath.

Grazing administration on the southern half of the Plumas developed rapidly and in a few days the permittees had called a meeting of their protective association at Bangor, inviting us to come and "take our medicine." Chief of grazing Chris Rachford, Dave Rogers and I attended. This was a memorable meeting in that I finally produced a plan which met with the universal approval of the people concerned. If I remember correctly, almost every permittee there took the floor at one time or another. For about two hours they berated the Forest Service management. Finally, however, they ran out of material, and there had not been what I could call a constructive suggestion from the entire group. Whereupon, I merely presented my plan, described the class of livestock they owned, the reason the stock were of that class and told them I was not particularly eager to put my plan into effect but that if they had a plan which would improve the condition, especially benefiting the small permittees in the foothills, and improve the quality of breeding, I would have no objection to trying it. I immediately got a favorable response and before leaving the meeting, they had agreed to at least the major features of my plan and had ordered three carloads of registered bulls to replace the scrubs then in use by most small owners.

In February 1920, I transferred to the Inyo Forest, relieving Chester Jordan as forest supervisor.

In my first approval of grazing applications on the Inyo, I made a decision which was to plague me for some time to come. Mr. M. Q. Watterson, a member of a leading banking firm in Inyo County, called at my office to insist that I reconsider my disapproval of a permit to grant range to a Basque sheepman from Kern County who was clearly a Class "C" owner without prior right. He was very insistent and pressed me for a reversal of decision. I refused in plain but polite language. Three years later it was necessary to render a similar decision. This time I was attacked through a petition charging me with all manner of both unfairness and competence. A copy of this petition was mailed to the district forester, claiming support by the community. An investigation by Jesse Nelson, who was then chief of grazing, disclosed the fact that they were unable to get a single signature on the petition, and the lawyer who had drawn it up and carried on correspondence in connection therewith, tried to deny authorship. I quote this as the most complete vindication of the grazing regulations ever to come to my attention.

I arrived at Mt. Shasta on March 6, 1926, to take over the administration of the Shasta National Forest, one of the worst fire protection jobs in the region. In the next few years, firefighting was made more effective through a substantial increase of fire equipment and intensive training of the personnel. Road building also contributed toward an improved system of control. In 1927, we started definite progress on road construction with heavy machinery.

Prior to 1926, it had been necessary to close to hunters and campers that portion of the forest embracing the private ownerships of Long Bell, McCloud and Red River Lumber Companies, in cooperation with the owners. In anticipation of again closing the area, the landowners had employed patrolmen and were posting against trespass. The sportsmen of Northern California were showing hostility to the closure and disclaiming responsibility for fires. It occurred to me that we did have fires for which hunters were not responsible and therefore, we might make arrangements to use hunters as firefighters and minimize risk due to their presence in the forest. So, I instructed that all hunters' conveyances be equipped with a shovel and axe and that hunters should agree to fight any fire occurring in the area while camped there, in return for the privilege of hunting. Before putting this into effect, I called the regional forester on the telephone and received his consent for a trial. I found the sportsmen's groups 100 percent for it. The landowners were doubtful but willing to try it. I then arranged a system of registration, and since there was no regulation covering the procedure, drew a written agreement in which the registrants each signed as a party thereto. That the plan worked 100 percent is history and this was the origin of the shovel and axe regulation.

In the spring of 1928, Frank Meyers developed the first headlight for use on the fireline. I assisted Frank, and we found that by placing the lamp on the forehead, a man could look at his work with the light naturally directed where it was needed. These lights were not made available in sufficient quantity for two or three years, but they then increased the efficiency of night crews by about fifty percent.

Fire hazards were greatly reduced in August 1927, when the Southern Pacific Railroad Company was first prevailed upon to screen their locomotives. It has seemed to me that both equipment and technique of attack on fires developed most rapidly during this period of the late '20s and early '30s.

The CCC program provided several difficulties, one of which was divided authority with the Army. The Army was to approve all campsites and construct all buildings. This frequently resulted in delays. The building program demanded the efforts of both services to accomplish completion within a reasonable time.

I received a telegram at 11:00 p.m. one Saturday night that the Sims and Callahan companies would both arrive early Monday morning. One company had to cross the Sacramento River and the other the Scott River. There were no bridges and the streams were bank full. Frank Meyers was called out of bed to start work on these bridges with two nights and one day to build them. Log pontoons were the only possibility and they were completed on time. They stood every test of high water and served until permanent structures could be built.

On April 1, 1934, I was transferred to the Klamath. It would be impossible to describe the feeling with which I took charge of the forest on which I had entered the Service about twenty-six years earlier. I knew most of the people of northern and western Siskiyou County, as well as the forest force. There was but one employee remaining who was on the forest when I entered the Service, or who was even in the Forest Service at that time. That was Perry Hill, with whom I had taken the examination.

In taking over the administration of the Klamath, it was obvious to me that it would probably be my last assignment in the Service and I was especially eager to accomplish everything possible for the forest. The work was almost wholly construction development.

A LOVE AFFAIR—THE AIRPLANE AND ME

Andy Brenneis

Like others who started in the Forest Service in the 1920s, I began to think about possible uses of aircraft in Forest Service work. I didn't have a chance to work on the idea until I was sent to the Angeles Forest in Southern California in June of 1936, to be the deputy to Supervisor Bill Mendenhall. I found Supervisor Mendenhall and a radio technician Jack Moon extremely anxious to improve the communication system. One day I was surprised when Jack Moon said, "How would you like to ride in an airplane and see if we can start air to ground radio communication?" Of course I was delighted and found myself, not too long from then, sitting in the open cockpit of a Curtis Robin aircraft. After I got strapped in, Jack Moon put a TE radio set on my lap, a set that was about 20 inches long, 10 inches wide and 10 inches deep. This was the big receiving set with a big output for Forest Service radio at that time. We started off, and needless to say, in an open cockpit plane like the Curtis Robin, the motor noise was just a bit loud. I had on a good set of sponge rubber earphones, he had an established pattern, and I communicated with the ground. The reception on my end was not perfect because the coughing of the motor interrupted it. Nevertheless, with a TE set and antenna stretched out to the strut of that aircraft, we did have air to ground communication. To me that was an amazing accomplishment. It opened up an entirely new world. It meant that we could talk to people on firelines. We didn't have to string emergency telephone wires over endless miles and we could get authentic information by just riding up there and talking to the people who were on the fireline.

I didn't stay on the Angeles long enough to pursue this. In January of 1937, I was transferred to the Cleveland National Forest as forest supervisor. My appetite in the use of aircraft, however, had been whetted and it didn't take long to establish good working relationships with the United States Coast Guard which had a large air base at San Diego. The Coast Guard said they would be very happy to fly work missions for us. We flew some aerial patrols in the open cockpit Corsair and I recall being flown one day over Palomar with the chief aviation machinist mate pilot. As I looked out over the side I got on the intercom and said, "Chuck, I just don't think I could bail out of this thing if we got into trouble; we're too high." And his answer was, "Son, if I tell you to bail out, you'll bail out, because I'm just going to flip it over on its back and you'll go."

At the end of May in 1938, I was transferred back to the Lassen Forest where I had started my career, now as forest supervisor. The Lassen was not well equipped with radio at that time. It was the summer of 1938, at our annual fire school that we tried out some of the smaller S and SX radio sets for the first time. To most of the people on the ground, these were more of a novelty than they were a means of communication.

Johnny Nagel flew in from Red Bluff one day on a mission that he completed on the Trinity. I picked him up in Chester and asked him if he had plenty of fuel in the tanks. They were on the wings. He said, "Ya, I had the tank topped in Red Bluff." I said, "Well, just for the ducks of it, Johnny, why don't you check both tanks because we have about a three and one half hour flight. You've only got a total of about four hours in this thing. It's going to be over tough country and let's take a look." His right tank was topped and his left tank was almost empty.

I decided that we had to get messages to the ground. In many cases, the best place to drop them was either right on a fireline, where a crew was working or, into a lookout with telephone communication. My solution was to tie a bright orange streamer to a little block of a two by four with the message attached. The procedure was to fly over the target, the pilot would say bingo and I'd dump it out the right door. The cream of perfection for this particular technique was achieved at Antelope Lookout on the Bogard Ranger District when we dropped one on the catwalk of the tower.

During the time I was on the Los Padres starting in 1946, things were happening in the aircraft industry and the Forest Service was in a position to benefit from them. I recall when I got a chance to take my first ride in a helicopter with an excellent pioneer, Knute Flint. The first models, of course, lifted the pilot and one man; no cabin. There was a windshield and you sat up there "naked as a jay bird" with a safety belt holding you in place. There were several helicopter operators in the business, but the helicopter had yet to establish itself in fire control.

In September of 1948, we had the Wheeler Springs Fire just over the mountain from the City of Ojai. It was a bad time. Everything was dry. The fire started from a faulty butane valve in the Wheeler Spring Resort and flaming gas was pouring out into the oak trees which overhung the building. Needless to say, the fire went from there. From about two in the afternoon until dark that evening my estimate was that it had burned 7,500 acres. It was climbing the hills like a pack of greyhounds. This was tough country with only one or two roads in it after you got out of the bottom land. The brush was extremely dense and the opportunities to burn up people were real. I mention this fire because of two items. First, I think it turned out to be the largest helicopter operation carried on in the California Region up to that time. If I remember correctly, we had either five or six small helicopters operating and Cal Ferris had gotten a Sikorsky out of March Field. Second, this fire was really a going concern and in one afternoon, I think it was on the third day, it blew up on three corners at about the same time. In a matter of four hours it had picked up another 10,000 acres.

We made some decisions on strategy. We had to keep the fire south of the Sespe River or we'd have another 200,000 acres Matilija Fire (1932) on our hands. The fire had sneaked up one of the main ridges on Chief Mountain and was headed out towards Topa Topa Lookout. All wild country and all foot work and hand work. The next morning we noticed at dawn that there was a wisp of cloud in the sky. The weather people had told us that this day was to be more of the same. We knew that we couldn't stand more of the same. Early that morning I took Assistant Regional Forester Frank Jefferson up to the fire camp at the base of Chief Mountain. We noticed that one of our men, Doc Davis, was trying to get some brush to burn and it just wasn't taking. He was following our plan to burn out a line the first thing that morning. We talked it over with the zone boss for that area, Stan Stevenson. We wondered if it might be possible to put men up onto that ridge line on Chief Mountain where the fire was sort of "stinking around." It didn't seem to be going anywhere. If we could put the men in there, we might be able to "hot spot" it and keep it from going down the other side of the hill and heading for Topa Topa. This was rather a chancy decision to make. After using the old trick of wetting your finger and holding it up into the wind, then looking at that little stringer of cloud that was in the sky with another one trailing it, we all agreed to try it. But we would have to muster all our helicopters and concentrate them on that particular area. We didn't use the Sikorsky because it was plainly too big to get into a tight squeeze. However, we did concentrate all the small helicopters and Stan put the men in there one at a time. And by golly, they did the job. We dropped cargo, put in another fire camp across the Marcos Highway and we supplied by air another crew under Stub Mansfield that just wouldn't abandon its position. We also put a fireline of defense along the front line fire lane behind Ojai and held it. About nine that night I received a message over the horn from Lee Thomas, intelligence officer on the staff for the operation in the back country, and he reported that all lines were secured. He also reported that the line that had been put in by the men who went to Chief Mountain by helicopter was secure. Thanks to the ability of helicopters to put down men in a crucial part of the line, we had won the battle.

It was in the summer of 1951, that I was involved in a large aerial operation on the Lems Summit Fire, Six Rivers Forest, south of Gasquet on the Smith River. Several of the fixed wing aircraft operated out of a relatively new airfield on the Klamath River at Happy Camp under Cal Ferris. Their primary mission was to drop supplies to men and animals on a line which was strictly back country.

Two incidents stand out in my mind about this air operation. One was serious and the other had a humorous angle. The serious one occurred about dusk. A radio call came in from the top of the mountain. They had an injury. Fred Bowen was on hand with some other chap who happened to be next in the rotation. I asked this pilot if he felt like tackling it. He says, "No, thanks." I'd known Fred a little and had flown with him. I asked, "Fred, is there a possibility of getting that guy out of there tonight or is it just too chancy?" He thought it over for half a minute and said, "Well, if we're going to do it we have to do it right now." I said, "Well, your choice." He said, "Will you put a few cars up at the lookout to the north of this field and have them turn on their lights? There is a small clearing up there and I can set down in it." The answer was affirmative. He left and it was only about 20 or 30 minutes before he was back out with the injured man.

The second incident had a funny twist to it, although it didn't seem so at the time. This was a tough fire and two strings of mules had been trucked all the way from the Missoula, Montana, Remount Depot to do the pack job. This may seem far out but there weren't many pack animals left in the country. The mules were taken up on top to do a lot of work right on the firelines and they needed feeding. We'd been having rather stinky weather in the early mornings. This one fly camp, up on one of the ridges, was in need of groceries and water. In addition, somewhere down the list was a request for hay and grain for the pack stock. The fog and smoke cleared out about 11:00 a.m. and the aircraft started moving out of Happy Camp with the larger loads. We couldn't see them from our position because we were still socked in, but we could hear the aircraft droning, evidently making drops. Then, they went away. All of a sudden a voice on the radio called me and said, "Andy, call that fire camp and tell them that if we've got to eat this shredded wheat then they'd better drop us some milk to eat it with."

What had happened was that the first load went out with hay for the mules and nothing for the men. You can laugh about this now, and we could the next day, but at the time, it really wasn't very funny because in about two hours the smoke and fog filled in again and that was the end of the day for drops.

MY LIFE WITH HELICOPTERS AND FOREST FIRES

Carl Wilson

I met my first helicopter on the Bryant Fire on the Angeles Forest in August 1947. My Lassen sector team and I had been sent to help control this ugly monster which had boiled out of Big Tujunga Canyon. Although we weren't important enough to warrant flying us up from the canyon bottom fire camp to Mt. Lukens, hundreds of other firefighters and thousands of pounds of freight were flown to the spike camps and firelines on top of the mountain by that fabulous Bell 47-B-3. In addition, that one ship evacuated sick, injured and trapped firefighters, flew reconnaissance missions and hover jumped firemen. It was truly a spectacular operation. This was not a "first use" for a helicopter on the Angeles, because earlier in 1946, a ship scouted, mapped and dropped freight on the Red Rock Fire near Castaic. However, the Bryant Fire was the first time that extended operational fire use of a helicopter had been made in the United States.

Interest in the helicopter as a firefighting tool really accelerated during Firestop in 1954. Experiments during Firestop included using large and small helicopters for laying fire hose and delivering water. In addition, pumping equipment and the hover jump were further tested. I remember doing my own testing of the helitanker concept while zone fire boss of the Jamison Fire on the Cleveland in 1954, stealing the idea from Firestop. I borrowed a five gallon paint bucket from fire camp and Hugh Ramsdell (the pilot) and I made several passes over some hot spots in an oak thicket. The water did reach the ground and the flareups were quieted.

Experiments with helicopters continued in 1956, after I transferred from the Angeles to the PSW Station. Shortly afterwards, the "Helitack Program" was formed. With the strong support of the WO, the station and the region, the national program was set up in cooperation with the Corps of Engineers at Fort Belvoir in Virginia. The overall objective was to "make helicopters an integral part of firefighting forces" nationwide. By the way, M. M. "Red" Nelson deserves special credit for the word "helitack" because he nagged Keith Arnold and me into coming up with an acronym or something that could be used to represent the project. So, the word was born about mid-span on the San Francisco Bay Bridge one evening en route to the RO.

Early results of the program in California were encouraging. In a firefighting test operation on the San Bernardino Forest in July 1957, 1,000 feet of linen hose in a suspended tray were laid across heavy brush by helicopter in less than one minute. Experts on the fire said that it would have taken six to eight men about two hours to do the same job on the ground.

Tests made of the long necked 35 gallon neoprene helitank were completed in 1955, and the accessory was ready for operational use in 1957. Although 35 gallons is a relatively small amount, when it is accurately placed at the right time, it can be worth 35,000 gallons later. "Helipump" was another innovation, along the same line, and the unit was developed in 1957. The pumper consisted of 10 gallons of water, an eight and one half gpm pump and 200 feet of one inch lightweight hose. This accessory proved to be of great value at critical spots on fires, particularly with a well trained helitack crew.

The hover jump, which had been developed on the Angeles in 1949, was further explored in 1956 as part of the helitack program. Jim Murphy used mannequins for some of the early tests and later firemen were put into protective smokejumpers' suits. Results showed that the "helijumps" were best made into open spots to avoid rocks or logs. Jumps made into brush were best made on those types which were a natural "mattress" of densely intertwined branches, such as buckbrush and scrub oak.

In 1958, the number of helicopters used on fires greatly increased and there was some experimental use of large commercial and military ships on forest fires. This was the year that the light turbine, Alouette II, was used in firefighting. Also, I believe it was about this time that our regional pilot Cal Ferris, coined the unforgettable term, "omni directional termination facility," meaning of course, a heliport.

The use of helicopters in forest firefighting in California continued to grow. Light and medium size turbines appeared in 1975, and their use increased exponentially. In the past decade or so, increased payload to altitude performance, coupled with speed, proved to be a significant development. The fire services now had the capability to move thousands of firefighters, many tons of freight and drop 100 to 1,500 gallons of water or retardant from buckets or fixed tanks. In addition, the technique of rappelling, instead of hover jumping firefighters, was perfected and used in some Forest Service regions.

First tests were made of nocturnal use of the light helicopter during the 1965-1967 period. This research and development effort was stimulated by the management analysis of the 67,000 acre Coyote Fire disaster which occurred in and adjacent to Santa Barbara in September 1964. The fire review showed that use of either airtankers or helicopters during the first night might have stopped the fire while it was relatively small. The major problem, according to fire experts like Norm Farrell, was lack of safe access for men and equipment during the critical hours when the fire was "lying down" under more favorable wind, temperature and humidity conditions. As I remember, Regional Forester Charlie Connaughton asked if there was something that research could do about testing the feasibility of night operations with aircraft. PSW Director John McGuire, who later became Chief of the FS, said, "We'll give it a try."

After two years of intensive field tests and study, we (in particular, Jim David and Ralph Johnston) learned that it was technically feasible to fly the light helicopter at night.

LAND ACQUISITION IN THE LAKE TAHOE BASIN

Andrew R. Schmidt

Editor's Note: The following is an abridged excerpt from an unpublished document prepared by Andy Schmidt under contract with the Pacific Southwest Region (R5) titled "A History of the Forest Service Role at the Lake Tahoe Basin."

From the original Lake Tahoe Forest Reserve in 1899, the progression into the present three legally proclaimed national forests—the Eldorado, Tahoe and Toiyabe followed. The Lake Tahoe Basin Management Unit was administratively formed in

April 1973, to provide uniform management of national forest lands in the basin, but these lands remain legally a part of the three proclaimed national forests. The reserved public domain base of about 68,150 acres included only two shoreline tracts of approximately 92 acres and about a half mile of shoreline. The national forest land base had grown to 132,772 acres as of March 1, 1979, and the length of shoreline to 10.69 miles.

The first land exchange at Lake Tahoe was made with William Kent, former U.S. Congressman, by a special Act of Congress of April 11, 1922, in which he obtained 2.03 acres of national forest land in exchange for 1.56 acres of land he owned. Kent also donated some 22.7 acres of land to the United States in two donations in February, 1920 and June, 1922, which became the William Kent Campground and Recreation Area with some 300 feet of shoreline near Tahoe City.

The first major addition to the public domain base occurred by land exchange in 1936, when some 7,776 acres of land in the Upper Truckee River and Trout Creek areas were acquired from the Carson & Tahoe Lumber & Flume Company and the El Dorado Wood and Flume Company. Although much of the area was old cutover lands with an attractive, thrifty stand of second growth timber, about 3,000 acres of virgin timber in the more remote areas remained. The companies were seeking to liquidate their assets during the late "depression" days and accepted $3.00 per acre or $23,328.60 for their lands. The estimated timber volume of 32.7 million board feet was valued at 50¢ per thousand board feet for Jeffrey pine and 10¢ per thousand board feet for white and red fir. The exchange was a so-called tripartite exchange in which the companies received money derived from national forest timber cut by third parties. In this case the national forest timber was cut by the Fruit Grower's Supply Company on a sale dated March 10, 1936, on the Lassen National Forest and the Clover Valley Lumber Company on a sale dated October 6, 1930, on the Plumas National Forest. Title to the lands passed to the United States in 1937.

One of the most fortunate land acquisitions at Lake Tahoe was the Pope and Baldwin tripartite exchanges at the south end of Lake Tahoe in 1950-51. The two, handled generally as one,

FS Lineman Lee Morford constructing telephone line from Orleans to Somes Bar, Klamath NF, California – 1934.

called the Pope-Baldwin Exchange, resulted in the acquisition of some 3,503 acres of excellent recreation lands including nearly 10,000 feet of prime beach on Lake Tahoe and 17,800 feet of shoreline including that on Fallen Leaf Lake.

In the Anita Baldwin Estate case, 2,501 acres for $768,100, the timber was cut by the Caldor Lumber Company on the Eldorado National Forest and the White Pine Lumber Company on the Modoc National Forest. The timber for the George Pope Estate case, $178,364 for 1,022 acres, was cut by Winton Lumber Company on the Stanislaus National Forest. Some parcels of land were also donated by the two Estates.

Interesting background to the two exchanges is furnished by a copy of a letter dated November 3, 1950, from Ray Knisley, then acting as trustee for the Anita Baldwin Estate, to State Senator H. E. Dillinger of Placerville, who had some question concerning the acquisitions. "Lucky" Baldwin, who died in 1909, wanted the property kept in a natural state. The property was kept that way by Anita Baldwin until her death in 1939. The heirs were mindful of the desires of both "Lucky" and Anita Baldwin but needed money to settle the vast estate. Ray Knisley had been acquainted with Anita Baldwin and family since 1915. According to the Knisley letter, he contacted Frank Dougherty, President of the Los Angeles Chamber of Commerce, to see if he could interest the State of California in acquiring the lands for a park. The state indicated the area was too large and that money was not available. Knisley then turned to the National Park Service, Washington, DC, in 1944, knowing that they were interested in Lake Tahoe. Knisley proposed, in addition to the Baldwin property, a package including, with permission of the owners, the George A. Pope lands and the Brigham (Ebright) lands at Cascade Lake and combine these with the Desolation Valley Wild Area as a national park. The Park Service officials said the area was too small and suggested he contact the Forest Service.

According to Ray Knisley his reception at the Forest Service office in Washington was very cordial—"they rolled out the red carpet for me and said that we'd love to do it. We don't know how but we'd love to do it." Knisley waited until after World War II for the return of B. M. Baldwin, one of the heirs, from the

Service in the summer of 1948, before contacting Ed Smith, Supervisor of the Eldorado National Forest. Ed Smith was in full accord with the idea, having been a supporter of acquisition of the property through channels of the Forest Service earlier. Soon after, Ed Smith took the Chief of the Forest Service, Lyle F. Watts, and Regional Forester Pat Thompson and several of their assistants to see June Lake, near Bishop, as a horrible example of what could happen to the Baldwin property. Knisley remembers that the party sat out on the porch of the old Baldwin cottage and talked about what they had seen and the possible exchange. He recalls Chief Watts saying, "I'm going to put you fellows on the spot. Do you think we should do it?" The answer was a unanimous "Yes." Watts said, "I agree. And furthermore having had the opportunity to do this, if we do not complete it, posterity should write after our names—'He lived in vain'." In the meantime, the Brigham property became unavailable at a reasonable price, so Knisley, acting as the agent for the Baldwin heirs and as a friend of the Pope heirs, worked with the Forest Service on acquisition of two estates.

Through Ed Smith's close working relationships with the El Dorado County Board of Supervisors, he was able to gain their support for the tripartite exchange, but to offset loss of the county share of national forest receipts, Smith agreed to establish a summer home tract (Spring Creek) on a piece of the Baldwin property. The Pope heirs had planned as an alternative to selling the property to subdivide into home lots and commercial areas including a hotel. County officials had become somewhat concerned about too rapid development at the south end of the lake and were more amenable at that time to Forest Service acquisition.

Additional support for the exchange came from the California State Legislature in the form of a resolution endorsing the exchange. Earl Lee Kelley, former head of the California State Department of Public Works, whom Earl Bachman had known since he was supervisor of the Shasta National Forest, appeared before the legislature in support of the resolution. It took common effort of many Forest Service people at the forest and regional levels to gain sufficient support to make the proposed acquisition a reality. Through this exchange the recreation complex of Pope Beach, Baldwin Beach and Fallen Leaf Campground came into being.

Opportunities to acquire lands through tripartite exchange procedures became difficult following World War II and virtually ceased in the 1950s. The acquisition program turned to land-for-land exchanges. A series of such exchanges took place in the Eldorado National Forest in the late 1950s and early 1960s in which the Forest Service exchanged isolated pieces of national forest land engulfed by rapid development at the south shore for private lands in the so-called green belt area east of Pioneer Trail to prevent undesirable scattered development in the forested area. Similar exchanges were made in the Tahoe National Forest area of the basin.

By mid-1960, the Forest Service had about run out of isolated national forest lands to be used as land base for land-for-land exchanges at Tahoe. Fortunately about this time enactment of the Land and Water Conservation Fund Act of 1965 provided possible source of funds even though the fund was nationwide in scope and included grants to states for purchase and development of recreational lands.

The first large purchase case in the basin under the Land and Water Conservation Fund (L&WCF) was a part of the George M. Whittell Estate on the Nevada side in October 1970. Purchased were some 4,752 acres at a cost of $1.5 million.

The next large purchase of backdrop forested lands occurred in October 1974, when the Forest Service acquired some 10,121 acres of land in the northwest sector of the basin from Fibreboard Corporation for $9,925,000. Another 63 acres in Nevada at the stateline between the two states was donated by the corporation.

Before the Fibreboard purchase case started, acquisition of the Meeks Bay Resort property was underway. Macco Corporation, a subsidiary of the Pennsylvania Railroad, had bought the property in late 1968, with intentions to redevelop the property into a 1,700 unit condominium and resort complex. Macco had been in contact with the Forest Service shortly after about a land-for-land exchange to round out its ownership to fit its ideas on development, but the Forest Service would not give up key access to Desolation Wilderness. When the

Pennsylvania Railroad got into financial difficulties, Macco's source of investment money also "dried up." John T. Williams, project manager for Macco, let it be known to Robert Rice and Andrew Schmidt in late 1970, that the corporation would be receptive to an offer to buy the property. Later that year in a "chitchat" session following a meeting, Schmidt and Rice were talking to Don Ellis, President and Steve Brandt, Executive Secretary of the League To Save Lake Tahoe. In an offhanded way, Schmidt and Rice suggested to Ellis and Brandt that the league buy the property to save Meeks Bay from development and hold it for eventual Forest Service acquisition. The suggestion was immediately picked up by the league executives saying they had people with money in their organization who would be willing to do so. The "germ" of acquisition was followed up by a letter on March 29, 1971, from Regional Forester Doug Leisz to Don Ellis expressing the desire of the Forest Service to acquire Meeks Bay for public use but since no money was presently available, he promised to work to get the necessary funds. This was the go ahead signal for the league to act. Later that year, William R. Hewlett of the electronics firm of Hewlett-Packard Corporation acquired the property at the request of the league. In 1974, L&WCF money became available and Meeks Bay was acquired from Mr. Hewlett for $3.1 million in December. The 645 acre property included 3,328 feet of shoreline with 1,800 feet of excellent beach.

Another property shown on the "Public Lands Priorities" map was the Rabe Estate lands which lay behind the Forest Service's Nevada Beach Recreation Area. The Forest Service was not able to immediately act because of fund availability, so the Nature Conservancy purchased the property in July 1977, for later repurchase by the Forest Service which occurred in February 1978. The 236 acre tract was purchased for $3,859,139.28.

The ongoing acquisition program of the Forest Service has added many other significant lands for public use in the Lake Tahoe Basin. Between 1935 and 1965 approximately 12,000 acres were acquired by land exchange. From January 1965 until January 1, 1979, some 41,478 acres, nearly all in the Lake Tahoe Basin valued at $48,931,211 have been acquired.

Much effort and creative thinking by many people inside and outside of the Forest Service are contributing to the success of the program.

MEMORABLE PEOPLE

BROILED EGGS ARE INVENTED

Richard McArdle
(Submitted by Dave Bernstein)

Editor's Note: The following story about Bob Marshall was submitted by Dave Bernstein, who wrote, "Some years ago I wrote an article about Bob Marshall, focusing on his activities in the Northwest. I had noticed that he and Dick McArdle had been assigned to the PNW Station in 1924 so, on speculation, I wrote to McArdle to see if there had been any contact. In return, I received a long detailed letter, part of which is enclosed. I think it is a very interesting look at a couple of our country's leading foresters early in their careers. Among other things, it illustrates McArdle's legendary total recall and gives an insight on the man who later became the Forest Service's first Chief of Recreation and leading proponent of its Wilderness system."

When Bob Marshall was first detailed to the Douglas fir yield study our first job was to locate and remeasure some permanent plots near Wind River. These plots were on Panther Creek. My diary says that we left Hemlock RS (now Wind River RS) at 8 a.m. on September 16 and went up Panther Creek by the Warrens Gap Trail and were packed in by Elvin Blaisdell. I don't think Bob had had much camping experience. We did a full day of work before starting to set up camp. John McGinn (the other member of the crew) and I set up the tent and fixed up some sort of table while Marshall collected firewood. I thought he'd find a dead cherry or some other dry standing tree.

As usual, at that time of year it was raining. As usual also, it got dark early in dense timber. We couldn't figure what had happened to Bob. When he finally showed up he was dragging some very small green hemlocks. McGinn located a small standing dead tree and after some delay we got a fire going. But when I looked at the ends of the hemlocks that Bob had cut I could understand why Leo Isaac had told me that Bob had broken every thermometer the station had because he used a thermometer like an axe, and an axe like a thermometer. It was my mistake. I should have made sure that Bob knew how to make a fire in a rainy forest. All of us were wet to the waist because we had had to wade through high salal underbrush so two members of the crew were pretty upset. But Bob was not and never lost his good nature the whole time he was with me—except one time when he was measuring trees and stepped in a yellowjacket nest.

Marshall, who never had proper field clothes, at least in 1924, said he had so many half worn-out dress suits that he wanted to wear them out. When he stepped on the yellowjackets they swarmed inside his pant legs and stung him rather severely at certain tender parts of his anatomy. I can still hear Bob shrieking, "Oh dear, Oh dear..." as he ran around trying to get rid of the jackets. McGinn was a gent with a hair trigger temper and when he heard Bob's "cussing," he shouted at Bob, "If you have to cuss, for God's sake cuss like a man." I tried to catch Bob when he passed us on one of his frantic circuits so that I could rip his pants off but I couldn't catch him. I was really sorry about Bob because he was stung extremely badly. This incident was in the late afternoon of the one and only day Bob worked in Oregon. I tried to persuade him to let me get him to a doctor but he absolutely refused. I was sorry I let him talk me out of that. But he said he was leaving early for Portland next day anyway and would see a doctor there. I never knew if he did or not.

But to get back to the camp on Panther Creek. By mutual agreement, McGinn and I decided that perhaps we should not ask Bob to cook any of the evening meals. One crew member would leave early and have supper ready when the other two came in. Bob insisted, a dozen times a day, that he should take his turn in fixing the evening meal. He was that way, always willing to do his full share and more of the work. Finally, Bob beat on me so much that I told him OK it will be your turn Friday evening (we planned to pack out after work on Saturday, assuming we could finish).

McGinn had to have fish on Friday and had grandly offered to catch all the fish we needed but I thought I'd better be sure and took along two cans of salmon. Bob asked what he should cook for the Friday dinner. I told him that we hadn't much left so to cook whatever he could find but save enough for breakfast Sunday.

As near as I could reconstruct that Friday evening meal, Bob used every piece of a six-man mess kit, using a larger kettle when he had no space in the pan he was using. First, he told me that he had sliced our one remaining loaf of bread (this was before the days of bakery sliced bread) and toasted each slice laboriously over a big fire. Then he combined the toast with (as near as I could figure) two or three cans of evaporated milk, one or two cans of peas and the two cans of salmon, and I don't know what other ingredients.

When McGinn and I got in, Bob had what looked like gallons of what he told us was a salmon soufflé. Oh yes, he had found half a dozen big lemons and because he thought salmon needed a touch of lemon juice he added the juice of all the lemons. He was crouching over a raging fire with the biggest kettle sitting in a bit of water in the biggest frying pan (he said it was a rather makeshift double boiler). Bob thought the "soufflé" tasted fine but to McGinn and to me it tasted like lemonade with salmon and peas.

McGinn told me privately that he was going to get up early and beat Bob to the breakfast fixing job. About three in the morning I awakened by light on the tent flaps. First I thought we hadn't extinguished our fire but then I knew that wouldn't be true because it was raining. I looked out and there was Bob making breakfast. He had used the rest of the flour and had made hotcakes which he was stacking on a nearby log. Had a stack about a foot high. I was tempted to awaken McGinn and decided perhaps I shouldn't. I went back to bed.

Three hours later when McGinn and I got up Bob was making what he called "broiled eggs." He had hard-boiled all the rest of our eggs, shelled them (or maybe he didn't) and was frying the whole eggs in copious grease. Told me that he had this inspiration during the night and his ambition was to make this new dish "broiled eggs," the national dish of the U.S. The eggs were like golf balls. I couldn't cut one and when I tried to do it, the egg bounced off into the brush. It was typical of Bob that he took our dissatisfaction with his breakfast in good part, perfectly good natured.

Maybe I shouldn't have put that in my recollections of Bob Marshall. I am sure that he learned everything that he needed to know before many years. He wasn't stupid by a long shot. Far from it. His mistake was to try to do more than his share of the work. He wasn't a shirker at any time I knew him.

If you write up anything about Bob Marshall, I hope you will stress his many good qualities. He was generous to a fault; he would give the shirt off his back and throw in the coat for good measure. In fact, he would insist on it. He was smart, genuinely brainy. I never heard him say an unkind word about anyone. He never sought credit for what he did but went out of his way to give the credit to other people. If anything that I have written in this letter makes you think that McArdle didn't appreciate Bob Marshall's many good points, please scratch it out. Many times in the years between 1924 and 1939 when he died (I think of a heart attack) I laughed at Bob but he always laughed with me. In my book he was OK.

LEE EMERSON – LAWYER, LABORER, LEADER AND LOVER

Rudo L. Fromme

It was a customary warm, sunshiny July day in 1911, in Grants Pass, Oregon, when a tall, rugged, good-looking, Clark

Gable type of man in his early 30s stepped into the forest supervisor's office to ask for a job at day wages. He wished to work only for the summer and hoped it could be something a bit strenuous or venturesome, like backpacking supplies to firefighters or scouting out trails in rough country. Pay was of secondary consideration.

He went on to explain that he was a member of a law firm in San Francisco, held a college law degree but had also had a couple of years of civil engineering. He just wanted a few months of rough it and if he could make it pay its way, fine and dandy. "Aha," said I, "Maybe you would like to relocate some of the more abominable steep switchbacks on the Rogue River Trail."

The next morning he was with Ranger Jesse DeWitt at Galice, who got him going on that very job. Later that summer, he was in charge of a small trail construction crew, using blasting powder and causing Dutch Jake and Hathaway Jones to scramble down from their high perched cabins to see who was shooting up the country. However, he was a diplomat par excellence and soon became a bosom pal of these old, cranky sourdoughs. Probably, I should retract the term "bosom" pal as used for Dutch Jake, for I never saw him when his soiled shirt wasn't open to the navel and the wooley red hair so matted down with syrup, bacon grease, egg yolk and sweat as to be anything but inviting.

However, not everyone accepted the trail at first. The source of a complaint was an elderly, grizzled mining prospector whose shack on lower Mule Creek chanced to be a hundred feet or so below a steep slope. It was across this steep slope that the new Forest Service trail was being constructed under foremanship of Lee Emerson.

The complaint first cited carelessness on the part of the trail foreman in using dynamite, causing rocks to bombard his shack on the creek below. It then went on with the old complaint that the Service, by refusing to burn the brush off annually, was causing bugs, flies and mosquitoes to propagate freely and making good hiding places for rattlesnakes. Also, that the building of horse trails opened the back country to town and city travelers who were a pest to the mining prospector wishing solitude. Furthermore, it stressed that the mining prospector was "striving valiantly to increase the wealth and prosperity of this most wonderful country." Following the signature were the words, "A Brother Mason."

The letter indicated better schooling than was apparent to me from his whiskered, uncouth appearance. However, I got him and Lee Emerson together in an explanatory session and it wasn't long until apologies were clearing the muddled atmosphere. Lee was offering to help the old fellow with some mining law difficulties as I left.

Naturally, we all felt relieved that our trail foreman's efforts had worked out so satisfactorily, but his plan for following through seemed to be going beyond the call of duty. He wrote that he had invited the old miner to be his guest for a certain dated dinner at a quite popular Chinese chop house in Roseburg. No one, however, anticipated the finality of Emerson's excellent gesture.

Maybe it was some combustive combination of certain elements in the chop suey mix. Lee, however, thinks it was just plain overstuffing of a long, starved stomach. At any rate, the poor fellow passed away in his sleep at the hotel that night. Perhaps the bed was too soft and soothing, also. Lee had planned to eat breakfast with his newfound friend, but, after waiting until almost midmorning, he and the hotel clerk went into the room and found him as stated. Lee mentioned, however, that death must have been quite peaceful as there seemed to be a smile of quiet contentment on the old, grizzled face. He followed through with funeral expenses from his own pocket, refusing to accept any contributions from us in the office. This is the true unvarnished termination of the case of the Mule Creek mining man, who complained, by letter, of Forest Service policies and practices to his Brother Mason, President Howard Taft.

"Yup," added Hathaway Jones, "She was still a goin' Woof! Woof!" but kind o' faint like. Well, more like "Woof! Woof!" (given in a hushed whisper). That's the way Lee Emerson imitated this Mr. Jones in his "gospel truth" yarn about one certain exploit of his faithful old hound, "Queenie."

It seems that Queenie had been gone for about four long days and nights. Jones could hear "Woof! Woof!" pretty strong the first day or two, but "the damn brush was so tarnation thick" that he couldn't find "zackly whar the 'Woof! Woof!' wuz a'comin from." When he did, on the fourth day, finally locate old Queenie "she wuz a'layin' down, stretched out on her back, but a'lookin' straight up into a tall tree a'watching a big mountain lion, what wuz flopped down on the first big limb. Both of them looked awful tired." "I should think so," commented Lee. "Wasn't her tongue pretty thick, barking away all that time without any water?" "Oh, she had water," perked up Hathaway, "And the cougar did, too. Used the same water hole—right near by—only, of course, on opposite sides. I seen both their tracks. While Queenie was a drinkin' on one side, that big cat would slide down the tree, grab a few laps from the other side and be ready to scoot right back up the tree agin jest as soon as Queenie started lickin' her lips."

According to Lee's further understanding, Hathaway brought the cougar down with one shot. Lee thought he (the cougar) was probably too tired to care very much, anyway. Jones left the big cat to be brought in another day, 'cause he had to carry Queenie home in his arms. She was nothing but skin and bone, and probably on her last "Woof!"

I noticed a growing, overpowering interest on the part of our assistant stenographer to Lee's recurring visits to our Grants Pass office. His presence seemed to induce an increasingly stupefying action on her normally studious attention to her typewriter. She appeared definitely hypnotized and the wheels of forest industry slowed down perceptibly. On my last inspection hike to his trail building operation late that season, he surprised me with this query. "What's the matter with that girl in your office? Is she plumb nuts?" Then he showed me a note she had slipped in with time slips or instructions I had sent him. Of course, I read it. I was asked to.

And now I feel rather apologetic at broadcasting love's young dream. On second thought, it wasn't so young, and neither were the two principal characters of this drama. Correction: Tragedy.

So here goes, as near as I can now recall: "Dear Mr. Emerson: I hope you catch a big, wild cat this fall." (He had mentioned in the office to me, his plans for hunting, as our trail allotment ran out.) I also hope you'll remember the girl in the office with its hide. I'd just love to have it along my bed at night so that I might get down on my bare knees and run my fingers through its woolly hair and then dream of you." Instead of Lee's bringing a hide "in" he decided to hide "out." The steno also shortly departed our environs. Maybe she eventually caught up with him, but I imagine not. I never heard of either one of them after that. I expect love died ablooming and just withered on the vine.

AN OLD TIME FOREST SUPERVISOR

Robert W. Cermak

In 1905, at the tender age of 26, R. H. Charlton took over as Forest Supervisor of the Angeles Forest Reserve. His predecessor was Everett B. Thomas. Thomas was an effective supervisor under the General Land Office, but he ran afoul of financial regulations. He was removed by Gifford Pinchot, charged with padding his expense accounts, convicted and sent to prison. That must have been a daunting reminder for a new supervisor.

The Angeles consisted of two divisions, the San Gabriel and the San Bernardino and had the largest force of any forest in California. Most of both divisions was backcountry, patrolled by rangers and guards. Charlton soon recognized the importance of recreation in the Angeles and its effect upon fire control. Even before 1910, Charlton was campaigning for campfire permits on the Angeles. By 1915, the forest was receiving a half million recreation visits each year. He proved after the hot 1914 fire season that 87 out of 153 fires were caused by recreationists and the next year the Angeles became the first national forest to require campfire permits. By 1920, campfire permits were required on all Southern California forests.

Charlton was energetic and full of ideas. Sometimes the ideas backfired, such as when he arranged for two bonfires on top of Mt. Wilson to signal the winner of the hotly contested 1916 presidential election. He put chemicals into the brush piles to produce a green fire if Woodrow Wilson won and a red fire if Charles Evans Hughes won. His friends in the press made sure the stunt had plenty of coverage. California's voters decided the contest, but the vote count was not complete until early on the morning after election day. Charlton received a premature message and lit the red bonfire, thus sending word to all of Los Angeles that Charles Evans Hughes was the next President. Charlton's face was as red as his bonfire when he found out the truth.

Another idea involved goats and had plenty of promise but proved unworkable just as it did 60 years later when it was tried again. Probably the first firebreaks were built before 1900 on the steep ridges of the Angeles front by individuals and water companies worried about fire and flood damage to their property on the lowlands. Charlton had convinced local counties to deposit cooperative funds used to build and maintain these firebreaks. By 1912, there were about 60 miles of firebreaks snaking up and down the ridges and hogbacks of the Angeles front country. Charlton was glad to have the firebreaks and wanted more but the chaparral sprouts in the clearings were growing faster than his men could cut them. It occurred to Charlton that goats could be used to keep down the brush.

In 1912, he got permission from the Secretary of Agriculture to graze goats free of charge for firebreak maintenance. He negotiated a deal with an Arizona herder and brought in 700 goats for the experiment. He found what others have found since; goats not only eat sprouts on firebreaks, they also eat anything else in sight. After three years of trying to keep the goats where they belonged, he gave up. It apparently was easier to herd rangers than goats.

That wasn't the only experiment by the eager Charlton. Very likely the first aerial fire patrol was conducted by one of Charlton's Rangers, S. V. Parney. In the summer of 1913, Parney rode with pilot Howard Gill and observed a series of test fires along the San Gabriel front country. This was really an amazing feat. Today it would be akin to orbiting the earth in a homemade rocket ship. Six years later, the Army Air Patrol launched aerial fire control with daily flights over District 5 forests. While Charlton did not risk a ride in that rickety aircraft in 1913, he did go up in an observation balloon in 1918 from Arcadia and drifted along the same route taken by Gill and Parnay.

His enthusiasm for aerial fire control was put to good use in 1919, when Charlton was the first to scout a large fire by air. The 1919 fire season was the worst experienced by District 5 since the disaster of 1910. It was capped off by what was the greatest acreage loss in the history of the Angeles. The San Gabriel Fire began on September 12th, followed by the Ravenna Fire on September 19th. The two fires, driven by Santa Ana winds, burned until the 26th and blackened 135,000 acres. The fires attracted national attention and established Southern California as a special fire control situation with special needs.

Five years later it was another great fire on the Angeles that proved Charlton's undoing. The worst fire season in District 5 history was capped off in 1924 by another San Gabriel Fire. This one burned for a month, destroying trees and brush on 50,000 acres and bringing out a highly disturbed Chief Bill Greeley from Washington. The Angeles was subjected to the first national Board of Review, a process that was open to the public. The results of the board were to change the face of the Angeles forever and to move R. H. Charlton to supervisor of the Cauchita Forest in Arkansas, where in 1930 he died in an automobile crash.

Charlton had been supervisor of the Angeles for 20 years. Reading between the lines, one gets the feeling that he may have been the sacrificial lamb for a mismanaged fire. However, there was evidence that the San Bernardino division was in chaos. Charlton constantly strove for better fire protection and had many successes with the public but in the end, the forest was just too big and events were moving too fast. One of the most important results of the review was to break out a new San Bernardino Forest from the old Angeles and make the job of the Angeles supervisor half as demanding.

PACKERS: MASTERS OF THE TRAILS

Edward C. DeGraaf

On the Apache Forest that straddled the state line between Arizona and New Mexico, there was some unusual pack train travel by local ranchers. The Hudson Brothers had relatives in the Tonto Basin who made moonshine in quantity. They brought the moonshine by pack train from Arizona to Reserve, New Mexico, where a brother-in-law, Allred, operated a saloon throughout prohibition. The pack train consisted of several small Spanish mules, each one packing two medium sized oak kegs of moonshine. The charred oak kegs also served to cure the moonshine en route. The lead mule was a rusty red color with an all white head by name of "Hereford." Often when coming down the trail, Dick Hudson might be riding ahead and singing some popular song while brother Jerd brought up the end of the string without singing.

For normal packing it was quite common to use large panniers made from stretched cowhide that held all kinds of camp gear very well. The bedrolls were thrown over the top and lashed down.

In the Northwest I got to know an old packer who was a regular aristocrat. He was a Nez Perce Indian who was very fond of horses but who refused to handle mules. He had grown up herding Indian ponies on the bunch grass ridges of the Okanogan country. Roy Beeman had a great reputation as a packer of the north Cascades having handled large strings of horses and packed all kinds of difficult cargo, timbers, dynamite, etc. Roy always wore deerskin gloves in packing and handling ropes. He was an expert in making up cargo of fire tools and supplies and was a very able firefighter in his own right.

Another packer on the Wenatchee, who was considered one of the best in the Northwest, was Slim Hollingsworth who had the Icicle Horse Ranch outside of Leavenworth, Washington. Slim worked as many as 30 or more horses with a few mules. He had no prejudice regarding mules except for animals that were a cross between a jack and a "hot blood" mare; they could be very "spooky." However, he could handle them all. In the fall he often packed hunting parties in the Cascades. On one occasion I saw him pack out a large black bear. He choose a spooky little mule named "Blam." Slim took a gunny sack and smeared it with bear blood and hair. He then rubbed this over Blam's head and neck while he held him up short. The mule shook with fright as Slim tied a blindfold on. The bear was then loaded on and securely lashed in place. The mule was then led successfully down the trail, shaking as they went. Slim could make a trail camp in short time and when he made coffee he measured by the handful.

On one occasion on the Mission Ridge Fire out of Cashmere, Fremont McComb, Lloyd Olson and I started out of fire camp at 2:00 a.m., scrambling up the steep canyon slopes. Slim was loading up with cans of water and prepared food expecting to climb out after daylight. Olson, McComb and I made it straight to the top and when we stopped to rest at the fireline, Slim was right behind us with his pack animals loaded with water cans and food. How he worked his way up the canyon wall with those horses in the dark, I'll never know.

A pair of good packers in the Cloe Elum country of the Wenatchee were Dave and Alfred McDonald. Dave was district assistant for the Forest Service while Alfred was a freelance packer. On one occasion they contracted to pack a large water control valve to a stream outlet at a lake back in the mountains for irrigation control. They hit on a travois arrangement with the valve lashed to the poles that were in turn secured to the pack saddle. In this manner they dragged the travois up the trail, Indian style to its destination.

On the Olympic Forest the road foreman, Wallace Hopper, was a converted packer. He had grown up as a bronc rider and packer in northern Arizona where it was a way of life. On the Olympic he had packed over all the trails crossing the forest and parts of the park. When the forest road system developed the need for road and drainage repair and maintenance, Hopper left packing and took up tractor, grader and truck work. However, his first love was packing and he enjoyed riding the forest trails.

One experience with an old pack mule lingers with me. The Entiat District on the Wenatchee used pack and saddle stock several of which were raised on local ranches. The district had an old pack mule named "Playboy" that loafed in the corral all summer as a free boarder. I checked on the mule's age and it was nearly 30. So I suggested to the ranger they dispose of Playboy since they no longer used him. The ranger agreed and "dry gulched" old Playboy. As time went on I realized I had urged the disposal of an old Entiat District pet who was popular with the locals, ranchers and school kids. It took time to regain the local goodwill.

SOME FOREST SERVICE COOKS I HAVE KNOWN

Slim Hein

As the years roll by, many of the details of my years with the Deschutes National Forest get kind of fuzzy in my memory. Not so with memories of the good camp grub we used to have and the cooks that prepared it under less than ideal conditions. I will eulogize some of them here and hope that in so doing fond memories of old times will return to the reader.

In October 1929, the Deschutes set up a bug camp at the foot of Fox Butte to treat the 1926 burn. The cookhouse was a tarpaper shack used by the Fox Butte guard after the cabin on top of the butte had been burned up by the fire. The dining hall was a large wall tent set against the cook shack and heated with the usual jumbo sized Sibley stove. Depending on where you sat at the table, you either roasted or froze while eating.

Our cook was a Mr. Tripp, who I believe, had cooked for Fred Matz's survey crew during the summer. Tripp had a wooden leg which affected his ability to cook not a bit. With the aid of one flunky he put more high quality grub on the table than the 20-man crew could eat.

Particularly, I remember his generous supply of pies and light bread rolls. While the men were eating these and other victuals, Tripp would stand in the cookhouse door grinning and sending more of the same to the rustic table.

I can't remember when our deaf cook, Jimmy Doyle, started cooking for the Deschutes, but for years he was the standard cook for the road crews and fire crews. He was an all around good cook and firmly believed that pie was the cheapest food to make contented workers. Plenty of pie exposed when the men sat down at the table would surely curb their appetite for roast beef. He usually made "double strength" coffee which he "cut" before serving. Woe to the digestive tract that got a charge of the uncut stuff. This discouraged men hanging around the cookhouse drinking coffee prior to mealtime.

In spite of his deafness, he seemed to hear what he wanted to hear and ignored the rest of the idle conversation. He could always hear an uncomplimentary remark about the food and would invite the talker to eat elsewhere.

Our one-armed man, Price Garlington, wasn't always a cook. He admitted, without boasting, that he lost his arm as the result of a poker game. He started as a cook's flunky in a Forest Service WPA camp at Fall River, where he became the second cook and eventually became a first class cook. By using ice picks to anchor spuds for peeling and meat for cutting, he did every chore in the kitchen. He became a good pastry cook and could make light bread rolls second to none. I never heard him ask for help. An old tomcat was quite a nuisance around Fall River Camp. While Price was left alone in the camp one weekend, the tomcat was neutered. With one hand! Believe it or not.

In the winter of 1936-37, we had a big bug camp at the old Brooks Scanlon Water Tanks Camp, 40 miles southeast of Bend. Most of the 75 men required quite a bit of "antifreeze" to combat the 30 below temperature. They replenished their supply on weekend R&R trips to Bend, which lasted normally 'til about midweek. Eventually their hangovers began lasting well through the week. What has this got to do with camp cooks?

Well, it turned out that our cook, John Gomer, saw an opportunity to bolster his income and was stocking booze by the case and retailing it to the crew in the latter part of the week at a neat profit. I never found him addicted to consuming his own stock and he was a first rate cook.

Another unforgettable cook, though not on the Forest Service payroll, was Mrs. Sannerud, the wife of the Brooks Scanlon Co. steel gang foreman. The 15-man gang boarded with Mrs. Sannerud at old Camp Two of the Fort Rock District. Here she plied her talent in two of the company's old portable cabins, one as the kitchen and one as the dining room. Often times three of us boys on pine beetle survey and spotting work boarded with the steel gang. You can't imagine the amount of grub that disappeared from that table. The Sanneruds were old country Norwegians and it is well known that they excelled in the kitchen.

Mrs. Sannerud put up a lunch box for the three of us that would easily suffice for four. Always a whole pie, and to avoid hurting her feelings, we never took the fourth piece of pie back. At times we had lunch at the cookhouse if we were working close by. Again a whole pie for the three of us. After we had devoured our first piece, she would make the rounds offering the fourth piece left on the pie plate, first to Les Hunter then to Cal Francis. After they had refused she made her last appeal to me, the youngest of the crew, and no doubt having the biggest appetite. She always said, "Mr. Hein, you are my last hope. Won't you have another piece of pie." I never failed her.

One weekend shortly before Thanksgiving, when boarding with her, I went to a turkey match at Redmond. I was a better than average marksman in those days and came off with four turkeys. Being single and with no need of the birds, I gave one to my brother, one to my sister and one to the last man I had beat in a match. I took the fourth one back to camp and gave it to Mrs. Sannerud who prepared Thanksgiving dinner for the steel gang Tuesday evenings as everyone scattered Wednesday evening for the holiday.

Well, her old dining table fairly sagged under the load. Roast turkey, baked ham, white potatoes, sweet potatoes, vegetables, salads, hot rolls—you name it. It was there piping hot and Mrs. Sannerud prepared it all herself.

When about done with the main course you could hear the egg beater whirring in the galley and in came the jumbo sized pieces of pumpkin pie smothered with a mountain of whipped cream. Oh Lord! Give me strength. I gave it my best effort. I didn't marry Mrs. Sannerud's daughter, hoping to get a good cook. She had no daughters.

When I came back from WWII, Mr. Sannerud had died and Mrs. Sannerud had left Bend. About 1965, I saw in the Bend paper that she had died and there was to be a graveside service in Bend. Out of respect for an old friend, I attended and felt honored to assist in carrying her to a resting place beside her husband. The few in attendance were grey haired widows of the long gone steel gang men.

Here's to the camp cooks of long ago! Unsung heroes.

MAURICE A. BENEDICT

Roy H. Blood

At long last I have been given the chance to speak my piece about M. A. Benedict, the Sierra's third supervisor—"Ben" or "Benny" to all of us who know and love him. Looking back over some 44 misspent years of knowing Ben, the memories that stand out most strongly are of his daily kindnesses and consideration and understanding, consistently and without a single mean or petty act to mar the picture.

I like to remember my first meeting with Ben—January 8, 1912, at North Fork, a mountain village of some 100 souls, and a surprisingly small number of heels. The trip from Friant by horse stage had been long, but quite rewarding. At six mile intervals, there were oasises where mail was dropped and where refreshments were available for man and beast. After taking full advantage of the opportunities offered at Friant, Sesame, and O'Neals, the driver and horses were in sorry need of replacement. During the short periods when I was not required to walk because of steep and muddy roads, I was permitted to ride above with the drivers, and both were full of sympathy for me, a youngster who was going to work for a lost cause, a "bunch of college dudes," who were trying unsuccessfully to "buck the cowmen" who had first rights to grass and land and water. North Fork lived up to all advance billings—a hotel, a store and two friendly saloons. Friendly in that when business was too dull for both, the personnel and clientele moved en masse across the street from one to the other.

The headquarters office was located on a knoll on the banks of the North Fork of Willow Creek and its flag made a brilliant splash of color against the forest green. The rangers from all outlying districts were in for the winter and I was lucky to be able to meet in one group the fine bunch of men with whom I was destined to spend a great part of my life—men of vision and integrity and an almost fanatical love and reverence for the trees and soil and water that made up their universe. Men with one or more degrees from our leading universities working happily and proudly with those who were almost illiterate, sons of the plains and mountains who had left their ranches or logging operations or other pursuits to do what had to be done to protect our remaining resources from ourselves.

Ben was deputy, second in command. A renowned athlete from the University of Nebraska. Selected by Lou Barrett as guard on the Plumas and promoted to ranger before getting a chance to take the examination. Ben was tall, blondish and handsome, but already running short of hair on top. He had been completely and immediately accepted by the entire group and was affectionately called "Benny" then and throughout all the years that followed. He was the willing recipient of all our troubles and joys, the helping hand that was given without question or hesitation. His advice and suggestions were respected by all, regardless of age and experience, and to him alone we owed our constantly improving work conditions and the better living situations for wives and families. A man with a beautifully developed but kindly sense of humor, and a grin that came from within. Those were the fuller years with short funds and long hours and the knowledge that your own small niche in the picture was of tremendous importance.

I remember a time at Kaiser Meadows when he cured Gene Tully's toothache with a pair of side cutting pliers. Three times they slipped off while Gene was dangling in the air at the end of Ben's powerful right arm. In talking with Gene today he told me that the tooth never ached again, that he still has it and that he was later instrumental in selecting his older brother's career, who now lives in San Diego and writes his name, Chas. F. Tully, D.D.S.

I assisted Ben one night in saving from a watery grave one of our North Fork residents by name of Patrick Alexander Christian Williams, a carpenter. During the early fall of each year we had a heavy run of salmon in the San Joaquin River. The regular procedure was for a group to wade the river shallows, the leader carrying a wire basket of burning pine knots for a torch, with the others wielding their spears at will. Williams, with considerable inventive genius, had devised a coverall suit which he painted with a solution to make it impervious to water.

He undertook the torchbearer job, stepped into the only deep hole in the river and his solution worked in reverse, permitting water to enter but not to leave. Weighted down by a ton of water and a ten syllable name, he was on his way out when Ben caught him and operated on the seat of his suit.

Ben had one of the early Fords—the "gray goose," which played a large part in the lives of many. The roads were too steep and muddy during much of the year to permit speeding the engine to the point where the lights would properly function. There are still roadside scars where Ben had miscalculated. When the San Joaquin and Eastern (said to provide the cheapest known rail transportation by the hour) was completed to Big Creek, Ben had a Ford auto equipped with flanged wheels for patrol purposes and placed on a daily run from Auberry to Big Creek. Somewhere in the Sierra files is a picture of him and Chester Jordan, each looking like King Tut before the unwrapping, after the car had failed to negotiate one of the 24 degree curves. So far as I can remember, the Ford is still at the bottom of one of Fresno County's deepest canyons.

Through a gentle understanding of Indian problems, Ben won over the local tribes and individuals and became their counselor and medicine man. Named "Ben Dick" by Charlie Daulton, he became one of the few favored "whites" and a potentially serious fire problem was averted. Some very real friendships were developed and maintained throughout the years. Many Indian children were later named "Ben" but none was called "Maurice."

One other good deed which must be credited to Ben! One of the grazing inspectors arrived to take a back country trip. He had brought with him properly chaperoned, his fiancee who perhaps wished to see her lover in his own environment before the final knot was tied, some months hence. Ben, in his part as matchmaker, pled a shortage of pack stock to carry the extra bedding, etc., for a single woman. His argument was so convincing that our local Justice, Con Short, performed a ceremony that night that converted the trip into a honeymoon and resulted in many happily married years.

Summing up, I find that I have said practically none of the things that I feel about the man who has done so much for so many with no thought except for the other fellow. In all these years, I have never known what religion Ben favors. Whatever it is, more of us should have more of it. I would like to dedicate this to the small group that was with Ben on that memorable day in January of 1912—Paul Redington, Audie Wofford, Ernie Dudley, Ted Shaler, Jack Farley, Ed Mainwaring, Chester Jordan, Roy Booth, George Cabin, Billy Brown, Hale Mace, Mal McLeod, Boot Taylor, Jess Roberts and Julia Shinn, each earnestly offering you friendship and loyalty and a chance to share their love for all nature.

FRED WEHMEYER

H. C. "Chris" Chriswell

"How in h___ did you get here?" Assistant Supervisor Jim Iler asked. "Every road into Pendleton is closed." This was my welcome to the Umatilla when at 8:00 a.m. Monday morning I reported for the assistant ranger job on the Heppner District in 1937.

I had left the Winthrop RS the day before and drove most of the day through a blinding snowstorm. The snow was drifting badly through the Cold Springs cutoff. The car would just barely make it through some of the drifts. I kept track of the distance from the last farmhouse lights in case a drift stopped the car. I wasn't really worried as I had complete cross country ski equipment with me. It was a great relief when I finally drove up against a big pile of snow in front of the Pendleton Hotel about midnight. I thought this was normal winter weather in eastern Oregon.

That morning every road into Pendleton was really closed. The highway to Milton-Freewater wasn't opened for a week. I watched the dozers breaking out the grey cement-like snow, colored by dust from the wheat fields.

There were two assistant ranger positions on the Umatilla at that time. Tom Brown was assigned to the Walla Walla District and I was at Heppner. We moved twice a year on these seasonal jobs and spent November through March in the supervisor's office.

All the work nobody wanted to do had been carefully saved for me. Those who worked on the old status records and the fire protection tax rolls will know what I mean. I soon discovered that I was at the bottom of the totem pole. No sympathy was wasted around that office. Besides, wasn't I soon to disappear into the vast Heppner country which was the domain of one district ranger named Fred Wehmeyer.

He was a complicated man. I never met a person in the Heppner country who didn't like him. The year 1937 was a time when most of the ranger districts were administered by the "Old-Time Rangers." Young foresters were replacing them as they retired. They had come under the Civil Service via the ranger's exam. Greatly varied in talents, roughhewn in personality, they were men we would tell tales about around our campfires. They were the greatest and most interesting men it has ever been my privilege to know. All six districts of the Umatilla were administered by them.

Fred was raised largely by Hazard and Zora Ballard at the mouth of Lost Creek in the upper Methow Valley. He was a wild kid. Fred lied about his age when he was 14 and joined the Navy. He became a Spanish-American War veteran when mustered out of the Great White Fleet. Passing the old ranger exam, he received an appointment on the Chelan Forest. He was ranger on nearly every district of that forest when P. T. Harris was forest supervisor.

When Fred was at Chelan, he was plagued with grass fires that burned up onto the forest. He became expert at handling them. One dry summer day a fire broke out on the east side of Lake Chelan. Fred called the supervisor at Okanogan and asked P. T. for 200 men. Remember, a ranger was supposed to "calculate the probabilities." The next morning P. T. sent 30 men to Chelan as the fire crowned up the mountain. By evening it became evident to Fred that P. T. considered 30 men more than enough to man the fire. He went back to town, marched into the theater, shut down the show and drafted every able bodied man to fight the fire. It wasn't long before he was transferred to the small Touchet District on the Umatilla. One of the drafted men was a small town politician. There were two rivers on this new district, the Touchet and one other. When Fred was asked what his new district was like he would say, "Well, it ain't very big, you can touch it on one side and pat it on the other." Later on Fred was transferred to the Heppner District.

In those days there was much discussion about the "Eastside" and the "Westside." Some maintained there was no difference in managing a ranger district, regardless of which side of the Cascades it was on. No so, the old-time rangers say. My discussion got quite heavy one day with Fred. He finally stopped me cold. "At a ranger's training session at Wind River we were all having a look at some of those young pole-sized Douglas fir forests in the Wind River area. The Hemlock Ranger described the history of the area. Experts from the RO and the experiment station talked at great length about natural pruning, soils, growth and site quality. Others brought out yield tables. Finally, a bowlegged ranger in back, who had been admiring the young trees, pushed back his Texas style Stetson and exclaimed in a loud voice, 'My God, look at all the corral poles'."

My training as an assistant ranger was: "With 720,000 acres of protective area, this thing is too big for one man. You take the east end and I'll take the west. Later we'll trade." Fred was of the school that you learned by going out and doing things. He felt this weeded the men from the boys. It was successful with me. I've never worked so hard before or since, or had so much fun doing it. "You get all the fun of being a district ranger without the responsibility," Fred used to mutter.

Oh yes, there was one other bit of training. "This ol' horse's name is 'Nuts'. He's been in the Forest Service family so long he probably knows more than we do. Watch him. He'll fool you goin' out but he'll go h___ bent when you turn him towards the barn. Give him his head and he'll take you any place on the district you want to go." Nuts had been ridden by forest officers for several years. I could hang on the saddle all manner of maintenance tools that dangled and jangled. I would stand up in the saddle to replace a low hung split insulator, dropping cutoff No. 9 wire about his ears. He wouldn't move a muscle. Upon meeting a bear, Nuts would calmly stand there watching the bear until it finally ambled away. With a grazing map in one hand I would point Nuts towards where a sheep camp was indicated. Before long there would be the camp.

Fred was an atrocious driver. Like some people of those times, a team of horses would have been simple but driving a car safely just wasn't understood. He received a brand new Plymouth panel as an official car. We soon noticed him driving his personal car. About a week later he drove the panel to the office. He was proud as punch as he pointed to the fenders and entire side of the vehicle. We were appalled! The new car couldn't have looked worse. Fred had run off the Heppner grade coming in from Tupper one evening. He had gotten a rancher to pull him back on the road with a truck. He had spent a week of evenings and two Sundays in his backyard pounding out the dents with a peen hammer. When we timidly asked him why he hadn't taken it to a garage, he glowered at us and said, "Cause I don't want anyone in the SO to know about it."

In the early spring Fred couldn't wait to see what effect the winter snows had on the district. He would drive up the road as far as he could. Sometimes the snow stopped him but usually it was the mud. Except for about three main roads, including the Heppner-Spray Highway, there wasn't a graveled road on the district. When a snow bank melted it took a couple of days for the frost to leave the ground. For a very short period of time the soil would not support a vehicle. Every year, without fail, Fred would drive onto one of these innocent appearing places and sink up to all four axles.

"Now you know those Chevy pickups. A bunch of engineers worked for years to design 'em so the bed was almost on the ground when you got stuck. The only thing to do is to lay down in that mud in your clean uniform and slowly worm your way under that low box. You work and struggle to get your shovel in beside the wheel. You rassel and you huff and puff. Finally, you carefully work the shovel back out and there is a little bitty piece of mud on it. You take a stick and scrape off the shovel, then take a stick and scrape off the stick." Fortunately for our piece of mind, the ground wasn't long in this muddy condition. All summer it was deep dust. Fred always said, "The difference between mud and dust in the Heppner country is about five minutes."

We used to take tick shots each spring. Fred always had a reaction to them. One day he was nursing a red, swollen upper arm. He swung around in his revolving chair. "Those damn ticks climb out on the tips of the bitterbrush and all clap hands in glee when they see me coming. If one of 'em dares bite me now he'll curl up his toes with the spotted fever."

One day Fred was cussing forms (one of his pet peeves). He was mad at what he called those "nasty meemos." He kept referring to the good ol' days. Suddenly he swung around in his chair and glared at me. "What in h___ am I talking about! The good ol' days! I'll tell you what the good ol' days were like. You start up the trail dragging the pack stock behind you. About 3:30 p.m. you start looking for a place to camp with good horse feed. You locate a place and unpack the stock. Then you peel off the saddles and put bells and hobbles on the horses, turning them out to feed. Then you start making camp and getting firewood. You drop everything and bring the horses back to camp. While cooking dinner you notice the bells stop ringing and you go after the horses again. Getting them back, you restart the fire and finish dinner. Just before dark you bring the horses close to camp. Finally, bone tired, you crawl into your bedroll. Just as you drift off to sleep the bells become silent. Nothing will waken you quicker. You pull on your boots and chase horses half the night. When dawn cracks, you're up and chasing horses again, sometimes walking three to seven miles. Hobbles didn't slow them nags. You feed them a handful of grain and then get breakfast. After currying and brushing, you saddle up. By the time you break camp, make up the packs and throw a few squaw hitches, the sun is pretty high in the sky. Then you start all over again. That's what the good ol' days were like. Thank God, they're long gone."

Fred's diaries were something. He had boxes of them under a steel cot in the alcove. Every day was faithfully recorded. The entries were almost all the same: 8:00 started work—5:00 quit—8 hours. These entries did not endear him to his supervisor or the administrative assistant. Neither did his stock answer.

"H___, if I put down everything I did in my diary I wouldn't have time to do it." Fred worked from dawn to dusk and most of his Sundays. His memos to his supervisor were eagerly awaited by the office force. The girls usually read them before routing them to the boss, and they enjoyed his humor. The supervisor often threw them into the basket.

When I introduced my new wife to Fred he said brightly to Iris, "We certainly need you in the Forest Service. Maybe you can help us solve a serious problem." Iris answered, "What is that, Mr. Wehmeyer?" He replied, "Why, reproduction in the virgin forest." Her face turned beet red as she look at Fred's wide grin.

Some of the older rangers had their wives travel with them. They were as much a part of the Forest Service as their husbands. On horse trips they would camp tend, cook and take care of the stock. Iris went with me whenever she could. After guard school most of my work was follow-up training of guards and lookouts, followed by range inspection and "other duties as assigned." When spending a day with a guard, Iris would visit with his family. The guards became our best friends; they were wonderful.

We often got home late Friday night from a week's trip. Iris would shop Saturday morning while I worked in the office. We would load everything in the car Sunday evening and head out again to the district to make camp. It was a busy and eventful life.

There was a shelter and fenced pasture at Long Prairie. We camped there and turned my rented horse out in the pasture.

While we were camped at Long Prairie, Fred brought the new supervisor, Ralph Crawford, out to meet us. As we visited, Ralph kept looking at our bed that Iris had made under a big "yeller pine." The mattress was made of fragrant white fir boughs and Iris had topped everything with a thick wool-filled quilt of her mother's. Finally, Ralph walked over to the bed and after admiring it for a moment, related the following: "Back at Leavenworth, I was getting married and had planned a pack trip up the Icicle for a honeymoon. I had ordered two fancy sleeping bags from Sears. When it came time to pack up, only one had arrived. We couldn't postpone the trip. Being just married like you two, I thought how great one sleeping bag would be. That was the most miserable week I ever spent. Don't let anyone kid you. You can't make love in one sleeping bag."

The last summer at Heppner we got a soaking rain in September. It rained all Thursday night and all day Friday. Iris had been getting mighty homesick, so coming in from the district Friday afternoon we decided it was a good chance to get back to Winthrop and visit her mother. No one was in the office. I couldn't find either Fred or the dispatcher. Finally, leaving a note on Fred's desk telling him my plans and asking for annual leave for Saturday morning (Monday was Labor Day), we left Heppner in a hurry as we had a long way to go that night.

On Tuesday morning I entered the office with a cheerful "Good Morning." Fred didn't even answer with his usual grunt. I thought his "Forest Service ulcers" were acting up again. Sitting down at my desk I noticed a small brown envelope. Inside was one of those one half letter sized memos. "You left your official post of duty without permission in the middle of fire season. Only the forest supervisor can approve such leave. You are A.W.O.L." /s/ Fred W. Wehmeyer, District Ranger.

I couldn't believe it! I asked Fred to turn around and waving the memo at him, asked what in h___ it meant. He replied, "If John Irwin were still supervisor, he would bring personnel action against you." I pointed out the soaking condition of the forest and told him it was about the silliest ruling I had ever heard of. After a lot of discussion and hemming and hawing, I finally cooled down and Fred reached across and crumpled up the memo. "Maybe things are finally changing for the better." And there under those moving bushy eyebrows was that wonderful grin.

LIFE WITH OSCAR EVANS

Editor's Note: For many years Oscar Evans was chief of timber surveys in R5. Oscar stories were numerous. A few of them follow, with the name of the author separating each story.

FROM ANDY BRENNEIS:

I wasn't one of Oscar's boys, but I did have an interesting experience in 1931 at the ranger school at Quincy. Oscar took us on for a bit of timber cruising which we obviously needed badly! We all had our starting points so "off we went," happy as larks and wondering what was so new or different about cruising for Oscar. We soon found out. Pretty soon strips were crossing one another and students were getting their tapes together like spaghetti. Why? How could this happen? Oscar had screwed up the works by resetting the declinations on all our compasses. Score one for Oscar!

FROM CARL WILSON:

The first night's meal at the Grizzly Flats Camp in 1937 was a shocking experience. I was sitting on a bench at the table waiting for the food to be served. I felt cold water running into my lap. To my dismay I learned about the initiation rite! Several boys rolled up the oilcloth tablecloth into a sort of trough. A person at the far end of the table poured a glass of water down the trough and the unsuspecting individual, me, was thoroughly drenched. I learned later that this happened to every new member of the crew.

One Sunday when one of our guests was Bob Burton's wife, several of us decided that she ought to be initiated into our crew. She was sitting there in her Sunday dress, enjoying the food and conversation. All of a sudden Mrs. Burton screamed and jumped up. Mr. Evans said, "Mrs. Burton what are you doing?" Bob Burton knew what was happening so he said, "Mr. Evans this is a common occurrence. At least once during every meal my wife screams loudly, jumps to her feet, and sits back down, and we just have to accept it." Every member of the crew, excluding Oscar Evans, of course, roared with laughter. I'm sure that this kind of initiation rite continued for many years. I suspect that Oscar Evans never learned what was going on.

FROM TOBE ARVOLA:

The time was late in the election year of 1940, when Roosevelt was pitted against Republican Wendell Wilkie, and the place was an USFS cruising camp way out on the Forest Hill Divide on the Tahoe National Forest. As could be expected, we had spirited discussions about the presidential candidates after each supper, with Oscar trying to dominate us with his conservatism. He plumb hated FDR and didn't hesitate to show it. The crew was mixed—some for FDR and others for Wilkie, but not too openly to side with Oscar.

One day on the sly, we decided to foil Oscar. We cooked up a scheme to have a rigged straw vote after the usual political session in the evening in order to settle the issue. Oscar fell for it, so we had our secret balloting. The agonizing result was one lone vote (Oscar's) for Wilkie and all the rest for that damned Democrat. Oscar was really mad. He put on his hat and stormed out of the cook's tent. This ended all the political talk at the supper table for the rest of the season.

FROM MYRON KRUEGER:

I was busy in my office at the University of California when Oscar dropped in at about 6:00 p.m. He told me he had just been assigned to a conscientious objectors' camp in the redwood region. It was his job to give the inmates information about forestry in the evenings after work. He asked for a kit of wood samples, such as Professor Fritz had prepared. I obtained such a kit for him. At that point I said, "Oscar, what kind of gooks are these COs?" Oscar replied that some of them were truly conscientious in their viewpoint but that all of them failed to realize that man, in keeping with the rest of the animal kingdom, is forced to fight for everything that he requires; in nature there is constant strife. Oscar went on like this for two or three minutes and then said, "Why am I talking to you about this? You're married too."

FROM LEO "SPIKE" SLATTERY:

In 1934, Oscar Evans' cruise camps were located at several sites on the Oak Knoll District of the Klamath NF; Beaver Creek, near Dry Lake LO (headwater of Doggett Creek) and Horse Creek, about a mile or so downstream from Baker's cabin. In mid-May Oscar set out from Berkeley in a Forest Service pickup loaded with gear and supplies. He took two passengers, Bob Howden and Spike Slattery. Both lads were on their best behavior and tried to answer respectfully to Oscar's constant stream of questions on forestry. At one point as the pickup chugged past a gulch full of willows and alders Oscar beamed, "My God, think of all the photosynthesis going on down there."

Oscar's arrival at camp resembled the return of dad from town, loaded with mail, fresh milk, fresh meat and plenty of jovial jabs at individuals. Out in the woods on the cruise line, Oscar was a stern perfectionist, checking the compassman's map particularly carefully. I recall Oscar inspecting my map, one corner of which was blank. "My God, man," he exploded. "You'll have that go down in history as flat when it's steep. Can't you see that contour? It's as plain as a goat's ass going uphill!" He made his point. At another time I heard of a compassman

West Boundary Ranger Station, Willamette NF, Oregon - 1936.

Fort Jones Ranger Station, Klamath NF, California - 1936. Left to right: "B" Guard Residence, "E" office, water tank tower, warehouse and "A" Ranger's residence.

Oscar was checking. "What's your foresight, boy?" he asked. "That big white cloud straight ahead, sir," was the reply. They say Oscar muttered to himself for hours afterward.

Toward summer we moved camp to Horse Creek on a bar shaded by alders and bordered by a broad shallow stretch of clear water. It seemed idyllic until our first hike upstream revealed several residences with toilets built on stilts over the stream. By now yellowjackets, hornets, beetles and wasps were enjoying camp life right along with us. One day Oscar brought Mrs. Zink, our super cook, a flit gun to rid the cook tent of the flying beasties zooming overhead. Her first blast of insecticide dropped several score, but her elation was short lived. Most of them landed on her head and lodged in her curly locks. This brought out a scream of horror and so ended the Horse Creek pesticide project of 1934.

FROM KERMIT CUFF:

One of the more memorable incidents of my forestry career occurred in 1938, when I was a member of Oscar Evans' timber survey crew working on the Eldorado. Our camp, consisting of tents, was situated at Gerle Meadows which at that time was pretty far out in the boondocks. It was a typical Evans arrangement with tents lined up in near military fashion with crew quarters at one end, the cook (kitchen) tent in the middle and Oscar's situated at the far end about 100 yards away.

The youngest member of the crew, Wayne Gray, had just completed his freshman year and had signed on as forestry student trainee at the princely salary of $65 per month. He was a nice kid but quite shy and reserved, no doubt due in part to the fact this was his first experience in the woods. Presumably for this reason, he rated rather low in Oscar's esteem.

Every member was assigned certain camp chores in addition to his daily stint on the cruise line. There was generally a pretty good correlation between the nature of one's camp assignment and his position on Oscar's totem pole. At about midsummer when Oscar decided the lady camp cook needed help, poor Wayne was delegated to arise at 4:45 each morning to light the cook stove fire, fetch water for coffee and otherwise assist in preparing breakfast.

After a week of this, Wayne reported with some concern he had sighted a bear at the edge of camp. This news was treated as a joking matter by Oscar who merely asked if it was papa, mama or baby bear, while laughing at his own wit. He continued this facetious treatment by either repeating the same question or asking, "How many bears did you see this morning?" I am sure Wayne was not too happy with this situation but before that week ended fate intervened to render some justice. It was around 2:30 a.m. a few nights later when all was quiet except for the faint croakings of a lonely bullfrog or two far down in the distant meadows. The sylvan tranquillity was shattered by loud and excited words from the solitary tent, "Bear in camp! Bear in camp! Oh, boys, he's here beside my tent! Come drive him away!"

This sudden ruckus resulted in an immediate stir of activity among the rudely awakened crew members as they fumbled around in the dark trying to get oriented and organized for the rescue. The only weapons available were simple tools of their trade—Jacob staffs, Biltmore sticks and hand axes. Thus armed and wearing only the odd pieces of clothing they could grab in the dark, they must have made a most unusual looking party of bear hunters. There was an understandable lack of enthusiasm as they proceeded towards the scene of the cries, nervously flicking their flashlight beams every which way in search of their quarry they hoped not to find. Any inclination to hesitate was banished by the continuing exhortations from the man most concerned. By making loud assurances to Oscar that help was coming and by noisily clattering their weapons together they succeeded in causing the bear to withdraw without bloodshed.

From this time on the bear was no longer treated as a joking matter. However, the crew felt otherwise. When further paw prints were discovered around camp it became feared the bear was interested in our food supply. In order to discourage the unwelcome visitor, Oscar ordered a Coleman lantern be kept lit all night in the cook tent. On it he attached a placard, "Do Not

Turn Out This Light." It was only, a day or two before some wag had added, "The bear has poor eyesight and needs the light."

There was bedlam in camp the first night Oscar was absent, as mimics kept up a din of the now famous words "bear in camp." The episode is still regarded by crew members as the outstanding event of that year. The remainder of the season whenever Oscar was away, some late retiring smart aleck could be expected to sound off "bear in camp!" Life for Wayne Gray was also more bearable (no pun intended). Never again was he bothered by belittling remarks about bears.

FROM JOHN HALL:

Oscar was almost always cheerful whether things went his way or not. For the job he had developed a complete how-to-do-it mapping and timber cruising handbook. He had developed and applied the "4-step" method of training long before fire control ever heard of it. Oscar ran a tight ship. He was a good disciplinarian and knew what was going on. He maintained excellent control records and made frequent field inspections, showing up unexpectedly in the damnedest places. He got a good day's work out of all his crews; no goof-offs tolerated. On our day and one half off each week he saw to it that we had something to do. He enjoyed taking us to town and to dances to meet the local girls. In 1932, while on the Modoc, he took the dozen of us in the old Reo, caged in the back like a bunch of yowling (singing) wild animals, to Crater Lake for a long weekend and later to Mt. Lassen, which we all climbed. There was song and storytelling around the campfire at night. Good food to eat and a festive table and a summer of good fellowship and purposeful work.

One of Oscar's delights, as the working day began, was to drop us off on the road near our starting corner with the exhortation "get up the hill like a striped-ass ape," his only vulgarity. He would then distribute the rest of the crews and double back to see if he could beat us to our corner. When he beat us, we would find him sitting on it and we got a good lecture about using the hand compass and pacing to find our objective.

Having worked the summer of 1932 for Oscar, being a forest supervisor's son and studying at Cal that fall, I was invited by Mrs. Evans to have afternoon tea with her at their home in Berkeley. It was English and formal, and I the only guest. Mrs. Evans was a gracious lady, discreet, well-informed and self-possessed. Their home was tastefully furnished. I presumed that she had learned to accept Oscar's bumbling ways, and saw through to the kind, good hearted gentleman and faithful husband, which he most certainly was.

In closing, no story about Oscar ever reflected on his dedication or professional competence. His maps and cruises represented the state of the art in its day and were used and appreciated by a generation of foresters.

JOE RICHARDS—SAWMILL MAN

Edward C. DeGraaf

Any story of the Hiawatha should include a small sawmill and its operator. When the Manistique District was activated a small sawmill was found tucked away in the south edge of the Iron Creek swamps to the north of Kitchitakipi—the Big Spring which now is Palms Book State Park. This mill was not an unusual one since nearly every small mill is generally a mechanical monster of ingenuity that would amaze Rube Goldberg. The story lies with the operator—Joe Richards.

To get to the small mill, one had to drive up an old rut road, leaving the good road behind at the Big Spring parking area. From there on the alder thickets brushed the car on both sides until one reached the mill clearing. The mill was a low elongated building with the owner's residence in one end. Here Joe Richards and his wife lived in harmony with the swamps with a few scattered neighbors. The small mill that included a shingle weaving attachment, plus a generous number of dogs. Joe and his wife both loved their old sawmill, and between producing a little lumber spasmodically, occasional custom sawing, turning out a few shingles and selling a few logs they made enough income to meet their needs.

Joe was a short heavy set man in his sixties who walked about the mill yard with a rolling gait and with a friendly grin on his face. He had a great shock of grey hair that would have shamed the present generation of shaggy hairstyles. Setting off the friendly grin was a pair of sharp shrewd eyes that hinted at a keen mind. The general appearance was of a bareheaded Santa Claus in old ragged clothes.

Mrs. Richards was a natural helpmate for Joe. She was a short, heavyset individual, and with no reflections intended, she was a natural "five by five." She was generally dressed in work clothes, well tucked into roomy bib overalls. When working at the mill she wore a bright bandanna about her head. When the mill was operating she was the stationary engineer and fireman, while Joe was the millwright and head sawyer.

The mill was powered by a horizontal boiler with the usual gauges, steam pistons, drive shafts, pulleys and belts. When Joe wanted to run the mill for a day or two he would get up a head of steam in the morning and go about his millwrighting. At the same time, one of them would pull the steam whistle a few blasts which would be heard for a couple of miles.

This whistle would alert the scattered neighbors that the mill was going to operate as soon as they could get there to work. They completed their chores and hiked into the mill site, and as soon as a crew was mustered the sawing and shingle weaving would start up. The mill would really get under way when Mrs. Richards would step from behind the boiler, where smoke and steam were rolling out and shout to Joe at the sawyer's stand, with a wave of her arm—"Give her hell, Joe!!" She then returned to her firing and watching the steam and water gauges. Joe would operate the carriage and saw while the neighbor workmen handled the lumber at the edger and trimmer.

Earlier mention was made of the dogs. These consisted of many noisy Pomeranians raised by Mrs. Richards. At one time I remember counting 27 of them. They had a fenced run outside that connected with the house part of the mill, with a small door opening out through the wall. When one stepped out of the car in the yard he was met at the pen fence by 27 or more little red dogs that looked like red foxes—all yapping as loud as they could.

Once I would head for the house they ran pellmell through their doorway to greet me in the house. The dogs would be all over—on top of the table, the sink, chairs—anything that brought them nearer to my height. All the time they were yapping away in fine form and creating a perfect bedlam, Joe would just chuckle while Mrs. Richards would scold the dogs and finally drive them outside via the small door which she closed. Then normal conversation could begin. To this day I never see a red Pomeranian but I think of that noisy horde of little dogs. Joe explained to me one time that his wife loved her little dogs so much she found it very difficult to part with them although they had ready sale.

Joe's source of timber was on the land adjacent to the mill. There was a large tract of land that had quite a scattering of hemlock, mixed hardwood and cedar logs on the ground. The original operation had shut down before the logging had been completed so there was enough timber available to keep the small mill busy intermittently for small scale operation. Joe had acquired the mill and had expected to continue a salvage operation. However, before long the land was optioned under a FS acquisition deal so no more salvage logs and timber were available until the title was eventually passed to the U.S. Joe wanted to buy the salvage timber and tried to persuade me to make a sale. I tried to explain to him that the FS did not own the land and timber until the transaction was completed and title passed, and only then could we sell the timber when finally owned by the U.S. Government.

Joe put up with my explanation for bureaucratic delay until he finally undertook a solution of his own. As this was in the early days of the New Deal, Pres. F. D. Roosevelt was constantly in the news. So Joe wrote the President a personal letter, stating his problem and asking why he could not buy the timber that was wasting on the ground. My first knowledge of this "end run" was when a copy of his letter came down through channels with a brief letter from the President's office, plus one from the chief's office, to look this man up and solve his problem.

By that time the regional office was busy on a solution, as I'm sure this problem repeated itself throughout the region.

Very shortly a regional letter of instruction came out that set up a temporary policy for sales on land under acquisition. The money from such sales was put into an escrow account to be processed and distributed later when title was cleared. This enabled me to get out to Joe's and initiate a salvage sale. Joe told me, with a twinkle in his eye, about his letter to Pres. Roosevelt and the letter of acknowledgment he had received from the White House. He was convinced that I was making a sale to him at the direct order of Pres. Roosevelt. He got a great chuckle out of it each time I stopped by to scale him out or make a new salvage sale. We became good friends, especially when I learned he came from Evart in Lower Michigan where the town doctor was a schoolmate and friend of mine.

CRONEY'S MINUTE MINDER
(first two paragraphs)

Joe Ely

A MAN FOR ALL SEASONS
(last two paragraphs)

Phil Lord

Editor's Note: Fred P. Cronemiller was the assistant regional forester in charge of wildlife management in R-5 when he retired. He was known far and wide as Croney. –E.J.

Most people are acquainted with the homemaker's "minute minder" by which she times the eggs or meat or whatever is cooking. Fred Cronemiller had an innovative minute minder that he had devised. We would make camp in the evening and get a little fire going and then put on some small pieces of fir bark, which after while would burn with a smokeless blue flame. I always had a piece of tin that I carried in a sooty burlap sack. After this had been set over the fire, Croney would get out his minute minder. This consisted of a bottle of wine, two man size.

He would slice the potatoes into a piece of foil, add milk, butter and salt, wrap the foil and put it over the fire. Each of us would have two drinks, interspersed with the appropriate amount of conversation. Salt and pepper the meat. One drink. Put meat on fire. Another drink. Turn meat. Another. Potatoes done. Meat done. Bottle empty.

Croney was one of the best camp cooks I ever knew and I knew some good ones. Anyone who never saw Croney flip a pan of trout has missed something. He could also rattle the lid on a Dutch oven with the best of them. In my estimation he was a complete man of the camp, as the Mexicans say.

You couldn't lose Croney in wild country; he was a good horseman and packer. He taught me a way to tie off a diamond hitch that never came loose but which could be untied with one quick jerk. I used it as long as I packed. Croney could speak enough Spanish and French to get along with most any sheepherder. He could, if necessary, drink wine with sheepherders, whiskey with loggers or buckaroos or lemonade with temperance people. You know, I don't think it mattered who they were, Croney just liked people.

WILLIE'S WILD RIDES ON THE WHITE MOUNTAINS

Jack A. Godden

Over a ten year period, 1958-1968, Willie Haynes our R-7 radio electronics technician probably established a record for the number of exciting trips in performance of duty as a radio technician on the White Mountain National Forest in New Hampshire and Maine.

A radio specialist by the name of Biggerstaff from the Beltsville Lab designed the elaborate high band radio system for the White Mountain National Forest in the late 1950s. Its implementation began in 1958, and it included the installation of two repeaters, one on Cannon Mountain on the western side of the forest and another on top of the 6,288 foot Mt. Washington. By the time I arrived on the forest in 1960, the system was in operation. The supervisor's office in Laconia could communicate with all ranger districts as well as any mobile units in the field. Although heavily financed initially from fire funds, engineering and timber were to finance many of the mobile and later, hand sets. It was an operable fire/administrative radio net, with multiple channels.

Willie's home base was in Hot Springs, Virginia. He installed the system and was the primary specialist involved in its maintenance. This meant a minimum of two visits to the Forest every year after the system was operational and the "bugs worked out." As all offices had electrical power, the reliable 30 watt desk models provided dependable service. Willie could drive to these locations in the comfort of his station wagon. The six fire lookouts maintained before implementation of aerial detection in 1966, were supplied pack sets, fixed mounted antennas and heavy duty batteries. Five were located on mountain tops; one was accessible by car. On one of Willie's early service trips in 1959, Stinson Mountain Lookout was having radio and antenna problems. Ted Gienty, then forest engineer worked out a helicopter ride for Willie. The Bell B-1 had the lift and power for the climb but in "owly" winds on the summit, the copter flipped over and Willie was separated from the pilot by a spruce top through the bubble. No reportable personal injuries.

Work on the Cannon Mountain repeater installation was more to Willie's liking. A cable car used year round for skiers and sightseers provided him with a ride up and return down from that summit. A helicopter ride to Mt. Washington, where the other repeater station was located, was not in the offering. There were three ways to the summit: 1. Hiking—the shortest distance was six point four miles, a five hours and 30 minutes time estimate. This was not a good choice. 2. Ride the Cog Railway built in 1869. 3. Drive the Carriage Road (or Auto Road) completed in 1861, its eight miles to the summit. Willie opted for the Cog Railway for one of his early trips. On this attempt the train broke down halfway up. There he sat for a couple of hours until another train rescued Willie and his equipment. The next day he made it to the top via the Carriage Road by four wheel drive vehicle supplied by the district. This eight mile stretch of road didn't give any light hearted people the sense of security in its upward climb or downward descent. Its gravel and rock ledge surfaces were known to be hazardous. But Willie made it to the top this time and enjoyed the association with the television engineers at the WMTW-TV transmitter. They supplied power and space for our radio repeater. Willie worked up a contract with them to do emergency repairs on our repeater and antenna.

Another year went by and Willie arrived on the Forest in early June and he decided to drive the road to the summit in his station wagon. His confidence built up when he successfully got to the summit. After a few hours maintenance he started his descent. Within two miles he lost control of his car and eased it into a ditch on the upside of the road just before the "outside curve" above the Great Gulf Wild Area (now Wilderness). If anybody had good radio communications Willie did. He roused up the Androscoggin Ranger Station to come to his assistance. They retrieved Willie and his station wagon. Little damage was done to the car but he had concern he might ever survive another trip to the tops of the White Mountains.

"Three strikes and you're out," the saying goes but not for Willie. The regional office let him hire an assistant—Bill Vanorsdale. Bill had worked his way up through the ranks as forestry aid, tech, general district assistant on the Monongahela National Forest. Together they made a great pair. Only Bill did the driving the next time and thereafter on the White Mountain National Forest trips.

A picture of Willie with his dark brown black fedora hat, heavy plastic, black framed glasses, matching his morning five o'clock shadow, leaning against his wagon and its tool boxes, is what's needed to really tell this story. Willie retired in the late 1960s and has gone to meet his maker and probably doing a little repair somewhere more peaceful than Mt. Washington and his White Mountain National Forest trips.

DAVE ROGERS

Ted Niehaus

As old-time Forest Service folks know, the early day honest supervisors were pretty much one man shows. On the Plumas, we had a theoretical hierarchy but in fact everyone reported to Dave Rogers. At the same time, Dave was known for supporting his people.

I recall an incident where I had offended a constituent through some arbitrary decision. Upon reflection, I knew I was wrong but did nothing to rectify the situation. Being very young, I thought a government man could never admit an error. In a few days the offended party came to see Dave and I was summoned to explain my action. To my surprise, Dave read the man off and sent him on his way. I got up to leave, thinking I must have been right after all. Then, in a stern voice Dave said, "Wait a minute. Now Ted, you were skating on damn thin ice and you know it and you had better not do that again."

MEMORIES OF SOME PEOPLE WHO EDUCATED ME

Jack Spencer

When I began my Forest Service career I judged the agency by the kind of person who worked for it. I suspect that many people in the small towns where Forest Service offices were located did the same, rather than by judging the organizations' policies or programs. Following are a few people from my beginning days in the Forest Service—people who had a profound impact on me and, more importantly, on the Forest Service and on how it was viewed by the public. Even today I call up the memories of these people when I need help.

THE DINOSAUR

My first summer job with the agency in 1950, was as a blister rust control laborer on the (then) Cabinet National Forest in western Montana. We worked from a tent camp set up on a ridge miles from the town of Trout Creek. In addition to the 75 college kids who dug up ribes plants (currant bushes) by the roots using a "hoedag," a small crew was assigned to the camp to spray herbicide by helicopter in other areas to kill the ribes, which was the alternate host of the white pine blister rust disease that was killing large numbers of this valuable tree. One member of the crew that mixed the herbicide, loaded it in tanks on the helicopter and serviced the craft, was an older man in filthy clothes, liberally soaked with diesel oil—one of the ingredients of the mix—who was unshaven, chewed snoose, swore with happy abandon like a logger and appeared to me to be a wino hired off the streets of a nearby town as a pick up day laborer. With characteristic youthful intolerance I sneered at his lack of "professionalism" and did not even consider that he could be a permanent employee of my chosen agency. When I asked an older crew member who he was, he replied, "He's the forest supervisor!"

Later I learned to appreciate that he was one of a soon to be extinct breed of old-time foresters who personally dove into whatever needed to be done—more comfortable with manual labor and field work than with office work and who were contemptuous of or oblivious to what was "politically correct." All of us "blister rusters" learned that summer that appearances can be deceiving, that working with your hands is not an indicator of one's contributions or station in life and that there is a rich tapestry of all kinds of people in the organization from whom to learn.

MRS. GREEN, AN EXTRAORDINARY WOMAN

Mrs. Green was a cook on the Kaniksu National Forest at the ranger station in Clark Fork, Idaho, when I was detailed there temporarily in 1952 to scale logs. She lived on a small ranch outside of town with an invalid husband who could do no physical work. The dirt road that was their lifeline to the outside world was not a county road, and therefore, was not plowed in the winter. Each morning after the first snow fell, when she had milked their cows, she pulled a toboggan loaded with filled milk cans to the county road two miles away so they could be picked up by the milk truck from the local dairy.

During the summers she worked for the Forest Service and ruled the kitchen and mess hall with what today might be called "tough love," borne of a life of hard work, and a sense of honesty based on a simple belief in the old fashioned virtues. Her physical strength was legendary, as was her cooking prowess. Once I offered to help her lift a half of beef from the butchers' block in a walk-in cooler to a meat hook near the ceiling. We both struggled with it getting nowhere, until she asked me to let go and get into a better position. Before I could get to my new position, she lifted the meat on the hook all by herself, unencumbered by my hindering efforts. Her uncomplaining acceptance of her hard life and her unflagging energy to do the right thing did not go unnoticed by those around her.

AL SAFFORD—MASTER OF THE WOODS

I remember Al Safford in 1952, the good natured, thin as a rail alternate ranger on the Clark Fork District of the Kaniksu National Forest, as the person who taught me and hundreds of others like me, some of the mysteries of woodsmanship. The title, alternate ranger no longer exists, but previously it was earned by men with wide forestry experience and exceptional skill, but with no college degree, and was meant to provide a bit more of a career ladder to key nonprofessionals.

Al approached any task with a minimum of rhetoric and a maximum of effort. His conscientiousness translated into a "stick to it until it's finished" attitude that extended from mopping up a stubborn fire on a steep slope miles from the nearest trail, to snow shoeing in to a back country drainage with a greenhorn student to set up a camp on eight feet of snow to cruise bark beetle killed timber so it could be salvaged the following summer.

After ten days in that camp, we headed out to the plowed road, some 12 miles distant and found that Lightning Creek, which we had crossed coming in by jumping from rock to rock, was now an impassable torrent of white water transformed by the melting snow of a few warm spring days. There were few trees that we could fall that were tall enough and close enough to the stream edge to bridge the stream but we did find one poor candidate. However, it was leaning slightly in the wrong direction to fall where we wanted it to go, or so it seemed to me. Al thought he could "finesse" the tree across the stream, but as much as I respected his judgment, I didn't believe he or anyone else could do it. With our two man crosscut saw, Al gave me a lesson in physics that was not contained in any textbook by demonstrating the use of homemade wedges, "holding wood" on one side of the back cut and the timely use of a double-bitted axe as a lever just as the tree began to fall, to place the tree exactly where it had to go. From that moment on I would have followed Al anywhere and tried anything he thought possible. Al died some 20 years later of a heart attack while fighting a fire in northern Idaho—a fitting departure for a man who loved the woods and spent his life working in them.

RALPH HANSEN—A COMPLETE MAN

Ralph Hansen was a professional forester, a woodsman-scholar, and one of the finest men I ever knew. In the early 1950s, as district ranger on the Plains District of the (then) Cabinet National Forest in Montana, he demonstrated uncommonly effective management skills without ever having read one of the current blizzard of popular books on the subject, when he fired a potential lookout who didn't measure up.

When told that he was being laid off after finishing guard school training, the college student flared up in anger, determined not to go without a fight, and loudly demanded to know why he wasn't just as good a lookout as the others who would keep their jobs. Quietly, unperturbedly and nonjudgmentally, Ralph outlined the lad's failures to benefit from the training that all lookout trainees had just received, as well as his questionable work ethic when left unsupervised, that led to the ranger's belief that he would be better suited in another line of work.

Ralph stressed that the young man may have worked at a level satisfactory to his former employers and to his own expectations, and that he probably wasn't aware that the standards of the lookout job were necessarily high. This was not necessarily the young man's fault, Ralph continued, since his experiences were not yet very broad. But if he were to benefit from his firing, he might consider a change in his work ethic. Subdued by Ralph's logic, but undiminished in spirit by the manner in which it was presented, the young man shook hands with the ranger and drove down the highway to find his next job opportunity.

Ralph subscribed to the *Congressional Record* and sat in his little kitchen on the ranger station in the evenings discussing with his wife, and sometimes with me, the latest goings on in Washington. From his vantage point in tiny Plains, Montana, with no television commentators or "spin doctors" to influence his thinking, Ralph realized long before most of the rest of the country that Senator Joseph McCarthy's celebrated hunt for communists in and out of government in the U.S. during the 1950s, was a sham. He knew from reading verbatim transcripts of McCarthy's committee hearings that many of the questions asked of witnesses were misleading, self serving and sometimes dishonest; and that the headline-grabbing charges leveled against most of his targets were without foundation. Armed with the facts as he knew them, Ralph influenced those around him to look beyond the shallow, conventional wisdom to find the larger, sometimes unpopular, truth.

Ralph told me that in his experience as a district ranger, subjects like sociology, public administration, history and political science were probably more valuable to him than pure forestry subjects like silviculture, forest management and mensuration, and suggested that I consider taking some of them when I returned to graduate school. Reluctantly, I did take some of his recommended courses out of the enormous reservoir of respect I had for him, although I would rather have taken the familiar forestry courses that beckoned so beguilingly to me.

Ralph understood that the path to an open mind was built on a foundation of acceptance of, or at least an awareness of, different ways of thinking about things. The intricacies of this novel (to me) philosophy were nearly wasted on my inflexible mind the next year as I struggled to adapt to unfamiliar and uncomfortable classroom situations where the answers to questions were rarely black or white as I had been used to, but rather, were wrapped in shades of grey and inextricably woven with other, equally murky questions, much like the situation in the "real" world.

NORMAN FARRELL

Carl Wilson

I was pleased recently, when my memories of Norm Farrell were jogged a bit. I remember him as a first class gentleman. There was more, too; Norm had the most thorough understanding of complex fire control problems of anyone I ever knew.

While he served as assistant regional forester and I was chief of fire research and assistant director of the PSW Station, he made sure that we operated as a team.

Norm had mixed feelings about his fleet of airplanes. At times he was proud, and at others dismayed. I recall a flight from the bay area to the Chester Airport in the old, reliable Twin Beech. There were about six of us on board and we were shouting to be heard. But, suddenly, there was a deadly quiet. It seems that the two engines had quit when Red forgot to switch to the second fuel tank. Suffice to say, Red heard Norm's "constructive comments" loudly and clearly!

On another occasion, in the 1960s, Norm, Ollie Holmes and I were returning from an Inyo board of review in the Beech Bonanza, which he called the "Vomit Comet." We started out in rather poor weather, lots of cumulous clouds in our path. We were flying VFR (Visual Flight Rules) from one potential emergency land spot to the next in hopes that we wouldn't have to use any. Finally, it appeared that we wouldn't make it over the Sierras and might have to land at Lake Tahoe. Norm kept reminding the pilot that he didn't want to fly into clouds filled with rocks. As it turned out, there was a momentary break in the weather and we made it safely but sweaty, back to the Concord International!

Norm had some basic fire control principles which he elucidated on the appropriate occasion. Some of the most important were:

> If you don't have a line right on the edge of the fire, you might as well lay red string around the perimeter.
>
> A fire "contained" is not "controlled," and a fire that is 50 percent controlled is like a woman being 50 percent pregnant.
>
> In suppressing large fires, it is mandatory that you put the main fire camp as close as possible to the perimeter of the fire. If you find an attractive campsite in a nice, green meadow and plenty of water, keep going until you are on the fire. You can come back later and have a picnic there!
>
> Don't over organize for a large fire with a top-heavy headquarters staff. As a rule, by the time the main fire camp is running smoothly, the fire should be under control.

These principles of fire control (now called "fire management") should probably be carved in stone, because all still apply.

OUR LOSS IS HER GAIN

Earl Niewald

The mid-sixties was an exiting time to be in the Forest Service. New laws such as the Wilderness Act and programs like Job Corps kept things humming. The one that excited me the most was the Land Water Conservation Act. Acquisition money was plentiful and land was available. It was just a matter of finding properties and negotiating a reasonable price. As a ranger on the Rolla District of the old Clark National Forest in Missouri, I spent a lot of time on land acquisition. In this period before NEPA and NFMA there was plenty of time for rangers to be in the field.

I heard of an old farm with Gasconade River frontage that would qualify for L&WCF acquisition. It was owned by an elderly brother and sister who lived on the property. On my first visit I left the State road and headed down the road to their house. After driving a few yards it was obvious that anything less than a four wheel drive would never make it. I walked the mile and half to the property. There, on the edge of the woods stood their small, tarpaper covered home. The river bottom fields in front of their home, which were once productive farm land, were reverting to brush.

I was greeted by Tennis and Mary Ernestine Decker. Both were frail and of small stature. Mary told me immediately that she went by the name of Ernie. I was invited into their home but before they offered me a chair, Ernie got some tools out and asked if I would fix their stove. The pipes were loose and a leg had fallen off. Tennis was upset that she would treat a guest that way. He mentioned that they had a visitor sometime earlier and Ernie did not ask him to fix the stove. Ernie replied that she did not ask for help because he visited on the Sabbath. Tennis replied, "If your ox falls in a pit pull him out." Throughout our visit Tennis frequently quoted verses from the Bible. After a friendly visit I brought up the subject of purchasing their property. Their attitude changed immediately and a sense of fear came over them. They told me in no uncertain terms that their land was not for sale. My offer of a lifetime occupancy reservation made no difference. Their land was not for sale, period.

A short time after visiting the Deckers I heard that Tennis had died. The major cause of death was malnutrition.

I continued to drop in on Ernie occasionally, usually when I was in the area with a four wheel drive vehicle. Ernie welcomed company and loved to talk. But before visiting she always had some chore she wanted done. It might be bringing in firewood or fixing a hinge. The biggest job came each fall when she asked that I take down the pipes of the heating stove and clean them. She wanted to be sure the stove was safe before the heating season.

Ernie had never traveled more than 25 miles from her home. She bought the supplies she needed in Richland, which was about 15 miles away. She had also been to Waynesville which was the county seat. She often asked what it was like to live in a big town like Rolla. I offered to come and pick her up some day and take her to Rolla. She declined saying "big cities skeered her to death." She said she loved the woods; she could never live in a town.

Every month or two Ernie would ride into Richland with the mail carrier to get her supplies. She had an arrangement for a ride back. Then she would have to get her purchases from the State road to her home. It would take her two or three round trips of three miles each. The last year or two before her death she did have a distant relative who began taking an interest in her and helped her with her shopping.

Ernie said she never was interested in boys and had never had any desire to get married. She said she loved her mother and had promised her when she was a girl that she would never leave her. Her mother meant more to her than any man could. She was proud of the fact that she had taken good care of her mother until her time of death.

To my knowledge Ernie had never gone to a doctor or had a prescription drug. She had her Doctor Book, published around 1900, that described ailments and gave remedies. Most of the remedies came from native plants or patent medicines. She had a lot of faith in that book.

Ernie often spoke of her religion. She told me several times about the time she was "saved." She had met Jesus face to face out in the woods and he assured her he would never leave her. After this encounter she was baptized in the Gasconade River. She said it was not necessary to go to church because once you are saved, you are saved. She did miss Tennis though, because he was so knowledgeable of the Bible. As she said many times,

"Tennis was well versed in the Bible. He could have preached the finest sermon that ever was if only he had not lost his voice."

While Ernie had no garden she did take advantage of the things that grew around her. Her favorite time was spring when the polk and dandelion greens were out and abundant. She also picked berries and collected nuts to supplement the few groceries she purchased.

Besides her property, the two things that gave Ernie the strongest feeling of security were a large woodpile and the old age pension check of eighty dollars that came each month. Each fall she would buy wood, even though she had not used more than half of the previous years supply. As a result she had half piles of wood from several years rotting into the ground. I suggested that she buy only half as much one year and use what she had left. She would not hear of it, after all one never knows what the winter might bring.

One day when I stopped for a visit Ernie asked me to read a letter. It was from an attorney threatening a foreclosure on her property if she did not pay a debt within a few weeks. I told Ernie that she really needed to see an attorney. Her reply, "I ain't going to see no crooked lawyer. All they want is your money." The debt was for $2,800, an impossible sum for Ernie to raise. Ernie said they had borrowed the money from an individual nearly 25 years ago. They had never made a payment The man who had loaned the money had been lenient and had not pressed them. Now he was faced with the statue of limitations on debt and he had to collect now or lose his money. While Ernie did not tell me, I learned from another source that the money was borrowed to keep Tennis out of prison. He had been caught moonshining and either had to pay a fine or do four years of time in the penitentiary.

While I was trying to figure some way out of Ernie's dilemma, a man stopped by my office with a proposal. He was a cousin of the person who held the note. He would buy the note from his cousin, make a deal with Ernie and sell the property to the Government. He would reduce the price enough to allow for a lifetime reservation for Ernie. We went and talked with Ernie and she reluctantly agreed. She realized that was the only way she could stay on the property. A few days later I went by Ernie's home to see what kind of a deal they had made. She had received cancellation of the debt and a few hundred dollars cash. At first I was upset that this person had taken advantage of the old woman. However, this was probably the best arrangement for Ernie. To receive old age assistance she could have only a limited amount of cash. If she had received more she would no longer get the monthly check. To her that check was security. The Government did acquire the 280 acre property for the appraised price of $16,000. Ernie was granted a lifetime reservation.

Sometime after this I was driving down the State road when I saw Ernie at her mailbox. She had just been to Richland and had been dropped off. I offered her a ride home as that day I had a jeep. Ernie told me she had been to town to arrange for her funeral. She had taken the dress that she wanted to be buried in and had also selected a casket. She had prepaid her funeral with the few hundred dollars she received for her farm. She seemed very pleased that everything was taken care of. When we arrived at her house she had a chore for me. She handed me a piece of paper and a pencil and ordered me to write what she told me. She proceeded to dictate her obituary. When she finished she asked that I read it back to her. It was rather flowery and complimentary. I could tell Ernie was very happy with it. She said she would mail it to the undertaker and everything would be complete.

One winter morning I was in the area and decided to check in on Ernie. It was very cold for Missouri, below zero, and there had been an inch or two of snow a couple of days before. When I got to her place there was no smoke coming from the chimney. I walked around the house and saw no footprints in the snow. I was standing in front of the house, knowing I had to go in and dreading what I might find, when Ernie knocked on the window and motioned me to come in. It seemed colder in the house than it did outside. As usual Ernie ordered me to get some wood and kindling and build a fire. The house never really warmed up, but we did sit by the stove for a visit. Ernie said she was afraid to build a fire on very cold days. It would have to be a big fire to warm her house and she did not like big fires. Instead, she stayed in bed under a pile of covers. She had not used her wood kitchen range for many years for fear of fire. She used the heating stove for what little cooking she may have done. She also said she had not lit her kerosene lamps for the last couple of years because of the danger of fire. She went to bed before dark and got up when it was light.

On New Years Day of 1971, I was driving with my family listening to the local news on the radio. The lead story was a report of an elderly woman who was found burned to death in the remains of her home. Before the name was mentioned I knew it was Ernie. A hunter had stumbled onto the burned out ruins. The fire had occurred several days before but it was never determined just when or how it occurred. I have often wondered how a fire could have started when Ernie was so careful she would not build a fire on a cold day or light a lamp to see at night. There was some speculation that someone had gone into her home to rob her and had set the fire to cover their tracks. There was never any evidence to support the theory.

Ernie had two big fears in life. One was fire, the other that someday she would be taken from her home in the woods and put in a "home." It seems ironic that culmination of one kept her from facing the other.

Ernie's funeral was at the Fairview Baptist Church on a cold, gray January day. She was laid to rest in the church cemetery. I did not attend the funeral. On that day I was at a meeting in Kansas City. I have always regretted not telling the forest supervisor that I could not go to Kansas City that day, there was something more important that needed to be done.

The next week the obituary was printed in the Pulaski County Democrat exactly as she had quoted it to me, with only the date of death added.

Mary Ernestine Decker—age 81—was converted and joined the church in 1906—she was a gifted musician and highly respected by all who knew her—she was a friend of all who knew her—she was an excellent neighbor—she leaves many relatives and friends to mourn her passing. Our loss is her gain.

Ernie would have been pleased.

POTPOURRI

THE FOREST SERVICE BADGE

From a Washington Office Letter

On the transfer of the National Forests to the Department of Agriculture in 1905, competition in designing a new badge was opened to all members of the Forest Service in Washington. The judges were Gifford Pinchot, Overton W. Price and E. T. Allen, who had been one of the first rangers in 1898.

No design received was considered both appropriate and representative of police authority. While another contest was being discussed, Mr. Allen traced and lettered a conventional shield and William C. Hodge contributed the tree drawn on a cigarette paper. The combination was accepted without change as the solution.

When the first badge struck from the die was submitted by the manufacturer for approval, it was kept by Mr. Allen who presented it to the Washington office nineteen years later.

FOREST RANGER AND SUPERVISOR EXAMINATIONS IN CALIFORNIA

From the Files of the
Forest Service History Center

The first forest ranger and forest supervisor examinations were held during October 16 and 17, 1905. This was the first and only one given for forest supervisors. Candidates were given a rigorous field test that included:

Shooting ability at targets with a rifle at 100 yards and a pistol at 50 yards.

Use of an axe in a dense stand of 12 inch to 16 inch dbh timber.

He had to chop such a tree down, limb it, and pile the slash so that it could be burned without damaging other trees in the stand.

Pack a camp outfit of blankets, cooking utensils, tools and barley and demonstrate his ability to load it onto a pack mule. He was allowed to use any of the following hitches: Diamond, Square, Double Diamond or Squaw.

Saddle and ride a horse through several gates and demonstrate putting on hobbles and picketing with rope.

Set up a compass and run a line over a prescribed course.

Estimate the volume of timber on a 20 acre plot and record all of the trees by size and species.

At that time, salaries for various positions were:

Forest Supervisor	$1,800 to $2,500 per annum
Deputy Forest Supervisor	1,500 to 1,700 per annum
Forest Ranger	1,200 to 1,400 per annum
Assistant Forest Ranger	700 to 900 per annum

FREE USE PERMIT—FOR PERSONAL USE

Fritz Moisio

In the Ozarks the free use permit system did a lot to gain friends for the Forest Service, especially during the initial years of management on these new forests. We issued free permits for wood, bee trees, board trees and for other personal uses. The most unusual permit that I issued was to John Pollard. Pollard lived by himself on a small acreage surrounded by national forest land. He was a good cooperator. It was on a fall like day when I last visited him.

At this time he had enough wood to take care of his needs but he wanted a permit to cut a good, white oak board tree.

Good, white oak was scarce and it would require a premium reason to qualify for cutting one under a free use permit, or even commercially for anything less than bourbon barrel staves. John said that he wasn't going to make it through the winter and he would just love to have a white oak board tree to build a coffin before winter set in. I told him to select the best white oak on the claim and issued the permit: "ONE WHITE OAK FOR PURPOSE OF CONSTRUCTING A COFFIN—FOR PERSONAL USE ONLY."

The following spring John's grandson informed me his grandpappy had passed away a few days after New Years and was buried in his white oak coffin.

MIRACLE MAN

George Morey

Here is an item received by Ranger Cooper at Galice years ago. It accompanied a job application from a young dreamer on the East Coast.

"I always wanted to become a forest ranger. The first reason for wanting to be a forest ranger is that a ranger is an adventurer, bronzed with the sun, his body is muscular and healthy from the constant outdoor life. His green uniform blends with the green of the trees, a forest ranger presents a striking picture. He is a symbol of that dynamic work that is forestry.

"Then too who wouldn't like a job in which the whole outdoors is his office, where the air is as sweet as a fresh cut apple, and where the lakes are glittering at his feet and with a roof of branches over his head? Sometimes the ranger gallops to his post on horseback. At other times he depends on his trusty legs. This modern guard of the forest may have wings also, that he may fly to an emergency. He can drive a tractor or a ten wheel truck. His is no rocking chair job, it's a job for rugged he-men."

I hope that this young man realized his dream.

BRIEF HISTORY OF FEEDING FIREFIGHTERS UNTIL WWII

Rupert Asplund

I experienced my first fireline feeding while I was living on the Cascadel Ranch above North Fork. I had a couple of pack horses and was hired to pack supplies to fires in the back country. The warehouse at North Fork was full of surplus canned food from World War I. I recall a fire in the Chiquita Creek basin. The fire boss (Ed Madsen) ordered steaks, along with other staples and received a pack load of canned corned beef. Needless to say, Ed was a bit upset with the forest fire control officer.

On a fire after I started to work for the FS during 1927, I was crew boss with a crew of North Fork Indians. We received a supply of food on the fireline, again surplus WWI canned food. The only problem was there were no labels on the cans. You opened a can hoping for fruit and it would turn out to be tomatoes. I overheard: "What's the matter? White man no put-um picture on the cans!" Later came the C-Ration, with canned fruit, canned beans and a can of "good" brown bread. We were all happy! Then, the K-Ration.

A common sight in fire camps for years was the ice can stove with beans and strong coffee. The 10 man and 50 man cooking outfits came into the picture early in the game. Fires on the Trinity were supplied with long strings of mules as late as the early 50s. To serve the Jim Jam Fire up New River on the Trinity NF, a load of mules was delivered from Idaho by plane.

Then, came the airdrops. My first experience with these was on the Little Bear Wallow Fire on the Trinity about 1938. Burlap chutes were used with no rip cords! The food came "raining" out of the sky, landing in the general vicinity of the camp. Many of the chutes did not open. The sides of beef were well pounded and tender but the cases of eggs did not fare so well! The natives remarked that the woods animals lived fat that winter. So, with airdrops uncertain, there was a period when it was still more efficient to use pack mules or perhaps, build a road to the fire camp.

The first use of the rip cord that I recall was about 1944, when Major Bennett, a WWI pilot, flew an overhead crew from Idaho. He would not drop any supplies from the Redding Service Center until the chutes were repacked with rip cords. After this, the airdrops became very efficient, especially when the Ford Trimotor or other slow flying aircraft were used. During the war, the Air Force DC3 was used but it was difficult to get the pilots to slow the plane down to a speed that permitted chutes to open without breaking the shroud lines.

During World War II on the Trinity NF, one of the most important facets of the fire camp supply was the women's motor corps and fire camp crews. The motor corps did most of the delivery of camp supplies, including the pack stock. The fire camp crews, including cooks, took over operation of the camps.

RETRIBUTION

John "Ray" Bruckart, Sr.

During the period of more than 16 years as a district ranger on the Darrington District of the Snoqualmie NF, it was my privilege to become acquainted with many of the early settlers, trappers, loggers and miners who first settled the beautiful Stillagaumish and Sauk valleys. It was only after I had moved that I became aware of the somewhat unique character of the area and high drama in which some became involved.

One such epic, and there were others hardly less dramatic, I have decided I might evaluate at this time, as all of the principals have long since passed from the scene.

The time covered a period of more than 30 years and had its beginning about 1895. At that time, the Sauk Valley east of the town of Darrington was being settled. Although this was a heavily timbered country, most of the 15 or so residents were homestead claimants. Thus, the majority of these people were interested in the land and not in the timber. Some made substantial improvements and established permanent homes. The original wagon road, that was built to serve the Uncle

Custer and Goat Basin mines at the headwaters of the tributaries of the Sauk, had fallen into disuse with the closing of the mines and all but a few settlers gave up and left their homesteads but not before obtaining patent to the land.

One of the settlers remained on his homestead claim until his death about 1930. Jim Bedel had married an Indian woman and raised a family of two girls and one boy. He was to play a prominent part in this story. Sam Strom is the principal character. Sam, a young Scandinavian, had a homestead about 15 miles east of Darrington and about a mile west of the Bedel place. He was a hard rock miner and worked part time at the Monte Cristo mine when not being busy on his homestead. Sam built a good sized two story house on his claim from native split cedar boards. Although a bachelor, he built the building large enough to house a large family. Actually, the house was intended as a stopping place for travelers going to the mine in Monte Cristo.

During the Klondike gold rush (1897-1901), Sam Strom decided to seek his fortune in the northern gold fields. He did not want to abandon his homestead on the Sauk, however, and he arranged with a close friend of his, John Brown, another homesteader, to look after his place. Brown was an older man and had a 160 acre homestead a few miles west of the Strom claim.

During Sam Strom's absence in the north, John Brown lived in the Strom house at least part of the time. Anyway, while living in the, Strom house he became sick and died. As I recall from Forest Service records in the district files, the party to whom Brown (purportedly) willed his homestead completed the necessary steps to have the claim patented. When the non-existent will was probated, witnesses claimed they had seen the will but it was lost and not available to be submitted for probate. The court accepted the witnesses' testimony that there had been a will witnessed by two or more persons, and Brown's property passed to one of two brothers who were among those who claimed to have seen the lost will.

When Sam Strom returned from the Klondike, he learned that his friend Brown had died in his house and was buried in his front yard.

Mr. Strom continued to maintain his house on the Sauk River although in the meantime he married, raised a family and had a yearlong residence in Darrington. He visited his homestead on the Sauk several times each year at which time he made needed repairs on the house and generally kept the house and grounds in good condition.

On one of these trips, while making repairs on one of the rooms of the house, he noticed a loose board and upon investigation, later stated he found a sheet of paper on which was written a statement about as follows: "I, John Brown, am alone, sick and dying, and hereby will to my friend, Sam Strom, my gun and homestead, providing that I am given a Christian burial."

Mr. Strom proceeded to carry out the provisions of the alleged will. He asked me to witness the removal of Mr. Brown's body from the grave on his Sauk River property. The local packer provided the transportation to Darrington where the remains were shipped to a funeral home in Arlington. Mr. Strom then arranged for a funeral and burial in the Arlington cemetery.

Later, his attorney arranged for a date when the will he found would be presented to the Snohomish County Court in Everett. On the appointed day, necessary witnesses, of which I was one, were assembled in the court room where the attorney attempted to have the will accepted for probate. Unfortunately for Mr. Strom, the judge refused to have the will recognized because there were no witnesses.

There was nothing for Mr. Strom to do but give up his efforts to obtain Mr. Brown's property on the Sauk.

There were four individuals involved in the successful effort to circumvent John Brown's wish for the disposition of his property on the Sauk River. There were two brothers; one was killed in a saloon brawl in Granite Falls—the other was involved in a shooting incident in which his adversary lost his arm. For his part in the shooting affair, he was hauled into court, convicted and served five years in the penitentiary. The third person, a packer, was killed in a snow slide at Monte Cristo. The fourth, Jim Bedel, suffered a stroke and was a helpless invalid until he died about 15 years later.

Thus, retribution was visited upon the four persons involved in the alleged plot that later resulted in a tragedy for all of them.

WHEN BIGFOOT VISITED THE LASSEN FOREST

Phil Lord

I read in the paper that someone had discovered some "Bigfoot" tracks in the Lassen Volcanic National Park. As far as I could ascertain nothing more ever happened.

It did take me back many years, however, to one fall when a couple of men riding after cattle in the vicinity of Straylor Lake on the Lassen Forest saw something they couldn't explain. They caught only a fleeting glimpse of it as it disappeared into a dense mahogany thicket. Fearing ridicule they told no one until somebody else came up with a story of having seen in somewhat the same vicinity, an ape like creature that ran on its hind legs and was covered with brown hair. This was essentially what the two riders had seen so they told their story. They had tried to find tracks but there had been no fall rain and the ground was too hard to really show much. Then a few reports of other sightings were heard and the paper in Fall River Valley came out with an editorial or two concluding that it might possibly be a Bigfoot. After all, it wasn't too far to the Klamath and Trinity country, supposed to be the Bigfoot headquarters. Anyway, there was a lot of talk and conjecture about it in the Fall River, Little Valley and Dixie Valley areas.

There were several schools of thought. Someone remembered a traveling carnival that had gone through the country and said maybe an ape had escaped. Others thought the whole idea was phony and advised people claiming sightings to change their brand of refreshments. Some few Bigfoot buffs traveled around hoping to see it.

Ned Bognuda, owner of the Little Valley Ranch and proprietor of the Empty Holster Bar at the small settlement there, said, "Well, Bigfoot, Littlefoot, ape or whatever, these poachers that live around here will have it eaten up by spring, anyway." Then one day Fred Haynes, Superintendent of the lumber company with a mill at Little Valley made a trip to Susanville. It so happened that the Lassen was having some kind of a party that night and Fred was invited. Some business the next day made it imperative that Fred return to Little Valley that night. I remember that he left the party early as the road to Little Valley was long, rough and dusty.

As Fred was bouncing along over the lava rock and through the red dust near Poison Lake his car lights picked up the Bigfoot running down the road. He got a real good look at it before it dodged off into the timber. It was a good sized bear that seemed to have withered or crippled front legs and ran on its hind legs.

I don't think that Bigfoot would have liked the Lassen anyway—too dry and flat.

THE BAILEY MEMORIAL

Don Bauer

We are at the site of the Bailey donation on the Snow Slide Road west of Fawnskin, on the San Bernardino National Forest. There is a story behind this unique piece of national forest land. Margaret Bailey was a very prosperous attorney in Reno, Nevada. She had lost her husband several years previous to this. This would have been about 1959 or 1960. She was alone and very wealthy and owned this full section of land which runs pretty well up the mountain towards Butler Peak Lookout and embraced a microwave station. We were involved in timber operations in this general vicinity on national forest land and she thought we were cutting or marking on her land and became quite adamant to get off.

The survey had been run by Earl Nichols, who was district ranger at that time, and Miles Julik was the assistant ranger. Our survey was right. Out of that chance acquaintance and resolving of that problem, it developed that she had a deep sentimental attachment to this area which she didn't want to share with us but let us know that she didn't want any trespassing going on on her land whatsoever. We told her we would honor her request.

Then a few months later I got a call from Ivan Sack who was supervisor of the Toiyabe Forest. Ivan and I had served together in the Navy during World War II as photo interpretation officers in the Pacific Theater. Ivan said he had been approached by Mrs. Bailey on what the national forest might be able to do to acquire her section through purchase, but with a covenant on it that would keep it as a natural nature reserve forever and that she was interested in disposing of the property. We were committed at that time very strongly to multiple use. We would be agreeable to negotiate for acquisition of the property either through exchange or purchase but on the basis that it would be managed for multiple use purposes.

That was the beginning of an interesting series of events. The following spring Mrs. Bailey approached the University of California and inquired as to whether they would be interested in accepting her section of land, the full 640 acres, as a donation on the condition that they would manage it as a natural area without any disturbance. The University's response was they could consider this but they would have to be assured of at least $50,000 a year to administer the donation in perpetuity. She also wanted to donate it to some other society but every one of these agencies had a problem with administering donations. So then she approached Ivan Sack again with the idea of donating it to the Forest Service. We told her we would be agreeable to accepting a donation but with the understanding that the land would be managed in accordance with the laws of Congress.

That led to a field trip during which she explained to Ivan Sack, myself and Earl Nichols that she was very concerned because her husband's ashes were buried on this site and she wanted to be sure they would be respected. I asked her why he was buried here and the story unfolded.

As young people in love in Southern California, I guess in the early 1920s, they got married and went on a honeymoon. Their destination was the hotel in Fawnskin which was quite a resort. In their little Model T Ford as they putted across the country after two flat tires, following being married in Los Angeles that morning, the car broke down at this location. They pushed it off the road and fortunately had some blankets and things and they came to this tree and put up a tarp and spent their honeymoon night right here. She described the meadow when the sun came up the next morning and the water was running, and to her this was the most beautiful spot in the world. It had been her husband's wish that when he died he would be buried near this site. She had his ashes brought down here and buried without benefit of any official clearance.

As an attorney, she was quite aware that there are very stringent laws about burying people or their remains. She pledged us to secrecy and we assured her that as long as she lived we would treat that with respect. Then we negotiated the donation. She asked if the Forest Service would accept that and treat this place with reverence and respect. I told her that I could not document it but we would accept the donation with the covenant that the land would be managed in accordance with the laws of Congress and in keeping with the management direction for the Big Bear Ranger District, which was pretty much the management that the surrounding national forest was getting. That was my point, to show her that this was the way we managed this kind of country and if she liked that, that's what we would do in the management of her section.

The other part of it, which was not documented, was that I promised her that as far as her husband's resting place was concerned, I would personally see as long as I was alive, that each succeeding Big Bear district ranger would personally know where that spot was and know what the story was and the need to treat this little area with unusual respect and reverence.

They were both attorneys and had no children. The monument says he died in 1955. Apparently, she had developed a very outstanding reputation as an aggressive, effective divorce lawyer. But in her own marriage and the sentiment in her life, it was interesting to see that she had a deep reverence for the joy and love of their honeymoon trip. It is beautiful here and I think she would be very proud of the management she is getting here.

I described to District Ranger Jere Mitchell the actual location of the monument. The metal casket that has the ashes of Mr. Bailey was originally buried about 100 yards to the west, next to the road because when she brought the ashes down from

First truck logging in Beaver Creek, Klamath NF, California - 1937.

Oak Knoll CCC Camp, Klamath NF, California - 1937.

Reno there was a little storm going on and it was nighttime and she buried it at that location, thinking it was her land. When she saw our logging activity around there, she was disturbed. We pointed out that it was not her land, so we moved it to land that was hers at that time. We agreed to move the remains to land that was hers in order to mitigate a possible trespass on national forest land. We needed additional tools so Ranger Nichols got on the radio. All of this was top secret and he requested Miles Julik to bring a shovel and a crowbar and some other earth moving implements as fast as possible. Miles asked what he wanted them for and Nichols said he would tell him when he arrived. So Ivan Sack, Miles Julik, Earl and I moved the little box which was only about 18 inches by a foot square and maybe six inches deep. It's at the bottom of the memorial.

OVER THE FOREST HE LOVED AND SERVED

Walt Puhn

There is no road to Shuteye Peak and its lookout was one of the first in the national forest system, so it takes a hike, a horse or an overland vehicle to visit this shrine to the old-timers. Pancake Rock at the summit of Shuteye Peak (El. 8,351) on the Sierra NF, above North Fork is well named. In fact, no one can look at this distinctive boulder without thinking of a huge "stack of hots." If it was not for this distinction, it could now be appropriately named "Old Foresters' Rock," because installed at its base are bronze plaques memorializing a group of old-timers who worked together and made history on the Sierra and in the California Region.

The plaques read:

Chester C. Jordan
1880-1934
Ranger-Supervisor-Inspector
Loyal Friend and Forest Officer

Maurice A. Benedict
1882-1959
Supervisor of Sierra National Forest
1916-1944

Roy H. Blood
1890-1964
Administrative Officer
Sierra, Lassen and Cleveland National Forests

Eugene "Gene" Tully
1875-1973
Last of the First Forest Rangers
Sierra National Forest
1903-1919

Earl A. Morrow
1895-1973
Sierra & Plumas National Forests
1922-1954

On a smaller companion rock is a plaque which reads:

Samuel J. McFarland
1867-1933
Forest Service Lookout
Shuteye Peak 1925-1929
His Mortal Remains Lie Beneath This Plaque
But His Spirit Still Watches

THE HISTORY OF THE "K" TAG NAME

Doug Miller
(Submitted by Chuck Smay)

Did you ever wonder where the term "K" tag came from? I did and thought for awhile that it was some early day forester or engineer's jest at spelling corner tag. Somehow, I came across a

November 17, 1975, letter from Doug Miller, Research Forester to Bruce Summer in the regional office, which provides at least one version of how the term "K" tag came into existence. I was willing to accept this version of historical fact since I was the Ranger on the Georgetown Ranger District and familiar with the Davis Cabin site. I was further prone to believe since my first job in the Forest Service involved blister rust control spraying and checking. I have spent considerable time since then looking for and dealing with those "K" tags!. In any event here is the letter and history of the term "K" tag.

United States Department of Agriculture
Forest Service
Klamath National Forest
Yreka, California 96097

Reply To: 1600 Information Service November 17, 1975

Subject: "K" Tags

To: Bruce Summer
U. S. Forest Service
630 Sansome Street
San Francisco, CA 94111

In reference to our phone conservation of 11/13/75, I thought you might be interested in the following information on the term "K" tag.

The term "K" tag referring to our present day location posters by "old timers" originated at Davis Cabin on the Georgetown Ranger District, Eldorado NF during the summer of 1938. Johnny Crowell was the Blister Rust Control Checker foreman at the Davis Cabin Springs Checker's Camp for Reed Miller.

Beside the door of the office tent was posted a check list of all items a Blister Rust Checker was required to have for the job. The first item was listed beside the letter "A." It so happened that item "K" was cruiser tags or today's location posters. The term Cruiser tag that summer was shortened to the letter "K" and the Blister Rust Checkers started placing the symbol "K" on their eight inch to the mile work maps, indicating "K" tag locations on roads, trails, streams, corners, etc.

Since quite a few of the checkers were students and the main Blister Rust Control Office for California and Oregon was in the Bay Area, the term quickly spread through Forest Service Regions 5 & 6 and is still in use today. Just think, it could have been a Z tag!!??

Doug Miller
Research Forester

TWO FOREST OFFICERS

Joe Elliott, Jr.

Editor's Note: Joe Elliott, Jr. has given us a copy of an account of the careers of two Forest Service people, Joseph E. Elliott, Sr. and John Ray Bruckhart, Sr., written seven days after the latter's demise. The original version went to the "Thirty Year Club" of Region 6. There are earmarks of the unique in this story.

What follows below is an alteration in format of Joe's account in order to highlight career parallels. The whole has a special flavor since the author is the son of Joe Sr., and Edith his wife, is the daughter of Ray Bruckart, Sr. A second daughter of Ray's also married the son of a forest supervisor. –E.J.

Joseph E. Elliott, Sr.	John Ray Bruckhart, Sr.
Born: Grizzly Flats, California September 6, 1885	Born: Lancaster, Pennsylvania June 23, 1887

Formal education: A few years of grammar school.

Youthful years: California, Oregon, Washington

First job with U.S.F.S.: Fire fighting 9/1/08, Stanislaus NF, D-5 (R-5)

Career: Appointment from ranger exam. Later Deputy Forest Supervisor, then Forest Supervisor. Supervisor of Cleveland, San Bernardino, Modoc, Lassen, Sequoia NF's. Entire career in R-5.

Marriage: Married school teacher, Anita Zimmerman, Dec. 18, 1911. Honeymoon in San Francisco.

Children: Three. A son holds BSF from, University of California. (A son died 3 days after birth in 1914.)

National Forests & National Parks: Joe was Supervisor of Sequoia NF when a generous slice of the Forest was transferred to the NPS - Kings River Canyon N.P. - following a Presidential visit.

Impact on FS Policy: Left to FS historians to describe.

Impact on NF Personnel: Guided many newcomers into Ranger, Staff, Supervisor, Asst Reg. For. positions. Aided early development of 3 Regional Foresters, a Deputy Chief, & a Chief of Personnel.

Formal education: A few years of grammar school.

Youthful years: A southern State, State of Washington

First job with U.S.F.S.: Fire Patrol 7/6/09, Snoqualmie NF, D-6 (R-6)

Career: Appointment from ranger exam. Later Deputy Forest Supervisor, then Forest Supervisor. Supervisor of the Columbia (Gifford Pinchot), Olympic, and Willamette NF's. Entire career in R-6.

Marriage: Married school teacher, Fay Batdorf, Dec. 18, 1914. Honeymoon in San Francisco.

Children: Three. A son holds BSF from University of Washington.

National Forests & National Parks: Ray was Supervisor of the Olympic NF when the heart of the Forest was transferred to the NPS Olympic N.P. following the same Presidential visit.

Impact on FS Policy: Left to FS historians to describe.

Impact on NF Personnel: Guided many newcomers into Ranger, Staff, Supervisor, Asst Reg. For. positions. Aided early development of a Chief of TM WO, a Deputy Chief, and Chief of FS.

After retirement: Served as resident manager for a lumber company. Later served as manager of a city Chamber of Commerce. Died at the ripe old age of 83 on Aug 2, 1968 at Antioch, California, within R-5.

After retirement: Served as consultant to a lumber company. Served on a city water board. Died at the very ripe old age of 92 on Oct 31, 1979 at Eugene, Oregon, within R-6.

OFF THE WALL

Ronald E. Scott

Editor's Note: Ronald E. Scott, Planning Staff Officer for the Huron-Manistee National Forests submitted a 36 page publication that consisted primarily of "quotes of the day" from members of the planning team. From almost 200 printed we are including 30 of the best. His introduction to the publication is below.

"Off the Wall Comments," is a composite of humor and vitality that was captured during development of the Huron-Manistee National Forests' Land and Resource Management Plan, November 1981 to July 1986.

The humor and spirit of these comments kept many of us, me particularly, going, through the many hard and difficult hours of management problem identification, analysis of the management situation, prescription writing, formulation and evaluation of alternatives, FORPLAN analysis and finally writing the draft and final forests' plan documents.

"Off the Wall Comments" are dedicated to all personnel of the Huron-Manistee and Region 9 who worked so hard and diligently over many long hours to complete these plans. We now all look forward to the appeals and plan implementation.

1. Planning is a game—the object of which is to discover the object of the game!

2. In the final analysis, there is no final analysis.

3. A planner with a psychosis thinks two plus three equals four and his work is important.
   A planner with a neurosis knows two plus two equals four and his work is important, but it bothers him.

4. Planner's dilemma: We can make it right or we can get done.

5. Nothing is timeless and when all time has passed nothing remains. (Jeff Pullen)

6. Quickly: Name one job with more job satisfaction and better pay? "Trash Collecting"

7. The baffling winds of planning will blow forever or fornever. (Jeff Pullen)

8. Name one planning team member who is presently eligible for food stamps?

9. We're looking further than we'll ever hope to see. It's taking us places we don't know. We're searching for something that will never be. Still onward we go, to where nobody knows. (Jeff Pullen)

10. We take inaccurate data, apply unreasonable assumptions and come up with hard facts. (Lance Lindwall)

11. Land management planning is a third down and long situation. (Gordon Joiner)

12. How do we shut this thing off? Because you like doing it we can't do it forever! (Ron Scott)

13. Until you get that twitch controlled I'm not telling you anything. (David Snyder)

14. If we start now, we can have a day's work finished the day after tomorrow.

15. After you run into the wall a few times, you don't feel like hitting it so hard anymore. (Jeff Pullen)

16. To hell with precision, we're gonna do it over anyway. (Jeff Pullen)

17. I don't know, I don't care, and it doesn't really make any difference (Phyllis Green)

18. Sometimes stopping is the most efficient way to go forward. (Larry Leefers)

19. Yes, though I walk through the valley of the shadow of planning, I shall fear no prescription for I know DE PLAN. (Lance Lindwall)

20. I'm not contradicting you, I'm telling you that you're flat-ass wrong! (Jeff Pullen)

21. "You" make mistakes.
Mistakes don't make "You" anything. (Jeff Pullen)

22. Planning is so complex that there is no room for reason. (Jeff Pullen)

23. LMP is like building a bicycle while you are riding it. (Floyd "Butch" Marita)

24. LMP is like driving down the freeway with the windshield blacked over while looking through the rear view mirror to see how you are doing. (Ron Scott)

25. I don't understand the simple stuff. (Ron Scott)

26. In LMP there is nothing so small that it can't be blown out of proportion. (Jeff Pullen)

27. If one "Plan" is good, more "Plans" would be better. (Lance Lindwall)

28. If everything seems to be going well, you've overlooked something.

29. Optimism indicates that the situation has not been clearly understood.

30. Worst case scenario: When everything goes as planned. (Jeff Pullen)

SAYINGS

Gary Munsey

One of the most intriguing things about working for the Forest Service was to hear sayings that I had never heard before. I have tried to capture a few of the more memorable sayings and have attributed them to the speaker as best as memory will serve.

"Hurry up, you're burnin' daylight"	L. Rivera
"I wasn't sleeping, I was checking my eyelids for pin holes"	R. Pipkin
"A good vet would have this cow back on its feet in 15 minutes" (Referring to a beef steak cooked too rare)	
"Cackle berries and hogs belly" (Bacon and Eggs)	H. Gale
"He looked at me like a cow looking at a new fence post"	H. Gale
"We can kill two birds with one stone if we don't drop the rock trying to line 'em up"	B. Neal
"If we had some bacon we could have bacon and eggs, if we had the eggs"	G. Garrett
"He didn't know if he was punched or bored"	J. Klump
"Bawlin' like a calf in a hail storm"	J. Klump
"He's raisin' hell and puttin' props under it"	H. Mostovoy
"He's so slow you have to drive a stake to see if he's movin'"	
"This shovel doesn't fit my hand"	
"I'm too light for heavy work and too heavy for light work"	R. Pipkin

"It was so hot I saw a garden hose crawl clear across the yard to get in the shade of a barbed wire fence"

"It scared me to death and I ain't afraid of nothin'"

"Generally speaking, so-and-so is generally speaking"

G. Garrett

POETRY

THE PARADE

Stan Tixier

Joel Frandsen was the district forest ranger at Duchasne, Utah, several years ago. He tells the story of the time they put an entry in the "Big Parade." What a time they had!

The District Ranger at Duchesne
    And his hard workin' crew
Had been a-chasin' lotsa smoke
    More than a month or two,
The forest fire season there
    Had surely put a strain
On all the District personnel,
    They really needed rain.

So when a soakin' shower
    Came through, at last, one night,
The Duchesne District workin' force
    Were thinkin', now they might
Complete some other projects
    And plans that they had made,
Perhaps they might participate
    In next week's town parade.

They said, "By gosh, let's do it!
    Let's show the people how
This Multiple-Use Forest
    Is workin' here and now

To meet their needs and wishes,
    Show how we manage land,
Our entry in the big parade,
    Let's make it really grand!"

They figured, rather than a float
    They'd use the pack-mule string
To carry illustrations
    Of all assorted things
This Forest land produces,
    Like water, grass and wood
And outdoor recreation
    And wildlife, if they could.

So all the pack and saddle stock
    Were fit with a display,
Lashed on to Decker saddles
    In every kind of way,
They tied on posts and fuelwood,
    Some Tall Wheatgrass was dug
And hung on wavin' upright,
    Sure hard to get it snug!

They slung big photo-boards on one
    Depicting recreation,
And hung some water jugs on too
    For further information.
The most unique and challenging
    Of all the Decker loads
Was wildlife, and they pondered long
    About what could be showed.

They asked the State game warden,
    An advocate/believer
In wildlife conservation, if
    He'd help live-trap a Beaver
That needed moving' anyway,
    And so they did, by golly,
And tied it up there, trap and all
    On their most gentle molly.

They had no time to practice much
    Until that very morning,
So when they got 'em all lined up
    Without a lot of warning,
Mules looked at one another's loads,
    They shied and started balkin',
They cargo'd tilt and turn and shift
    As soon as they were walkin'.

The crew pulled latigos up snug
    And tightened up the hitches,
Old Chet set out a-leadin' 'em
    From center road to ditches.
He had to keep 'em movin'
    And never let 'em stand,
Lest they get tangled up and spooked
    And surely out of hand.

This unique equine entry then
    Got lined out fairly well,
And started off down Main Street,
    It wasn't hard to tell
That those old mules were snaky,
    Especially when the pack
In front of 'em would wobble
    Or swim from front to back.

The Wheatgrass load in front of
    The molly with the trap
And nervous Beaver, swayin'
    On every single lap,
As Chet led them in circles
    Each time the High School band
In front of them would slow down
    Or play their tunes and stand.

The pack string turnin' circle
    Now wide and wider spun,
Around a power-pole went most,
    In fact, went all but one,

The pig-tail ropes were broken,
So Chet lost all his mules,
And they were loose and buckin'
And actin' just like fools.

They scattered wood and wheatgrass
And jugs and photo-boards
Along Duchesne's main thoroughfare
And then went headin' towards
The Ranger District pasture,
But Molly, buckin' hard
Went tearin' hell-for-leather
Right thru the mayor's yard.

She ripped up the geraniums,
Destroyed the pansy bed,
And then tore down the clothesline,
A nightgown on her head,
She shook it off and stomped it,
At last she stood around
The trap and dizzy Beaver
Were turned and hangin' down.

The Beaver's nose was bloody,
His eyes were red and glazed,
His big broad tail was bent somewhat,
I guess he was amazed
That yesterday he's chompin' trees
And splashin' in a stream,
Today he's part of a parade
And Ranger District team.

When prizes were awarded
For bands and floats and such,
The District crew had not-a-doubt
They weren't a-gettin' much,
But if there'd been a trophy
For tryin' hard to bring
Excitement to the town parade,
They'd sure have won the thing.

SIUSLAW FEVER

Stan Bennett

It begins when you catch a faint strange smell,
    And it isn't the Albany, Toledo, or Gardiner citadels,
Nor the heavy hot bracken fern pollen-laden air,
    That makes you sneeze and catch your breath with care;
You might say it's like an elderly coho salmon,
    That made its redd and lies forever deaden,
Or a trace of paranoid sarcasm that rhymes with "Pulse,"
    Just lying in wait to deliver the final convulse.

At times the fragrance is stronger than sassafras,
    Somewhat like resins, turpintine and even ashes;
It's not offensive, stale or irritating,
    But rather it's refreshing, invigorating and stimulating,
Generating such a drive, a vigorous stampede,
    That even the fir, hemlock and alder weed,
Wondering of their numbered days, begin to shake,
    Fearing the axe, in something less than the 80 yr. rotate.

And then again like a fever, it diminishes to a bleat,
    Resembling only the TREK with it's neat delicious meat;
Only to spring up in the form of an October Storm,
    Spewing destruction and catastrophe all over the norm;
And then to be subdued and corraled by those
    Who have the fever, the fragrance, the fire
To straighten out both nature and man's insidious desires.

What is this thing called the Siuslaw Fever
    That catches one up in this myrrh
Of drifting sands, salty winds, fallen trees,
    Silting streams and many other controversies?
It can't be told in a word or a book,
    But rather in living, eating and learning
With the folks of this Forest and what they foresee,
    From the songs and the scenes of this land by the sea.

FIRE IN THE VALLEY

Rusty Watson

The moon hangs low over the mountain,
The evening is hushed and still.
Life in the brush is restless.
An ember glows in the valley,
On the slopes is a sickening sight.
All the animal life is fleeing,
And there is fire in the valley tonight.
The moon down over the valley,
The sky with the light is streaked,
Then a roar from the far off horizon,
And the ranger siren shrieks.
The air vibrates with the roaring,
And lit by the sickening lights.
The animals for shelter are fleeing,
And there is a fire in the valley tonight.

HOLLOW LOG

H. Basil Wales

An old hollow log may come into the mill
But the old hollow log's got some good in it still,
And a sawyer who's smart will find more in the same
Than a faller will see—or a scaler will claim.
Yes, you never can tell what a sawlog'll cut
If you judge it alone by a look at the butt;
It may rot at the stump where it's close to the ground,
But a bit higher up may be thoroughly sound.

So we ought to take men, when we see 'em-log run;
There are very few men who are all Number 1.
Here's to a fellah who ain't, I'm here to admit,

and Perhaps—even you—may be rottin' a bit.
But there's good in us all, as I often suspect,
And there's something in all not exactly select.
Yes, a log's like a man, and a sawyer who's smart
In an old rotten log finds a little good heart.

THE WILDLIFE REPORT

Stan Tixier

Bill Hurst was ranger on the Ashley National Forest in the early '40s and later forest supervisor there. He told this story a few years ago on a visit to the Ashley and I altered it just a little with poetic license.

Some years ago, when Bill was Ranger down at old Dutch John,
Before the "reclamation" of the Flaming Gorge was done,
The lush Green River Valley was abundant wildlife space
For lots of native species, it was near an ideal place,
Before the dam and reservoir, before the N.R.A.*
The Ashley National Forest was some different from today.

An annual requirement was a big wildlife report,
The Ranger'd count or estimate the numbers of each sort
And kind of wildlife species and he'd write 'em on a form
And send it into Vernal, they'd compare it to the "norm"
And average for the past five years or maybe six or so,
They'd run a Forest total to submit to the R.O.**

So each year Bill would tally up how many elk and deer
And antelope and sage grouse and how many bears were here,
He'd estimate the predators with every coyote's howl,
He'd show a figure for the bighorn sheep and water fowl,
This form had space for every single duck and coot and goose,
And every year he'd list the fact the District had ONE moose.

Now this cow moose spent summers in the meadows of the
Green.
A chompin' sedge and willows, she was very often seen
Alone and solitary, but one year Bill had to laugh
When she showed up, by golly, with a nice big baby calf,
So that year on the wildlife form, unusual to see
The space for moose, where Ranger Bill reported he had three.

The Forest Supervisor's staff who handled the report
Cranked up the phone and called old Bill and asked him just
what sort
Of happening had caused the moose to jump from one to three?
So Bill said, "That one cow had had a calf that year, you see,"
The guy said, "Cow plus calf makes two," his voice reached near
a shout,
But Bill was calm, and he replied, "You figure that one out!"

(*National Recreation Area)
(** Regional Office)

ESSAYS

MEMORIES FOR OLD-TIMERS

William G. Weigle

What a joy it was to know that the bridge was completed and that you had swam old Dobbin across this dangerous stream for the last time.

And what a satisfied look was on your face when you got your first wagon and later on about 1920, when you needed better transportation so badly and you were looking with longing eyes at the Indians going by in a new auto truck and you were still driving the two mules and a wagon but better times finally came along and suitable equipment was made available.

You were again happy when you got a telephone line into Elk Canyon and, of course, the Forest Service had to be neighborly and accommodating so the settlers were given the privilege to tie on, and next August during a bad fire you wanted the use of the line badly and sister X says, "Get off the line!"

You were a new man in the district, 40 miles away from a blacksmith, shop, store, or anything that could give you help in time of trouble, and how well you slept that night when you learned about old Bill Fixit over in the next canyon who would teach you how to shoe a mule or splice a cable.

In the forests of the open country, horses were used everywhere from the start but the forests of district six in the early days had few roads and continuous trails. Hence, means of travel over the forests was to a great extent on foot and in going from one watershed to another no one thought that 20 to 30 miles a day on foot called for any overtime pay.

The development of air travel in fire protective work in the late '20s and early '30s was considered hazardous as the planes were small and poorly powered and had few landing places except in timber. Even though it was hazardous, you never thought about refusing to go; you would probably take out some insurance, kiss your wife and kids goodbye and you were off.

The progress in fire protection and timber work has resulted in most of the forests being opened up by roads and trails to all lookouts and timbered watersheds and equipped with landing fields and two-way radio service.

Although the early forester had many unfavorable conditions to meet, there was some recompense in addition to the small pay he received. While on a trip, if he was so inclined, he was free to benefit by the wonderful fishing conditions in most of the large streams. For example, anyone not familiar with trout fishing in the upper reaches of the Coeur d'Alene, St. Joe and many other streams of Oregon and Washington prior to the time they were opened up by roads and trails, would consider the truth a fish story.

In looking back over the Forest Service work of years gone by, the old-timer has the pleasure of many pleasing memories and a feeling that he is glad he made forestry his life work.

In looking back over the first half century of the activities of the Forest Service, the early field officers met up with many conditions that required the judgment of a seasoned statesman, the knowhow of every kind of mechanic and the physical endurance of a superman, but the periods of greatest difficulties furnish the most pleasing memories.

What old-timer in the Forest Service does not look back with pleasure to the time when he surveyed with pride the completion of a ranger station that cost six hundred dollars or less.

Again his stooped shoulders straighten up when he thinks of the drilling of holes in the solid granite, loading with dynamite and shooting off a pinnacle rock and listening to it go thundering down the mountain 1,000 feet or more to get a space 12 feet square to build a lookout house on a 8,000 foot high mountain; the stringing of two parallel wires fitted with trolleys from a point 3,500 feet or more down the mountain up to the lookout site on account of the absence of a trail; the trolleys being fitted with wire baskets carrying rock going down to furnish power to pull a load of lumber going up.

A LOOK AT OLD LOOKOUTS

Stanley Stevenson

There has been a need to detect fires for thousands of years. However, I won't go back that far. I'll stick to wildland fires since 1900.

Forest fire detection in California started with people using "high places," including trees. On a few of the highest mountains in the early National Forest Reserves of the 1900s, a number of trees had small platforms built in or near the tops to provide a daytime view over the surrounding forest.

As the population grew and corresponding higher risk of fires occurred, more intensive detection of fires, particularly in forested areas, was necessary to safeguard communities.

The primitive fire detection and communication methods at that time were inadequate to protect the forests of the Western United States. Awareness of the value of our national forests was highlighted by the demand for lumber and by the desire to protect unique stands of trees.

Primitive shelters were constructed on the highest accessible peaks. Shelters were provided with water and allowed both day and night detection, using the heliograph for communication by day.

One of the lookout stations built in this early period, about 1910, was on Sherman Peak at an elevation of nearly 10,000 feet, on the Sequoia National Forest in California. The trees were felled and whipsawed on top of the peak into rough lumber for station construction. Smaller logs were split for stairs and underpinning. The roof was made of shakes split from some of the pine logs.

On Sunday Peak there was a similar lookout station, also built during this early period. One of the young men stationed on Sunday Peak in the early 1930s was John Monday. His favorite response when answering the telephone was, "Monday on Sunday." In the two decades to follow, John was the district ranger on the Stonyford Ranger District of the Mendocino National Forest in Northern California.

As quickly as possible, a faster method than the heliograph for relaying the location of fires to the fire suppression forces had been provided. The heliograph required sunlight and a knowledge of the various codes that were used, whereas telephones transcended these limitations.

Telephone lines, consisting of a single No. 9 gauge galvanized wire, supported by poles in open land and trees in forested areas, were stretched between each lookout, fire guard and the ranger station, for dispatching firefighters to the reported fire. The ground was substituted for the other wire to complete a circuit. A fire guard making his rounds could temporarily connect his portable phone to the line for nearly immediate communication.

To reduce the danger of line breakage in the forest when limbs or trees fell across the line, a split ceramic doughnut shaped insulator was wired around the line. The insulator wire was fastened to a tree hook hammered into the tree at the height a man could reach while standing on his horse. Enough slack was left in the line for it to reach the ground without breaking when loaded with snow or fallen objects.

Jennie Abbott, my grandmother, was hired by District Ranger Ray Stevenson, my father, since he couldn't find anyone else who would go up onto this isolated peak. On several of the pages of her diary, she wrote that it was a lonely place. Since the lookouts and fire guard stations were all connected by telephone lines to the district ranger station, during rainy weather, my mother would ring up the isolated stations and play her Victrola over the telephone to them. One of the fire guards could play the harmonica, another way the isolation was relieved for an evening.

Ione Pinnell, when she was about five years old, would visit her grandmother Jennie several times during the two years

Jennie was stationed on Sunday Peak Lookout between 1922 and 1923. Robert Beard, District Ranger on the Green Horn Ranger District, received the fire reports. He was one of the first Forest Service rangers appointed on the Sequoia National Forest after it was created in 1910.

As a six year old, I caught a ride on a pack train in 1921, over the eighteen miles of trail from roads end up to the Sherman Peak Lookout to stay with Grandmother Jennie for a couple of weeks. I'm not sure how much company I was, as she said many times that, "A whistling boy and a furry worm make the butter fly." Butter was a precious staple at the lookout and I was very fond of grandmother's pancakes.

The fire guard from Bonita Meadows Fire Station would come up the trail about three miles with supplies. He also packed water up from the spring three-eighth's of a mile below the lookout.

There were fire guards at Cannel Meadows, Bonita Meadows and Beach Meadows. A trail crew and sometimes an erosion control crew, equipped with portable telephones, were dispatched to fires detected by Sherman Peak Lookout. At that time, the Kern Plateau had no roads and transportation was by horse or walking.

Fire guards had a saddle horse and one or two pack mules to carry shovels, axes, a five gallon backpack pump and provisions for two or three days. The stock were caught each morning and saddled, with packs sitting nearby for immediate use.

The Kern Plateau is a moderately rolling timbered area, with some peaks where the dispatched fire crews could climb to locate the smoke of the reported fire.

The area is subject to lightning fires after thunderstorms. Often the resultant rain would cause moisture to rise like steam and appear as smoke. Lookouts were sometimes fooled, since they had recorded many of the lightning strikes that did start fires.

The lookouts were equipped with a centrally located two and a half foot dais, on which was fastened a map of the forest visible from the station. A plastic 360 degree transparent azimuth circle was oriented with the station in the center and 0 degree on north. An alidade was used to sight on the smoke. The lookout took a reading of the degrees from the azimuth circle and phoned the numbers to the ranger station. If other lookouts could see the smoke, they too called in their bearings.

The ranger or assistant would use a composite map with azimuth circles of all the lookouts from which he could stretch strings on each corresponding bearing. If the strings crossed, the fire was located. Many lookout personnel soon learned the names of the country's features and could name the exact location of smoke.

In the early 1930s, the Osborne firefinder made of metal was installed on the dais. It was easier and more accurate to use. The Osborne firefinder also gave vertical bearings. A series of photographs were taken from the lookouts encircling the complete view of the country, with corresponding bearings printed on the photos. They created a photographic panorama of the visible area encircling the building just below the ceiling. However, this was seldom used.

A complete study was made of the peaks in and adjacent to the forests to determine the location for the best visibility of high fire hazardous areas with the least double coverage. This study resulted in eliminating some lookout stations, while adding others elsewhere.

By 1950, population growth in cities and rural communities generated higher fire risk but also efficient detection, radio communication and fire suppression facilities. However, lookout stations in many places were victims to smog, dust and smoke and were abandoned in favor of aircraft and local fire reporting.

INSECTS BOTHER EARLY TRAIL BUILDERS

Fred Wehmeyer

I often wonder if those who now use modern roads and trails ever fully appreciate the cost in pain and misery, besides dollars, of building them in the days long gone by. Every mile of trail was constructed by someone being stung by bees. Nearly every down log, if rotten, contained yellowjacket nests and the air itself was full of insects. Horse flies, deer flies, gnats, no-see-ems and mosquitoes were so thick that one breathed with difficulty. In those days, when exploring for future trail routes, we generally camped on a ridge and packed water. There we could clear a spot and build smudge fires for the horses. The horses would stand head to tail in the smoke and tears rolled down their cheeks. They would venture off to eat only when a breeze brought a temporary lull to the hordes of ravenous flies and mosquitoes. Keeping an eye on the horses was an endless job because, if they thought there was any chance of getting away, they would take off.

I remember a story of Ranger Jamison of Loomis, Washington, who was back in the Toats Coulee country, about his horses wandering off during the night. He started trailing them and soon met another person doing likewise. They struck up an immediate and casual acquaintance. Jamison introduced himself as the local forest ranger and the other fellow said, "I am a prospector, been prospecting these mountains for the past twenty years." He then added, "That ain't exactly the truth as I've hunted horses fifteen of those years!" Painful adventures added many laughs to the ribald humor of the camp. I remember Glen Mitchell telling of the man they had hired to help cut the hay on their ranch at Molson. He said the hired hand passed him with his arms flailing like vanes on a windmill, cutting wild patterns through the hay field. As he passed, the hired hand yelled, "Ole Yesus, yellowyackets!"

On the way to a fire, about November 1, I had a Britisher in my crew, a soldier recently released from a Lancastershire lancer outfit. We were climbing up steep ground carrying fire tools. Attracted by our perspiration, a large bald hornet lit on the Britisher and changed the shape of the poor chap's ear. He let out a war whoop, but feeling somewhat conspicuous as a true Englishman, he tried not to be emotional. He then said, in a matter of fact conversational tone of voice, "I do believe that nasty insect bit me." Somehow, we "Yanks" lacked much in manners, for the affair struck me as being highly hilarious and we fairly rolled in dirt in laughter.

But back to bugs. I have read that there have been authentic cases of suicide by wild animals who could no longer endure the torture of the millions of voracious insects. I have seen range cattle, no longer able to endure the torture, let out a bawl and take off through the brush without any regard to injury, in the hopes of dislodging some of the insistent insects. I have also seen men working on a trail sit down and cry, completely worn out from the continuous battle of the bugs. I used to think that the devil surely created the insect world and if the time of retribution came about, I hoped to see the rascal staked somewhere on the Similikameen River with no protective net. That will be the day we can join Tug Boat Annie in saying "Happy days are here some more."

RED IVES REMEMBERED

Patrick Harrison

I am a dirt forester, not a literary critic, but I know what "place" means to the bookishly inclined. There can be locations; a country, a valley, a town, or in this case, a ranger station, that possesses such a personality that it becomes a "place." It can be written about the same way you might write about an old friend. The Red Ives Ranger Station up on the headwaters of the St. Joe River, in Northern Idaho, is definitely such a "place."

I couldn't conjecture how many Forest Service folk have touched, and been touched by, Red Ives. Remote locales have a special role in Forest Service mythology. The further up some gawd-fersaken gulch, the more appeal it has when described to the uninitiated.

Red Ives, on the old St. Joe NF, unquestionably held a spot high on the list of such "back end of nowhere" places. It was common for Alaskan transferees to pick Red Ives as their first station on return to the "lower 48." It wasn't all that different from the places they had know up North. "A halfway house for returned Alaskans," was one of Red Ives' many nicknames.

An old prospector named Red Ives had a dugout where "his" creek met the emerald clear flow of the St. Joe River. Several histories of the country allude to Red Ives, as a young man, as having been part of the first surveying expedition up the St. Joe. Whatever the vague history of the site's naming, the Forest Service's involvement in this beautiful canyon bottom dates back to 1934, with the consolidation of the Ward Peak and Pole Mountain districts. In 1935, the new district was moved to the Red Ives site, when the Civilian Conservation Corps (CCC) constructed a two story frame office/kitchen/bunkhouse and warehouse very close to Red Ives' old dugout. In '37 a barn and powerhouse were constructed. A CCC camp was a feature of the site all through this early period. A messhall was added in 1963, and two barracks were constructed in 1964. During the 1960s and '70s many large trailer homes were brought to the location to house the district staffers. In 1984, the Red Ives Ranger District was merged with the Avery Ranger District, and became a "work center."

My family and I were fortunate to spend four years at Red Ives (1977-1981) when Bob Lehman was rangering there. Ranger Lehman was a perfect "fit" for Red Ives. He was a classic ranger; decisive, yet easygoing, a natural leader, yet comfortable with letting good staffers run their own shows and just simply "a good man." The big cookhouse was run by a wonderful Finnish woman, Esther Smith. Her huge, "baked daily" cookies caused more than one Forest Service helicopter, bound from some location back to Missoula, to drop down to the heliport for "engine checks" and, of course, a run into the cookhouse.

The station was like a small, self-contained city, replete with its own sewage treatment facility, a huge powerhouse, and of course, the cookhouse and hundred-person barracks.

The most memorable part of the setup, for the staff at any rate, was the way the house trailers had been arrayed up Red Ives Creek. Over the years, the various residents had landscaped their yards, especially the side that fronted on the Red Ives Creek. It was common to find carefully laid native flagstone walks leading to tree shaded picnic spots close by the tumbling flow of Red Ives Creek.

Some ninety miles of backcountry road separated Red Ives from the nearest real community at St. Maries, where everyone moved for the winters. Groceries were trucked over the Bitterroot divide from St. Regis, Montana, where a grocery store accepted orders over the radio-telephone. Many are the stories of hurried trips "down the Joe" to get overlooked supplies, see a doctor, or have a baby!

The sense of community that grew from being so isolated was special, and the geographic location from the benign oversight of the supervisor's office often allowed us to operate with an impunity and effectiveness reminiscent of an earlier, less regulation bound Forest Service. For a small district staff, I have never seen its equal in simply cranking out accomplishment. I suspect it had always been that way "up the Joe" at Red Ives.

RANDOM OBSERVATIONS

Hank Mostovoy

How does one know when he is a witness to or taking part in a historical event? The following documents some changes which took place while I was working for the Forest Service, off and on, from 1945-1979. (16 years to 50 years old.)

As the saying went in 1945, "We are no longer in a caretaker status." Due to the GIs returning home, the Forest Service started out on a fast track. I noticed:

1. At the ranger district level, all multiple uses started increasing at an enormous rate. Allowable cut on my district went from 10 million to 25 within a couple of years and continued on up to 55 million.
2. People started recreating. This they did not have time to do before or during the war.

3. Being a fireman, I noticed the biggest change here. I saw fire prevention become a national word, due to a bear. With the increase of logging, road construction access to fires almost did away with the mule train and lookouts. These roads, however, also brought more risk and hazard together, which made for fires in areas we never had before. Manpower increased. New equipment was designed and used, especially with the advent of the helicopter and aerial retardants and their application.

4. I watched foresters laying out timber sales in areas their teaching told them not to. I watched an organization called the Sierra Club bring our timber sales almost to a standstill. At this time politics really started to rear its head; the clear cut and all it entailed.

5. I sat in on meetings when politics decided to make a national park out of part of our district. The in-fighting. The lies put forth by a national TV news organization to show support for the park. The Sierra Club and National Geographic throwing all their weight toward the park.

6. I watched as names were applied to areas that were nameless and questioned why? Why didn't we give these areas their Indian names? I was later to learn that this was only a continuation of what has been called, "The European way." Something to do with manifest destiny.

Much, much more took place down through the years. Those subjects mentioned, if expanded upon, could give a person a good idea of some of the historical events that took place from 1945-1979, in a small pocket of land in the national forest system.

LOS PADRES LOOKOUTS FLASHBACK

Raymond Richart

In the early 1980s, I revisited La Cumbre Lookout. (I had retired from the Los Padres NF as communications officer). I was alone and had plenty of time to look around. I climbed the steep stairs up the tower. The trapdoor to the catwalk was unlocked. I lifted it and found the door to the lookout also unlocked. I entered. There was a musty smell. One shutter was down, hanging by one corner, allowing plenty of light to enter. Other than the pull down map and the alidade, there were no furnishings. All was covered with dust. I had the feeling of having entered a tomb. I left, closing the door and trapdoor behind me and stepped backwards all the way down the stairs, hanging on to the stair rail.

The sun was warm and I sat down on a handy rock, dropped my eyes and allowed myself to slip back fifty years in time. The lookout was freshly painted. The shutters were tightly wedged open. No slack could be left in the fasteners or the wind would vibrate them loose. A happy, elderly man greeted me. "Hi! Nellis, you certainly surprised me, I must of been dreaming." It was Nellis Binns on Santa Lucia Lookout. I had called him with an "S" type portable radio and asked if he needed anything. He wanted his mail brought up and I had it. He had just pulled two loaves of fresh, baked bread from the oven. He cut me a good, thick slice from the heal of one loaf. Nellis could see me with his telescope just as I left the Indian Guard Station and had the baking timed for my arrival. He captured his own yeast germs and was a master at making fine yeast. His other hobby was making telescopes with old eye glasses. It was higher powered than the FS field glasses. It didn't take me long to finish radio improvements and I left with saddle and pack for the Indian Guard Station, loaded the stock in the trailer and headed for another lookout. (I always called before packing in to a lookout and brought them supplies and mail.)

"Greetings, Maude and Kurt." "Welcome, Ray, sure glad to see someone. I thought there was no other life on earth except pack rats, buzzards and bald eagles." Maude was an extra good cook. She baked biscuits that would tempt a person with ulcers. And strawberry jam! They were Texans, and biscuits and gravy with ham, was standard for breakfast. Kurt, when not on lookout duty, was a packer. I liked to have him with me because he could pack as fast as I could unload a pickup. He was jovial and nothing excited him. This was Cone Peak Lookout overlooking the coast, also the Hunter Ligget Military Reservation. At this instant I was sitting on a rock, with my back to the sun, which was getting low in the sky.

I was still in a trance. I was on my way to Topa Topa Lookout, spring of 1944. I was following a pack string of ten pack animals tended by Blain and Sherm Mansfield. Packs were loaded with radio equipment, fuel oil, grain for the stock and regular supplies. One pack mule had smoke coming from the pack; the mule got excited and tried to buck off the pack. This excited the mules next to it and pandemonium was the order. Sherm was nearest and ran the mule down and got it under control. A metal stove pipe had shorted out the storage battery and caused a blaze. Next, we surprised a bear as we rounded a bend. It was a black bear and freshly shed. Its fur was black and shining, and it rippled with movement as the bear galloped away.

Topa Topa Lookout was the home of the Williams'. They had tame deer and small animals. They had thrown out scraps to the does with fawns. As the fawns grew up they had no fear of the people so the wilder deer followed them through curiosity.

I was back now to La Cumbre Lookout, after traveling every trail and visiting every lookout and occupant, sharing their griefs, gripes and happiness. As I search my memory for names, I find many names too dim to read. From La Cumbre, all the high back country is visible including many other lookouts. The sun was set and there was a pretty pink sky in the west. The islands showed up dark as I looked across the Santa Barbara channel. I could see the breakwater and all the little boats inside. I could make out the Granada Building and the Fox Arlington. Farther west, the Goleta Valley and all that used to be farms now covered with buildings. Soon the lights were on; as it got darker there were many colors. It was a beautiful sight. It got much cooler and I was reminded to go inside. The lookout was dark and cold. There were no happy friendly people to beckon me in. It was closed forever, along with thirty-five others. There was no longer need for back country guard stations, saddle horses and pack stock.

GOLDEN EGGS FROM THE GOOSENEST

Bruce Barron

Because of its harsh terrain and extremes in climate, the Goosenest Ranger District on the old Shasta Forest (back in the early 1940s) was sometimes dubbed "The Little Siberia of Region 5." Those hot, dry summers punctuated with fickle thunderstorms, often left the district ablaze like a Fourth of July celebration. In winter, frigid winds and icy snowstorms swept down from the Northern Cascades, plunging temperatures to below freezing for days on end. However, beneath its harsh exterior, the Goosenest held a hoard of golden eggs!

One such golden attribute was the fertile soils at the Mt. Hebron Ranger Station. The family garden was a joy to remember. The now defunct variety of "Rock Hill" strawberry plants yielded bushels of huge, sweet, deep red berries the size of apples. Root vegetables were prize winners. They found no bottom to that rich frangible lava ash soil. Secret forays by ranger station personnel to that huge lava cave east of town, yielded quantities of precious bat guano that held those magical qualities for promoting such luxuriant plant growth. Not only did that yield valuable fertilizer, a connecting "lava ice tube" held thick layers of clear ice throughout the summer. On a hot summer day, it was like finding another golden egg when a short safari to the cave provide ice for churning some old-fashioned homemade ice cream. Adding generous quantities of those plump, succulent strawberries gave the "piece de resistance" to that delicious ice cream.

The Goosenest was copious in many more ways. There was that early summer harvest of wild currents, gooseberries and wild plums, which ended up in delightfully flavored jams, jellies, pemican and mince meat. Then, in the fall, there were those trophy sized mule deer to the East, mingled with the blacktails in the West. Hordes of fat ducks and geese flew low over nearby mountain passes and settled on favorite hunting ponds. Sage hens flocked on the edges of the rimrocks in the sage country to the east. On those days when storms were howling and it wasn't a fit day out for man, beast nor forester, one just threw another log on the fire and enjoyed the aroma of roast goose, fresh baked pastries, popcorn and hot coffee, while challenging neighbors to a friendly game of pinochle, poker or gin rummy.

Yes, those were only a few of the many golden eggs (opportunities) that prevailed on the old Goosenest.

LETTERS

Joe Church

I have often wondered but never asked if other district rangers received many or even any of those sealed, long, white envelopes in the district mail; envelopes from the forest supervisor addressed to the DR and always, at least in my case, marked <u>PERSONAL</u> or even <u>P</u> <u>e</u> <u>r</u> <u>s</u> <u>o</u> <u>n</u> <u>a</u> <u>l</u>. Such letters were addressed in various ways, mostly by typewriter but once in awhile by hand, such as:

Joe - Orleans
Joe Church - Orleans D.R.
Joseph B. Church, District Ranger, Orleans R.D.
District Ranger, Orleans Ranger District.
And other ways too, and that includes the Mi Wok R.D.

However, no matter how it was addressed, it meant I was about to receive the personal attention of the forest supervisor on a subject or subjects important to him and about to become important to me, like it or not.

I received this sort of attention from Charley Yates first when I was acting DR on the Scott River District of the Klamath and he was the forest supervisor. Actually, Charlie was pretty gentle with me.

When I moved first to Orleans on the Six Rivers and seven years later to Mi Wok on the Stanislaus, I was introduced to experts in the field, namely Wes Spinney on the former and Harry Grace on the latter. Those long, white, sealed envelopes seemed to show up weekly though I suppose in reality it was more like monthly. But no matter what the timing, their arrival was greeted by the exclamation, "Oh, Oh!" I discovered early in this game that the contents would be delivering a dose of reality and almost always also delivering the message that I wasn't as smart as I thought I was. Also, it wasn't very smart to answer them except in the general tone of, "Yes, sir," either, when I felt the subject just couldn't be shrugged off. Such responses could and almost always did bring on an even stronger reply, at which time even I could recognize the meaning of the old saying, "discretion is the better part of valor" and lay off.

Searching through my Forest Service box of memories—a large cardboard box packed with my personal Forest Service stuff—I found a couple of copies of supervisor letters, and even copies of my responses, the reading of which jumped me back about three decades or so and caused me to wonder how I ever got away with what I did.

In the summer of 1962, Wes came to the Orleans District for a look see at a couple of things of interest to him but didn't let me, the DR, know of his visit. So I took umbrage at this "breach" and accordingly wrote him a letter of complaint. His reply was quick and to the point. After a brief description of why he was there, he proceeded to tell me in no uncertain terms that "he was the forest supervisor of the Six Rivers National Forest and that he would go on it wherever and whenever he wanted and didn't need my permission or anyone else's to do so, and that I better be careful in the future of what I said to whom!!" The miracle is that I got by with it without, as far as I ever knew, Wes ever holding a grudge. I continued to get <u>PERSONAL</u>

letters from him, even occasional ones extending praise for doing something a little better than just right. Those kind would make my day but I still find that when I reread his letter it can still raise a sweat, even after all those years.

Harry Grace's style may have been a little different than Wes', but again there was no mistaking the meaning.

The DR training plans on the Stanislaus included, at least for me, the requirement that the incumbent read some books on management with the number and titles, within limits, being negotiable. The forest library was well stocked with such books and the local public library served as good backup. So I read books, though I suspect he felt I wandered a little far afield occasionally. Early in this particular game I read a book titled "Tough-Minded Management," or something like that. Since he liked periodic updates on accomplishments, I wrote him a note advising of this one. I received in short order his hand written reply in that by now familiar sealed, long, white envelope marked PERSONAL.

It was a single sheet of yellow, lined note paper which began by saying that he couldn't read my writing so, in the future, please have it typed. That didn't start out too well for me because I didn't think my writing was that bad. I had, and have, been told by too many people that my handwriting is poor, but of course, they are wrong too. Anyway, a little further along in his note, Harry told me that I should take tough-minded management to heart or words to that effect. That raised my hackles, whatever they are, and created a storm of protest in my mind. I put that storm on paper, three pages and all in hand. I raved at being taken to task without any explanation of how or why I was deficient, with no suggestions as to how I could improve my performance. I then itemized all those things which illustrated, as far as I was concerned, the work which I and the district people had done which disproved his statement and finished up by saying that I thought it was dirty pool on his part. I was worked up. The interesting part is that I could find no evidence that I actually sent my reply. I suspect I didn't for I remained the Mi Wok DR through 1970, maintaining I felt, pretty good relations with Harry, even receiving a couple of good performance ratings and in general, had a fairly productive and enjoyable Stanislaus tour. That's not to say there weren't more of those sealed envelopes marked PERSONAL, for there were and they were as unsettling as ever. I do vaguely recall replying once in awhile but surviving.

FIRE WATCH

Jackie Johnson Maughan

It's early September and for days I've been on extended hours—this means I cannot leave the 30 foot tower except to use the latrine. Even then I must take my hand-held radio.

My days have been running to 12 hours during this fire burst. Fifteen miles to my east is the big 15,000 acre fire, to my west another smaller fire. I've been watching the air tanker deliver retardant to contain it. This is dangerous because the canyon is steep and narrow and the pilot must get low enough to do some good.

Tonight I sit out on the catwalk and watch the big fire. I think about the 600 plus men and women down there, remember the storm that caused it, how I called it in, watching as it took off—one acre, four acres, ten. It was a goner way before our initial attack forces could get there.

The moon blooms red and below the coyotes set into a dissonant howl, as if they knew. Then I think, of course they know. They can see, hear, taste, smell better than I can. It makes me think of that old hymn, Amazing Grace.

Morning comes and I'm socked in by smoke, can't even see the outhouse. I'm alone, more than just lonely. I miss my husband, my friends. After all, I'm 300 miles from home. I get down maybe once a week to pick up my mail, use the phone, buy groceries, refill my water containers.

Summit Point, where I've lived now for 70 days, looks into the Eagle Cap Wilderness. The nearest town is Halfway, Oregon, population 500. It's a 13 mile, 3,000 foot climb on dirt road, then two miles by trail. Visitors are few.

These are the times I go stir crazy. Normally I can hike up and down the mountain to check up on the mule deer, the resident redtailed hawk and a badger which has been busy all summer excavating a burrow right in the middle of the trail. There's also a bear I know, in a heavily wooded side drainage I've named "Spooky Hollow."

But days of prison like isolation can make me question my sanity. Why do I do this summer after summer? Have I come to value this time too much? Have I come to love this airy space of falcons and eagles and hawks too much? I have watched the birds all summer and know of their habits; how the kestrels hunt the noon air in groups of six; how the owls come silent in the color of moonlight to kill my friendly ground squirrels. Is it only here that my writing, my thinking, can achieve clarity? Words fall like predators on the page. Will I become, finally, one of them?

I think about all these things, too distracted to read or maybe write. I think about what we call creature comforts—running water, electricity, and know I don't need them. I'd live this way if it weren't for my husband and my teaching job. But then who would I be, some hermit with an old black dog and a shotgun?

Then the afternoon wind begins to pick up. I watch as it blows the smoke to the east, then go out on the catwalk to take a reading. I think about how it would feel to become air, a thermal over the mountains.

I radio that the wind has shifted to the west/southwest and is blowing at 35 miles per hour. A half hour later and the smoke's cleared and I can see the big fire making a run into the dry timber near Deadman Point. I radio again.

Today that fire will take out another 5,000 acres. It will send smoke and ash boiling into the sky like a nuclear bomb. I'll watch this country going up in smoke. It hurts to see. But I know, even as I feel the fire burning through me and the land, that this is the way it must be. The land must renew itself; so must I. The old forest must die for the new to live. I must help the young that will come behind, for someday I too will become of smoke and ash. A painful lesson, this ecology of the natural, but one to gather strength from.

RETIREMENT WORDS—1968

Al Crebbin

"The things that I enjoyed most of this life, probably, were in the Marble Mountains. There were thousands and thousands of miles I have ridden along these creeks, with their tumbling white water and deep pools murmuring to themselves and chattering on their long, lonely way to the sea. Then up over the mountain, see the forest with the gigantic trees, soft mossy glens, soft carpet of needles under the trees and the little trees growing up trying to be big ones, struggling for 50, 60, 70 years to get their heads up in the light. Then, along the ridges with the rock gardens, the flowers, the beautiful arrangements, just the rocks and the flowers. And overhead the blue, blue sky in the distance, ridge after ridge of almost seemingly endless mountains. Then down through the glaciated rocks carved out 25 thousand years ago across the meadows into the lakes. There were hundreds and hundreds of campfires, their wonderful companionship, the talks or just gazing dreamily into the embers, and then to bed under the stars. Thousands and thousands of bright stars, and black velvet. And later on, often the moon would come out, turning the whole mountainside into a vast fairyland where the tinkling of the horse bells in the meadows was a musical accompaniment. Soon light would hit the first peak, accompanied by the drowsy chattering of birds in the trees. Often as the sunbeams stole down through the trees, a squirrel would start scolding when he found his privacy had been invaded. Then, as the sunbeams finally came through the forest and struck my sleeping bag, I'd know it was time to get up and go out for another glorious day. It's been wonderful.

"Now I know it's time to quit, to go. The mountains are getting steeper the days are getting longer and the manuals are getting thicker. I know it is time to go and make room for younger men with more energy and new ideas. And I am not really sorry, but just a little sad. Thinking of the close association I've had with so many of you wonderful people and the beautiful forest. But I'm truly glad, now I can go and stay at the place I'd rather be than anywhere else in the world. Home with Helen. And Helen and I wish you just as much happiness and satisfaction out of your careers as we have had out of our service in the Klamath Forest. Thank you very much."

ABBREVIATIONS

ADR	Assistant District Ranger
AM	Amplitude Modulation
AO	Administrative Officer
ARF	Assistant Regional Forester
AWOL	Absent Without Leave
AWS	Aircraft Warning Station
BIA	Bureau of Indian Affairs
BLM	Bureau of Land Management
BM	Board Measure
BO	Body Odor
BPR	Bureau of Public Roads
BWCAW	Boundary Waters Canoe Area Wilderness
CCC	Civilian Conservation Corps
CDC	California Department of Corrections
CDF	California Division (Department) of Forestry
C&H	Cattle and Horses
CHP	California Highway Patrol
CJRL	Cotton Jacket, Rubber Lined
C&T	Condition and Trend (Range Plots)
DA	District Attorney
DBH	Diameter Breast high
DC	District of Columbia
DR	District Ranger
EJ	Everett Jensen
ERA	Emergency Relief Administration
FCA	Fire Control Assistant or Aide
FCO	Fire Control Officer
FDR	Franklin Delano Roosevelt
FEMA	Federal Emergency Management Agency
FFF	Forest Fire Funds
FM	Frequency Modulation
FRD	Forest Road Development
FS	Forest Service
GHQ	General Headquarters
GI	Government Issue
GLO	General Land Office
GP	Gifford Pinchot
GPM	Gallons Per Minute
GS	Guard Station or General Schedule

HP	Horsepower
IC	Incident Commander
ICS	Indident Command System
IWW	Industrial Workers of the World
JF	Junior Forester
JP	Justice of the Peace
LA	Los Angeles
LEO	Law Enforcement Officer
LMP	Land Management Planning
LO	Lookout
L&WCF	Land and Water Conservation Fund
MDM	Mt. Diablo Meridian
MED	Miscellaneous Executive Duties
M&LS	Manistique and Lake Superior Railroad
MPH	Miles Per Hour
NEPA	National Environmental Policy Act
NF	National Forest
NP	National Park
NPS	National Park Service
OD	Olive Drab
OEO	Office of Equal Opportunity
OWR&NCO	Oregon and Washington Railway and Navigation Company
PA	Protective Assistant
PNW	Pacific Northwest
PSO	Project Sales Officer
PSW	Pacific Southwest
PT&T	Pacific Telephone and Telegraph
R	Range
R&D	Research and Development
RF	Regional Forester
RO	Regional Office
ROTC	Reserve Officer's Training Corps
R&R	Rest and Recreation
RS	Ranger Station
S	Superregenerative (Radio Model)
SFO	San Francisco International Airport
SMUD	Sacramento Municipal Utilities District
SO	Supervisor's Office
SP	Shore Patrol (Police)

SPARS	Women's Coast Guard Reserves (based on Coast Guard motto "Semper Paratus— Always Ready")
SPF	Semiportable Fone (Radio Model)
S&W	Smith and Wesson
SX	Superregenerative Xtals (Radio Model)
T	Township or Ten Meters (Radio Model)
TBM	Torpedo Bomber Medium
TNT	Trinitrotoluene or Trinitrotoluol
TVA	Tennessee Valley Authority
US	United States
USA	United States of America
USDA	United States Department of Agriculture
USFS	United States Forest Service
USGS	United States Geological Survey
V-E	Victory in Europe
VFR	Visual Flight Rules
V-J	Victory Over Japan
WAC	Women's Army Corps
WASP	Women's Air Corps (Force) Service Pilots
WAVES	Women Accepted for Volunteer Emergency Service (USN)
WM	Willamette Meridian
WPA	Works Progress Administration
W-W	Wallowa-Whitman
WWI	World War I
WWII	World War II
Xtal	Crystal
YACC	Young Adult Conservation Corps
YCC	Youth Conservation Corps

Class A through Class G fires relate to acreage size

NAME INDEX OF CONTRIBUTORS

Abraham, Dan B.
Anderson, Hugh V.
Arment, Charles
Arvola, Tobe
Asplund, Rupert

Bangsberg, Ralph
Barron, Bruce
Bauer, Don
Beeson, Russell "Russ"
Belcher, Gilbert E.
Bennett, Stan
Bernstein, David A.
Bjornsen, Bob
Blake, Grover C.
Blood, Roy H.
Booth, Roy
Bowerman, Harold R.
Brenneis, Andy G.
Brown, Bruce
Bruckart, John "Ray" Sr.
Bryan, W. D.
Buck, C. J.
Buck, Shirley
Burgess, Dorothy

Campbell, Robert L.
Caples, James W.
Carson, Dick
Cermak, Robert W.
Chatham, R. A. "Dick"
Chriswell, H. C. "Chris"
Church, Joe
Colvill, Les
Cooper, Loran
Cousins, Albert H.
Cravens, Jay H.
Crebbin, Al K.
Cronk, C. P.
Cuff, Kermit

Dahl, Harold A.
Davis, Tom
DeGraaf, Edward C. "Ed"
Denham, Avon
Denney, Rex
DuBois, Coert

Elliott, Joe Jr.
Ely, Joe

Filius, Dave
Floe, Sanford
Fromme, Rudo L.

Gleason, Clark
Glover, Art
Godden, Jack A.
Gray, Bob
Groom, Jack I.

Hall, John
Hanson, Pete
Harrison, Patrick
Heilman, Donna D.
Hein, Corwin E. "Slim"
Horton, Jack
Hunt, Buel B.

Irwin, Robert L. "Bob"

Jensen, Everett
Jones, Ira E.
Jones, T. J. "Tom"
Jorgensen, C. Glen
Joslin, Les

Klubben, Barbara
Klubben, Lyle

Krueger, Myron

Lewis, Duffy
Lewis, John F.
Lewis, Melvin M.
Lilligren, H. M.
Longacre, A. M.
Lord, Phil

Mack, Harvey B.
Maughan, Jackie Johnson
McArdle, Richard "Dick"
McClain, J. Fred
McEwen, Carroll "Mac"
McEwen, Cleowa "Cleo"
McGuire, C. C.
McNeill, James D.
Merritt, M. L.
Miller, Doug
Miller, H. B. "Dusty"
Mitchell, Roy
Moisio, Fritz
Morey, George
Mostovoy, Henry "Hank"
Munger, Thornton T.
Munsey, Gary

Neal, Carl B.
Nebeker, Dwain
Nelson, Mildred G.
Nelson, M. M. "Red"
Newhall, George N.
Niehaus, Ted
Niewald, Earl

Paine, Philip L.
Parker, Alvin
Parker, Scollay
Parmele, Benny G.
Pence, F. Carl
Peters, Donald
Phelps, Howard T.
Pratley, James R.
Puhn, Walter "Walt"

Radigan, A. J.
Richards, H. R.
Richart, Raymond
Roberts, Ira W.
Rogers, Russ

Saarni, Roy
Schmidt, Andrew R. "Andy"
Scholberg, Maren C.
Scott, Ronald E.
Shaw, Helen D.
Slattery, Leo "Spike"
Smay, Charles "Chuck"
Smith, Edwin F.
Smith, Harold E.
Smith, Jack
Spencer, Jack
Steele, Foster
Stevenson, Stanley
Swift, Lloyd W.

Tikkala, Bill
Tixier, Stan
Tonnesen, John Ed

Waha, A. O.
Waite, David M.
Wales, H. Basil
Walker, David O.
Walker, Gordon
Watson, Rusty
Wehmeyer, Fred
Weigle, William G.
White, Harry
Wilson, Carl
Wofford, Audie
Worthington, Richard "Dick"

Ziegler, Bob

Gilbert "Gil" W. Davies worked for more than 30 years in the U.S. Forest Service, retiring in 1990. All except the first six months were spent on the Klamath National Forest in Yreka, California.

Florice "Flo" M. (Hebard) Frank worked for more than 23 years in the U.S. Forest Service, retiring in 1985. The first 15 years were on the Klamath National Forest in Yreka, California. The remaining years were on the Winema Forest in Klamath Falls, Oregon and the Lassen Forest in Susanville, California as the forest contracting officer in both places.

Davies and Frank founded **HiStory ink Books** in 1992. To date, they have either authored or edited 17 different books, all connected to history.

MEMORABLE FOREST FIRES

200 Stories By U.S. Forest Service Retirees

- More Than **75** Years of Forest Fires from **1910** to **1988**
- Fire Stories From All Over the United States by **57** Retirees
- More Than **200** Stories Written in the Retirees Own Words

A Perfect Bound 6" x 9" Book with 506 Pages and 80 Photographs

Stories about dispatching, demob, firelines, aircraft, fire camps, initial attack, equipment, communications, smokejumping, travel, crews, slash disposal, trespass fires, law enforcement, training, command, safety, weather and much more.

The book describes accidents, tragedies, humor, close calls, snafus, fire gods, characters, goof ups, drinking, food, friendships and unforgettable experiences.

Sale Price: $21.95 each book
Shipping Price: $2.50 for each book
(California residents add $1.59 sales tax for each book)

FREE SHIPPING AVAILABLE: Order two or more books to same address. Titles may be same or mixed

ISBN: 0-9634413-9-6

Payment:

Personal checks, money orders, Government third party vouchers all accepted.

Make checks payable to:

HiStory ink Books
P. O. Box 52
Hat Creek, CA 96040

Free flyers on our other Forest Service and history books if requested.